Encyclopaedia of Mathematical Sciences
Volume 132

Invariant Theory and Algebraic Transformation Groups III

Subseries Editors:
R.V. Gamkrelidze   V.L. Popov

Springer
Berlin
Heidelberg
New York
Hong Kong
London
Milan
Paris
Tokyo

Vladimir L. Popov (Editor)

Algebraic Transformation Groups and Algebraic Varieties

Proceedings of the conference
Interesting Algebraic Varieties
Arising in Algebraic
Transformation Group Theory
held at the Erwin Schrödinger Institute,
Vienna, October 22-26, 2001

Springer

Vladimir L. Popov
Steklov Mathematical Institute
Russian Academy of Sciences
Gubkina 8
117966 Moscow
Russia
e-mail: popovvl@orc.ru

Founding editor of the Encyclopaedia of Mathematical Sciences:
R. V. Gamkrelidze

Mathematics Subject Classification (2000): 14C05, 14C22, 14F22, 14F99, 14J17, 14J70, 14J99, 14L05, 14L10, 14L17, 14L24, 14L30, 14L35, 14L40, 14M05, 14M07, 14M10, 14M12, 14M15, 14M17, 14M20, 14M99, 14N25, 14N99, 14Q15, 14R20, 17B20, 17B70, 22E46, 22E60

ISSN 0938-0396
ISBN 3-540-20838-0 Springer-Verlag Berlin Heidelberg New York

Springer-Verlag is a part of Springer Science+Business Media
springeronline.com

Printed in Germany

Typeset by LE-TEX Jelonek, Schmidt & Vöckler GbR, Leipzig
Cover Design: E. Kirchner, Heidelberg, Germany and V.L. Popov, Moscow, Russia
Printed on acid-free paper 46/3142 db 5 4 3 2 1 0

Preface

These are the proceedings of the conference *Interesting Algebraic Varieties Arising in Algebraic Transformation Groups Theory* that was held at The Erwin Schrödinger International Institute for Mathematical Physics, Vienna, Austria, from October 22 through October 26, 2001.

The conference was made possible through interest and financial and organizational support of The Erwin Schrödinger International Institute for Mathematical Physics, Vienna, Austria. On behalf of all participants I thank this institution and especially P. W. Michor, one of its Directors, for this interest and support.

It is an empirical fact that many interesting and important algebraic varieties are intimately related to algebraic transformation groups. To name only some, the examples are affine and projective spaces; quadrics; grassmannians, flag and, more generally, spherical (in particular toric) varieties; Schubert varieties; nilpotent varieties; determinantal varieties, Severi, Scorza and, more generally, highest vector (HV-) varieties; group varieties; generic tori in algebraic groups; commuting varieties; categorical quotients of Geometric Invariant Theory and the related moduli varieties of curves, vector bundles, abelian varieties etc.; simple singularities realized as that of the corresponding categorical quotients and nilpotent orbit closures. The idea of the conference was to trace the new evidences of this relation.

For various reasons several talks given during the conference do not appear in these proceedings. Below a complete listing of all talks given is presented for the information about the conference. The talks which do appear are generally expanded and/or modified versions of those given during the conference.

November 21, 2003 *Vladimir L. Popov*

List of Talks Given at the Conference

"Interesting Algebraic Varieties Arising in Algebraic Transformation Groups Theory",
ESI, Vienna, Austria, October 22–26, 2001

Monday, October 22, 2001

10.30–12.00
DAVID J. SALTMAN (University of Texas at Austin, Austin, USA),
Invariants of Symplectic and Orthogonal Groups of Degree 8.

14.00–15.30
DENNIS SNOW (University of Notre Dame, Notre Dame, USA),
The Role of Exotic Affine Spaces in the Classification of Homogeneous Affine Varieties.

15.45–17.15
NIKOLAI L. GORDEEV (State Pedagogical University, St. Petersburg, Russia),
Branch Locus of Quotients of Finite Group Actions.

Tuesday, October 23, 2001

10.30–12.00
ALEXEI N. PARSHIN (Steklov Mathematical Institute, Moscow, Russia),
The Krichever Correspondence for Algebraic Varieties.

14.00–15.30
CLAUDIO PROCESI (Università di Roma "La Sapienza", Roma, Italia),
Diagonal Harmonics.

15.45–17.15
FYODOR L. ZAK (Central Economics Mathematical Institute, Moscow, Russia and Independent University of Moscow, Moscow, Russia),
Orders and Classes of Projective Varieties.

Wednesday, October 24, 2001

10.30–12.00
CORRADO DE CONCINI (Università di Roma “La Sapienza”, Roma, Italia),
On Semigroups Associated to Irreducible Representations of Algebraic Groups.

14.00–15.30
JOSEPH M. LANDSBERG (Georgia Institute of Technology, Atlanta, USA),
Deligne Dimension and Decomposition Formulas from a Geometric Perspective.

15.45–17.15
VLADIMIR L. POPOV (Steklov Mathematical Institute, Moscow, Russia),
Self-dual Algebraic Varieties, Lie Algebras, and Symmetric Spaces.

Thursday, October 25, 2001

10.30–12.00
LAURENT MANIVEL (Institut Fourier, Université Grenoble, Saint Martin d'Hères cedex, France),
The Singularities of Schubert Varieties.

14.00–15.30
SHIGERU MUKAI (Research Institute of Mathematical Sciences, Kyoto University, Kyoto, Japan),
Minimal Counterexample to Hilbert's Fourteenth Problem.

15.45–17.15
CIRO CILIBERTO (Università di Roma Tor Vergata, Roma, Italia),
Varieties with One Apparent Double Point.

Friday, October 26, 2001

10.30–12.00
JUN–MUK HWANG (Korea Institute for Advanced Study, Seoul, Korea),
Automorphism Groups of the Spaces of Lines on Projective Manifolds with Picard Number 1.

14.00–15.30
NGAIMING MOK (The University of Hong Kong, Hong Kong),
Holomorphic Vector Fields and Deformation Rigidity.

15.45–17.15
EVGUENI A. TEVELEV (Independent University of Moscow, Moscow, Russia),
Rank Stratification of the Tangent Space of G/P.

List of Contributors

Ciro Ciliberto
Università di Roma Tor Vergata
Dipartimento di Matematica
Via della Ricerca Scientifica
00133 Roma, Italia
cilibert@axp.mat.uniroma2.it

Corrado De Concini
Dipartimento di Matematica
"G. Castelnuovo"
Università di Roma "La Sapienza"
Piazzale A. Moro,
00100, Roma, Italia
deconcin@mat.uniroma1.it

Vincenzo Di Gennaro
Università di Roma Tor Vergata
Dipartimento di Matematica
Via della Ricerca Scientifica
00133 Roma, Italia
digennar@axp.mat.uniroma2.it

Jun–Muk Hwang
Korea Institute
for Advanced Study
207-43 Cheongryangri-dong Seoul
130-012, Korea
jmhwang@ns.kias.re.kr

Daniel Krashen
Department of Mathematics
University of California, Los Angeles
Los Angeles, CA
90095, USA
dkrashen@math.ucla.edu

Joseph M. Landsberg
School of Mathematics
Georgia Institute of Technology
Atlanta, GA 30332-0160, USA
jml@math.gatech.edu

Laurent Manivel
Institut Fourier, UMR 5582 du CNRS
Université Grenoble I, BP 74
38402 Saint Martin d'Hères cedex, France
Laurent.Manivel@ujf-grenoble.fr

Ngaiming Mok
Department of Mathematics
The University of Hong Kong
Pokfulam Road, Hong Kong
nmok@hkucc.hku.hk

Shigeru Mukai
Research Institute for Mathematical Sciences
Kyoto University
Kyoto 606-8502, Japan
mukai@kurims.kyoto-u.ac.jp

Vladimir L. Popov
Steklov Mathematical Institute
Russian Academy of Sciences
Gubkina 8
Moscow 117966, Russia
popovvl@orc.ru

David J. Saltman
Department of Mathematics
The University of Texas
Austin, Texas 78712, USA
saltman@math.utexas.edu

Dennis Snow
Department of Mathematics
University of Notre Dame
Notre Dame, IN 46556–4618, USA
snow.1@nd.edu

Evgueni A. Tevelev
Department of Mathematics
The University of Texas
Austin, Texas 78712, USA
tevelev@mail.ma.utexas.edu

Fyodor L. Zak
Central Economics Mathematical
Institute of the Russian Academy
of Sciences
47 Nakhimovskiĭ av.
Moscow 117418, Russia

and

Independent University of Moscow
11 B. Vlas'evskiĭ
Moscow 121002, Russia
zak@mccme.ru

Contents

Notice that, when $k \geq 72$, the saturated ideal of a set of $\delta \leq (k-1)^2$ points in general position in $\mathbf{P}^4$ is generated by polynomials of degree $\leq k/4$ [GM].

The Theorem provides (as far as we know) new examples of **Q**-factorial projective varieties [MP]. In principle, the method we use could prove that, more generally, hypersurfaces in $\mathbf{P}^4$ with few isolated singularities are, under suitable hypotheses, **Q**-factorial.

When studying *smooth* surfaces in V, we have a more precise result, i.e. the following proposition:

Proposition. *Let $V \subset \mathbf{P}^4$ be a reduced and irreducible hypersurface of degree $k \geq 3$. Assume that the singular locus of V consists of δ ordinary double points. If $\delta < (k-1)^2$ then any smooth projective surface contained in V is a complete intersection on V. The same is true if $(k-1)^2 \leq \delta < 2(k-1)(k-2)$ and V does not contain any plane, and if $\delta = 2(k-1)(k-2)$ and V does not contain either a plane or a quadric.*

Moreover, if $(k-1)^2 \leq \delta \leq 2(k-1)(k-2)$ and V contains some plane π, then π is unique, V does not contain any quadric (except when $k = 3$), and the subgroup $\mathrm{Cl}_0(V)$ of the Weil divisor class group of V generated by smooth surfaces is freely generated by π and by the general hyperplane section of V. If $\delta = 2(k-1)(k-2)$ and V contains some quadric q, then q is irreducible and unique (except when $k = 4$), and $\mathrm{Cl}_0(V)$ is freely generated by q and by the general hyperplane section of V.

One may hope that Proposition holds true for *any* surface contained in V, but we have not been able to prove this. Neither we do know whether $\mathrm{Cl}_0(V)$ is equal to the full Weil divisor class group of V.

Notice also that the Severi variety of hypersurfaces of degree k in $\mathbf{P}^4$ with $(k-1)^2$ nodes, has two irreducible components, corresponding to the family of hypersurfaces whose nodes are in general position, and to the family of hypersurfaces containing a plane.

In order to prove our claims, we need the following lemma:

Lemma. *With the same notation as in the Theorem, let $X \subset V$ be any reduced and irreducible projective surface. Let $S \subset V$ be the surface residual to X in the complete intersection of V with a general hypersurface of degree $\gg 0$ containing X. Then S is irreducible and smooth away from the intersection of the singular locus $\mathrm{Sing}(V)$ of V with the singular locus of X. The germ of S at any point $p \in S \cap \mathrm{Sing}(V)$ is isomorphic to the germ at p of a surface consisting of a finite number $i = i(p)$ of planes contained in the same ruling of a 3-dimensional quadric cone with vertex at p.*

Proof of Lemma. Using a similar argument as in [S] one sees that S is smooth away from the intersection of $\mathrm{Sing}(V)$ with the singular locus of X.

Let $\pi : \tilde{V} \to V$ be the blow-up of V along $\mathrm{Sing}(V)$. Since V has only ordinary double points, then $\tilde{V}$ is smooth and its exceptional divisors E_i,

$i = 1, \dots, \delta$, are smooth quadric surfaces. The strict transform $\tilde{X}$ of X meets each E_i in a curve of type (a_i, b_i), with $a_i \geq b_i \geq 0$. Let H_V be the general hyperplane section of V. When $n \gg 0$, the general surface Σ of the linear system $\mid n\pi^*(H_V) - \tilde{X} - \sum_{i=1}^{\delta} a_i E_i \mid$ is smooth and irreducible, and meets E_i in a curve of type $(0, a_i - b_i)$. This proves Lemma because $\pi_*(\Sigma) = S$. *Q.e.d. for Lemma.* □

Now we are going to prove Theorem with the hypothesis (A).

Let $X \subset V$ be a reduced and irreducible surface. To prove that X is a complete intersection, it suffices to prove that the surface S, residual to X in the complete intersection of V with a hypersurface of degree $\gg 0$ containing X, is a complete intersection.

This said, for any integer $i \geq 1$, denote by m_i the number of points of $S \cap \mathrm{Sing}(V)$ with multiplicity i for S (compare with the previous Lemma). We have

$$\sum_{i=1}^{+\infty} m_i = \delta(S),$$

where $\delta(S)$ denotes the cardinality of the set $S \cap \mathrm{Sing}(V)$. Let $f : \Sigma \to \mathbf{P}^4$ be the normalization of S. Notice that Lemma implies that Σ is smooth. Denote by $\mathcal{N}_f$ the normal bundle of the map f. By the double point formula we deduce

$$c_2(\mathcal{N}_f(-k)) = d^2 + dk(k-4) - k(2g-2) - 2[\sum_{i=1}^{+\infty} \binom{i}{2} m_i],$$

where $d \gg 0$ and g are the degree and the sectional genus of S. On the other hand, taking into account Lemma, a local computation proves that each point of $S \cap \mathrm{Sing}(V)$ with multiplicity i for S has the same multiplicity i in the 0-cycle representing $c_2(\mathcal{N}_f(-k))$, i.e. $c_2(\mathcal{N}_f(-k)) = \sum_{i=1}^{+\infty} i m_i$. Summing up we obtain $\sum_{i=1}^{+\infty} i^2 m_i = d^2 + dk(k-4) - k(2g-2)$, i.e.

$$\sum_{i=1}^{+\infty} i^2 m_i = 2k(G-g) + \nu(k-\nu)(k-1). \tag{1}$$

In the formula (1), ν is defined by writing $d = pk - \nu$, $0 \leq \nu \leq k-1$, and

$$G = G(d,k) = d^2/2k + d(k-4)/2 - \nu(k-\nu)(k-1)/(2k) + 1$$

denotes the Noether–Halphen bound of g in $\mathbf{P}^3$ (see [S], [H]).

Now fix a point $p \in S \cap \mathrm{Sing}(V)$ with multiplicity $i \geq 1$. Let H_p be the general hyperplane through p, and denote by C the reduced subscheme of the scheme-theoretic intersection $S \cap H_p$. The description of the germ of S at p given in Lemma implies that the geometric genus $p_g(C)$ of the irreducible curve C is equal to g. Then we have

$$G - g \geq p_a(C) - p_g(C),$$

where $p_a(C)$ denotes the arithmetic genus of C. Using Lemma again, we see that $p_a(C) - p_g(C)$ is equal to the difference between the arithmetic genus of the reduced curve consisting of i lines lying in a quadric cone in $\mathbf{P}^3$ and the arithmetic genus of the curve consisting of i mutually skew lines. This difference is $i^2/4$ when i is even, and $(i^2-1)/4$ otherwise. And so we deduce

$$G - g \geq (i^2 - 1)/4. \tag{2}$$

Let h be the maximal integer $i \geq 1$ such that $m_i \neq 0$ (put $h = 1$ if $m_i = 0$ for any i). Taking into account that $\sum_{i=1}^{+\infty} i^2 m_i \leq h^2\delta$, using (1) and (2) with $i = h$, we obtain

$$\nu(k-\nu)(k-1) \leq h^2(\delta - k/2) + k/2.$$

By our hypothesis (A) we deduce $h^2 \leq k/(k-2\delta)$. Hence

$$\sum_{i=1}^{+\infty} i^2 m_i \leq \delta k < (k-1)^2. \tag{3}$$

By (1) it follows that $\nu(k-\nu)(k-1) < (k-1)^2$ from which we get $\nu = 0$. If $g < G$ then, by [E, IV.2 and VI.3], we have $G - g \geq k - 2$. By (1) and (3) we obtain $2k(k-2) < (k-1)^2$, which is not possible because $k \geq 3$. Hence we have $g = G$. By [H] it follows that the general hyperplane section of S is a complete intersection on the general hyperplane section of V, and so also S is on V. This concludes the proof of Theorem with the hypothesis (A).

Next we are going to prove Theorem with the hypothesis (B).

To this purpose, fix an irreducible hypersurface $F \subset \mathbf{P}^4$ of degree $n < k/2$, passing through $\mathrm{Sing}(V)$, and such that $F \cap V$ and $F \cap S$ are irreducible. Such a hypersurface F exists by the hypothesis (B). Denote by D the reduced subscheme of the scheme-theoretic intersection $F \cap S$. Using the same argument as in the proof of (2), we have

$$p_a(D) - p_g(D) \geq \sum_{i=1}^{+\infty} m_i(i^2-1)/4,$$

where $p_g(D)$ (resp. $p_a(D)$) denotes the geometric genus (resp. the arithmetic genus) of D. Notice that $p_g(D)$ is equal to the geometric genus of the section of S with a general hypersurface of degree n, and so

$$p_g(D) = [n^2 d + n(2g-2-d) + 2]/2.$$

On the other hand, since by construction D is an irreducible curve in $\mathbf{P}^4$ verifying a flag condition of type (nd, nk, n) [CCD], we also have (assume $k > n-1$)

$$p_a(D) \leq G(nd, nk, n) := nd^2/2nk + nd(n+k-5)/2 + \rho + 1,$$

where ρ is defined as follows. Write

$$n\nu = uk + v, \quad 0 \leq v \leq k-1.$$

Put $A = k(n-1)(n+k)$, $B = (n-1)k$, $C = nk-1$, $D = nk(n-1)(n+k-1)$ and $E = nk(n+k-2)$. Then we have

$$\rho = (Au^2 + 2Buv + Cv^2 - Du - Ev)/2nk$$

(see [CCD, p.132, *Proof of the bound of Theorem*] for the definition of $G(nd, nk, n)$, and do not use the explicit definition given in [CCD, p.120] because there is a misprint there). Summing up we obtain the following estimate

$$2k(G-g) + \nu(k-\nu)(k-1) \geq -2k\rho/n + 2k[\sum_{i=1}^{+\infty} m_i(i^2-1)/4]/n. \qquad (4)$$

Now we notice that

$$2k\rho/n \leq 0. \qquad (5)$$

In fact, a direct computation proves that, in the u v plane, the locus $\rho = 0$ is a real ellipse Γ, and the vertices of the rectangle

$$0 \leq u \leq n-1, \quad 0 \leq v \leq k-1,$$

to which u and v belong, lie either on Γ or in its internal region. Taking into account our hypothesis (B), from (1), (4) and (5) we obtain

$$\sum_{i=1}^{+\infty} i^2 m_i \leq k\delta(S)/(k-2n) < (k-1)^2.$$

As from (3), we deduce that S is a complete intersection on V. This concludes the proof of Theorem with the hypothesis (B). *Q.e.d. for Theorem.* □

Now we are going to prove Proposition. We keep all the notation introduced in the proof of Theorem.

We begin by examining the case $\delta < (k-1)^2$. To this purpose, by the Lemma, it suffices to prove that any smooth surface S of degree $d \gg 0$, contained in V is a complete intersection on V. Since S is smooth, then formula (1) becomes

$$\delta(S) = 2k(G-g) + \nu(k-\nu)(k-1). \qquad (6)$$

By our numerical hypotheses on δ, it follows that $\nu(k-\nu)(k-1) < (k-1)^2$, and so, as from (3), we deduce that S is a complete intersection on V.

Now we examine the case $(k-1)^2 \leq \delta < 2(k-1)(k-2)$ (hence $k \geq 4$). Let $S \subset V$ be a smooth and irreducible surface which is not a complete intersection of degree $d \gg 0$. By using (6) we get $\nu(k-\nu)(k-1) < 2(k-1)(k-2)$. If $\nu(k-\nu)(k-1) < (k-1)^2$ then $\nu = 0$ and so $G - g \geq k-2$ [E, loc. cit.]. By (6) we get $2k(k-2) < 2(k-1)(k-2)$, which is not possible. Hence $(k-1)^2 \leq \nu(k-\nu)(k-1) < 2(k-1)(k-2)$, from which we obtain $\nu(k-\nu) < 2(k-2)$, with $1 \leq \nu \leq k-1$. Therefore either $\nu = 1$ or $\nu = k-1$. By the classification of the curves with maximal genus [H], it follows that if V does not contain any plane then $g < G$. By [E, IV.2, IV.3, VI.2 and VI.4] we deduce $G - g \geq k-3$ and then by (6) we get $2k(k-3) + (k-1)^2 < 2(k-1)(k-2)$, which is not possible.

In order to prove the assertion concerning the group $\mathrm{Cl}_0(V)$ in the range $(k-1)^2 \leq \delta < 2(k-1)(k-2)$, suppose that V contains a plane π. The nodes of V which belong to π are the complete intersection of two plane curves of π of degree $k-1$. Therefore, if V contains another plane σ, the intersection of π with σ contains at most $k-1$ nodes of V. It follows that $2(k-1)^2 - (k-1) < 2(k-1)(k-2)$, which is not possible. Hence π is unique. Since π is unique, the argument above shows also that $\mathrm{Cl}_0(V)$ is generated by π and by the general hyperplane section H_V of V. Moreover π and H_V freely generate $\mathrm{Cl}_0(V)$, otherwise $m\pi$ would be a complete intersection on V, for some $m \geq 2$. This is in contrast to the fact that the singular locus of V is finite. If V also contains a quadric q, then q is irreducible and contains all the nodes of V (see below). It follows that the nodes contained in π belong to a conic of π. This is impossible, except when $k = 3$.

Now we analyze the case $\delta = 2(k-1)(k-2)$. Let $S \subset V$ be a smooth surface of degree $d \gg 0$, which is not a complete intersection. As before, we see that if $2k(G-g)+\nu(k-\nu)(k-1) < 2(k-1)(k-2)$ then V contains a plane. Then if V does not contain any plane, we have $2k(G-g)+\nu(k-\nu)(k-1) = 2(k-1)(k-2)$. It follows that $\nu(k-\nu)(k-1) \leq 2(k-1)(k-2)$, i.e. $\nu(k-\nu) \leq 2(k-2)$. As before, one sees that the strict inequality $\nu(k-\nu) < 2(k-2)$ leads to a contradiction. Hence $\nu(k-\nu) = 2(k-2)$ and so $\nu = 2$ or $\nu = k-2$, and $g = G$. By [H] it follows that the general hyperplane section of V contains a conic, which lifts to a quadric on V.

As in the previous range, one sees that, when $\delta = 2(k-1)(k-2)$, if V contains some plane π, then π is unique, V does not contain any quadric (except when $k = 3$), and the subgroup $\mathrm{Cl}_0(V)$ is freely generated by π and by the general hyperplane section of V.

Finally suppose that $\delta = 2(k-1)(k-2)$ and that V contains a quadric q (we may assume $k \geq 4$). Then q is irreducible. Let (l, g) be the homogeneous ideal of q in $\mathbf{P}^4$, with $\deg(l) = 1$ and $\deg(g) = 2$. If $f = 0$ is the equation defining V, then we have $f = Al + Bg$, for suitable polynomials A and B of degree $k-1$ and $k-2$. The zero-dimensional scheme Z defined by l, g, A and B has degree δ and its points are nodes of V. Since the nodes of V are ordinary double points, then Z is reduced and equal to the set of nodes of V. It follows that q contains all the nodes of V and that the nodes are the complete intersection of the hypersurfaces defined by l, g, A and B. Hence q

is unique, except when $k = 4$. At this point, in a similar way as in the case that V contains a plane, one proves that $\mathrm{Cl}_0(V)$ is freely generated by q and by the general hyperplane section of V.

This conclues the proof of Proposition. *Q.e.d. for Proposition.* □

Acknowledgement. The first author wishes to thank A. Corti for having pointed out the problem of proving that hypersurfaces in $\mathbf{P}^4$ with few nodes, are $\mathbf{Q}$-factorial.

References

[CCD] Chiantini, L., Ciliberto, C., Di Gennaro, V.: The genus of curves in $\mathbf{P}^4$ verifying certain flag conditions. Manuscripta Math. **88** 119–134 (1995)

[E] Ellia, P.: Sur les lacunes d'Halphen. In: *Algebraic Curves and Projective Geometry: Proceedings, Trento*, 1988, Lecture Notes in Math., vol. 1389, pp. 43–65, Springer Berlin

[GM] Geramita, A.V., Maroscia, P.: The ideal of forms vanishing at a finite set of points in $\mathbf{P}^n$. J. Algebra **90/2** 528–555 (1984)

[EP] Ellingsrud, G., Peskine, C.: Sur les surfaces lisses de $\mathbf{P}_4$. Invent. math. **95** 1–11 (1989)

[H] Harris, J.: The genus of space curves. Math. Ann. **249** 191–204 (1980)

[MP] Miyaoka, Y., Peternell, T.: *Geometry of Higher Dimensional Algebraic Varieties*, DMV Seminar, Bd. **26** (Birkhäuser, Basel 1997)

[S] Severi, F.: Una proprietà delle forme algebriche prive di punti multipli. Rendic. Accad. Lincei, s.5, 2^0 semestre, fasc. II **15** 691–696 (1906)

Boundedness
For Low Codimensional Subvarieties

Ciro Ciliberto and Vincenzo Di Gennaro

Università di Roma Tor Vergata, Dipartimento di Matematica,
Via della Ricerca Scientifica, 00133 Roma, Italia.
cilibert@axp.mat.uniroma2.it, digennar@axp.mat.uniroma2.it

Abstract. We prove that for certain projective varieties (e.g. smooth complete intersections in projective space), there are only finitely many components of the Hilbert scheme parametrizing irreducible, smooth, projective subvarieties of low codimension and not of general type. We give similar results concerning subvarieties with globally generated tangent bundle.

Introduction

In [6] Schneider proved that, when $n \geq (m+2)/2$, there are only finitely many components of the Hilbert scheme parametrizing smooth subvarieties of dimension n in $\mathbf{P}^m$ not of general type. De Cataldo [3] extended Schneider's result to the case of smooth quadrics of dimension m. In the present article we offer a further extension, which we now will state.

Let V be an irreducible, smooth, projective variety of dimension m, over $\mathbf{C}$. Let $\mathcal{S}$ be a set of projective subvarieties of V. We will say that $\mathcal{S}$ is *bounded* if there is a closed immersion $V \subset \mathbf{P}^r$ such that

$$\sup\{\deg(X) : X \in \mathcal{S}\} < +\infty.$$

This means that the varieties in $\mathcal{S}$ belong to finitely many components of the Hilbert scheme. In particular, this definition does not depend on the closed immersion.

Refining the proof of Schneider [6], in this paper we will prove the following:

Theorem. *Let $V \subset \mathbf{P}^r$ be an irreducible, smooth, projective variety of dimension m. Let $n \geq (m+2)/2$ be an integer and put $k = m-n$. For $i \in \{1, k, k+1\}$, assume that any algebraic class in $H^{2i}(V, \mathbf{C})$ is a multiple of H_V^i, where H_V*

2000 *Mathematics Subject Classification.* Primary 14C05, 14M07, 14M10; Secondary 14J99, 14K99.

is the hyperplane section of V. Then the set of irreducible, smooth, projective, n-dimensional subvarieties of V not of general type is bounded.

By the Lefschetz Hyperplane Theorem and Noether–Lefschetz Theorem [1, Corollary, p. 179] we obtain the following

Corollary. *Let $V \subset \mathbf{P}^r$ be an irreducible, smooth, complete intersection of dimension m. Let $n \geq (m+2)/2$ be an integer. Only when $n = (m+2)/2$, assume that V is general in the sense of Noether–Lefschetz, with $h^{a,m-a}(V) \neq 0$ for some $a < m/2$. Then the set of irreducible, smooth, projective, n-dimensional subvarieties of V not of general type is bounded.*

Notice that for any smooth hypersurface $V \subset \mathbf{P}^{m+1}$ of dimension $m = 2(n-1) \geq 4$ and degree ≥ 3, one has $h^{n-2,n}(V) \neq 0$ [5, p. 173].

Next we turn to the cases $n = m/2$ and $n = (m+1)/2$ and we prove some partial results concerning abelian subvarieties (see Corollary 1 and 2 below), and subvarieties with generated tangent bundle (Corollary 3). Notice that all these subvarieties are not of general type, because their first Chern class is effective. It turns out that, in certain cases, the set of such subvarieties is empty (compare with [7] and [8]).

Notation

Let $Y \subset \mathbf{P}^r$ be any smooth, irreducible, projective variety over $\mathbf{C}$. We will denote by H_Y the hyperplane section of Y, by T_Y the tangent bundle, and by K_Y a canonical divisor. If $\mathcal{E}$ is any sheaf on Y we denote by $c_i(\mathcal{E})$ its Chern classes and by $s_i(\mathcal{E})$ its Segre classes. If $Z \subset Y$ is a subvariety, we denote by $N_{Z,Y}$ the normal sheaf of Z in Y.

If $\mathcal{S}$ is a set and $f : \mathcal{S} \to \mathbf{N}$ is a numerical function, we say that a function $g : \mathcal{S} \to \mathbf{N}$ is $O(f)$, and we write $g = O(f)$, if $|g(\xi)| \leq Cf(\xi)$ for all $\xi \in \mathcal{S}$, where C is a constant.

The Proof Of Theorem

We begin by proving the Theorem stated in the Introduction. To this purpose, let $X \subset V$ be a smooth subvariety of dimension n, of degree d. Taking into account our hypothesis on the cohomology of the ambient variety V, and using a Barth-Lefschetz type of argument (see [2, Theorem 1.1 and proof]), one has

$$K_X = eH_X$$

in $H^2(X, \mathbf{C})$, for some rational number e. Notice that if X is not of general type then $e \leq 0$.

Now let q and γ be rational numbers such that $K_V = qH_V$ in $H^2(V, \mathbf{C})$ and $T_V(\gamma)$ is globally generated. Then also $N_{X,V}(\gamma)$ is. Since

$$c_1(N_{X,V}(\gamma)) = (k\gamma + e - q)H_X, \tag{1}$$

by [4] we deduce $0 \leq k\gamma + e - q \leq k\gamma - q$. Notice that taking $\gamma \gg 0$ we may assume $1 \leq k\gamma + e - q$. Therefore we get

$$1 \leq k\gamma + e - q = O(1). \tag{2}$$

We need the following lemma. We will prove it below.

Lemma. *With the same assumptions as before, for any* $i = 0, \ldots, k$ *one has*

$$(\gamma H_X)^{k-i} c_i(N_{X,V}) c_1(N_{X,V}(\gamma))^{n-k} = O(d).$$

Using Lemma for $i = k$, (1) and the self-intersection formula

$$c_k(N_{X,V}) = (d/t)H_X^k,$$

($t =$ degree of V) we obtain

$$(d/t)(k\gamma + e - q)^{n-k} H_X^n = O(d).$$

By (2) it follows that

$$d^2/t = O(d),$$

whence the boundedness for d. This concludes the proof of Theorem. *Q.e.d. for Theorem.* □

Now we are going to prove Lemma. We argue by induction on i, the case $i = 0$ being true by (1) and (2). Assume then $1 \leq i \leq k$. From the formula

$$c_i(N_{X,V}(\gamma)) = \sum_{h=0}^{i} \binom{k-h}{i-h} (\gamma H_X)^{i-h} c_h(N_{X,V}),$$

intersecting with $(\gamma H_X)^{k-i} c_1(N_{X,V}(\gamma))^{n-k}$ and using induction, we get

$$\begin{aligned}(\gamma H_X)^{k-i} & c_i(N_{X,V}) c_1(N_{X,V}(\gamma))^{n-k} \\ &= (\gamma H_X)^{k-i} c_i(N_{X,V}(\gamma)) c_1(N_{X,V}(\gamma))^{n-k} + O(d).\end{aligned}$$

Hence in order to prove Lemma, it suffices to prove that

$$(\gamma H_X)^{k-i} c_i(N_{X,V}(\gamma)) c_1(N_{X,V}(\gamma))^{n-k} = O(d). \tag{3}$$

To this purpose, first notice that since $N_{X,V}(\gamma)$ is globally generated then by [4] we have

$$0 \leq (\gamma H_X)^{k-i} c_i(N_{X,V}(\gamma)) c_1(N_{X,V}(\gamma))^{n-k}. \tag{4}$$

On the other hand one has (use [4] and [6, proof of Proposition])

$$c_i(N_{X,V}(\gamma))c_1(N_{X,V}(\gamma))^{n-i} \leq c_1(N_{X,V}(\gamma))^n,$$

from which, by (1) and (2), we deduce

$$\begin{aligned}
&(\gamma H_X)^{k-i}c_i(N_{X,V}(\gamma))c_1(N_{X,V}(\gamma))^{n-k} \\
&= c_i(N_{X,V}(\gamma))\gamma^{k-i}(k\gamma + e - q)^{n-k}H_X^{n-i} \\
&= \gamma^{k-i}(k\gamma + e - q)^{i-k}c_i(N_{X,V}(\gamma))(k\gamma + e - q)^{n-i}H_X^{n-i} \\
&= \gamma^{k-i}(k\gamma + e - q)^{i-k}c_i(N_{X,V}(\gamma))c_1(N_{X,V}(\gamma))^{n-i} \\
&\leq \gamma^{k-i}(k\gamma + e - q)^{i-k}c_1(N_{X,V}(\gamma))^n \\
&= \gamma^{k-i}(k\gamma + e - q)^{n+i-k}H_X^n = O(d).
\end{aligned}$$

Taking into account (4), it follows (3). This concludes the proof of Lemma. *Q.e.d. for Lemma.* □

Some Partial Results In the Cases $n = m/2$ and $n = (m+1)/2$

Now we are going to state some partial results in the cases $n = m/2$ and $n = (m+1)/2$. We need the following

Proposition. *Let $V \subset \mathbf{P}^r$ be an irreducible, smooth, projective variety of dimension m and degree t. Let n be an integer with $n < m \leq 2n$ and put $k = m - n$. Assume that at least one of the following properties holds.*

(A) *any algebraic cycle in $H^{2k}(V, \mathbf{C})$ is a multiple of H_V^k, where H_V is the hyperplane section of V;*

(B) *k is even and $H^{2i}(V, \mathbf{C}) = \mathbf{C}$ for any $i = 1, \ldots, k-1$.*

Let $X \subset V$ be a smooth subvariety of dimension n and degree d. Denote by $c_i(T_V)$ the Chern classes of V and by $s_i(T_X)$ the Segre classes of X. Then we have

$$d^2/t \leq [\sum_{i=0}^{k} s_i(T_X) \cdot c_{k-i}(T_V)_{|X}] \cdot H_X^{n-k}. \tag{5}$$

Proof. Let $L \subset \mathbf{P}^r$ be a general linear subspace of dimension $r - n + k$ and put $X' = X \cap L$ and $V' = V \cap L$. We have $\dim(X') = k$ and $\dim(V') = 2k$. From the normal bundle sequence

$$0 \to T_{X'} \to T_{V'} \otimes \mathcal{O}_{X'} \to N_{X',V'} \to 0$$

we obtain

$$c_k(N_{X',V'}) = [\sum_{i=0}^{k} s_i(T_X) \cdot c_{k-i}(T_V)_{|X}] \cdot H_X^{n-k}. \tag{6}$$

In the hypotheses (A), using the self-intersection formula we get $c_k(N_{X',V'}) = d^2/t$ and so, in this case, we deduce (5) as an equality.

In the hypotheses (B), the Lefschetz Hyperplane Theorem and the Hodge–Riemann bilinear relations imply that the intersection form on $H^{2k}(V', \mathbf{R})$ is positive definite. Hence we have

$$(X' - (d/t)H^k_{V'})^2 \geq 0.$$

Using the self-intersection formula we get

$$c_k(N_{X',V'}) = (X')^2 \geq d^2/t$$

from which, taking into account (6), we obtain (5). *Q.e.d. for Proposition.* □

As a consequence of the previous proposition, we obtain the following boundedness result for abelian subvarieties of a complete intersection.

Corollary 1. *Let $V \subset \mathbf{P}^r$ be an irreducible, smooth, complete intersection of dimension m. Let $n \geq m/2$ be an integer. When $n = m/2$ and n is odd, assume that V is general in the sense of Noether–Lefschetz, with $h^{a,m-a}(V) \neq 0$ for some $a < m/2$ (e.g., any general hypersurface of even dimension $m \geq 4$ and degree ≥ 3). Denote by t the degree of V and let η be the integer such that $c_{n-m}(T_V) = \eta H_V^{n-m}$. Let $X \subset V$ be an abelian subvariety of dimension n and degree d. Then we have*

$$d \leq \eta t. \tag{7}$$

The equality occurs except when $n = m/2$ and n is even.

Proof. By the Lefschetz Hyperplane Theorem and by the Noether–Lefschetz Theorem [1], we may apply the previous proposition to V and so, taking into account that all the Chern (and Segre) classes of an abelian variety vanish in codimension > 0, we get (7) from (5). *Q.e.d. for Corollary 1.* □

By making formula (7) explicit, in the case V is a hypersurface we deduce a more precise result.

Corollary 2. *Let $V \subset \mathbf{P}^{m+1}$ be a smooth hypersurface of dimension $m = 2n \geq 6$, and degree t. When n is odd, assume that V is general in the sense of Noether–Lefschetz, with $h^{a,m-a}(V) \neq 0$ for some $a < m/2$ (e.g., any general hypersurface of even dimension $m \geq 4$ and degree ≥ 3). Let $X \subset V$ be an abelian subvariety of V of dimension n and degree d. Then*

$$n!(2n+2) \leq d \leq \sum_{i=0}^{n} \binom{2n+2}{n-i} (-1)^i t^{i+1} = f(n,t).$$

Remark. In particular, when $f(n,t) < n!(2n+2)$, there is no abelian subvariety of V of dimension n. This is the case when n is even and $t \ll n$ (e.g., $t \leq 3$ for $n = 4$, $t \leq 5$ for $n = 6$, $t \leq 6$ for $n = 8$, etc.), and when n is odd and $t \gg n$ (e.g., $t \geq 3$ for $n = 3$ and $n = 5$, $t \geq 2n+2$ for any $n \geq 3$ odd).

Proof of Corollary 2. With the same notation as in Corollary 1, when V is a hypersurface then $\eta t = f(n,t)$. On the other hand, by the Riemann–Roch Theorem, we have $d = n!h^0(X, \mathcal{O}_X(1))$ and, by a theorem of Van de Ven [7], if $n \geq 3$, then $h^0(X, \mathcal{O}_X(1)) \geq 2n+2$. *Q.e.d. for Corollary 2.* □

We conclude with the following result concerning subvarieties with globally generated tangent bundle.

Corollary 3. *Let $V \subset \mathbf{P}^{m+1}$ be a smooth hypersurface of dimension m, of degree $t \geq m+2$. Let $n \geq m/2$ be an integer such that $k = m - n$ is odd. When $n = m/2$, assume that V is general in the sense of Noether–Lefschetz. Then there is no smooth subvariety of V of dimension n and globally generated tangent bundle.*

Proof. Put $c_i(T_V) = \eta_i H_V^i$, $\eta_i \in \mathbf{Z}$. Since $t \geq m+2$, then $(-1)^i \eta_i \geq 0$ for any $i = 0, \ldots, k$. Now let $X \subset V$ be a subvariety of dimension n, degree d, and globally generated tangent bundle. From [4] we have $s_i(T_X) \cdot c_{k-i}(T_V)_{|X} \leq 0$ for any $i = 0, \ldots, k$. From (5) we deduce $d \leq 0$, which is not possible. *Q.e.d. for Corollary 3.* □

References

1. Carlson, J., Green, M., Griffiths, P., Harris, J.: Infinitesimal variations of Hodge structure (I). Compositio Math. **50** 109–205 (1983)
2. Ciliberto, C., Di Gennaro, V.: Boundedness for codimension two subvarieties. Internat. J. Math. **13**(5) 479–495 (2002)
3. De Cataldo, M.A.: A finiteness theorem for low codimensional nonsingular subvarieties of quadrics. Trans. Amer. Math. Soc. **6**(349) 2359–2370 (1997)
4. Fulton, W., Lazarsfeld, R.: Positive polynomials for ample vector bundles. Annals of Math. **118** 35–60 (1983)
5. Parshin, A.N., Shafarevich, I.R. (eds.): *Algebraic Geometry III*, Encyclopaedia of Math. Sciences, vol. 36 (Springer, Berlin, Heidelberg 1998)
6. Schneider, M.: Boundedness of low codimensional submanifolds of projective space. Intern. J. of Math. **3**(3) 397–399 (1992)
7. Van de Ven, A.: On the embedding of abelian varieties in projective spaces. Ann. Mat. Pura Appl. **4**(103) 127–129 (1975)
8. Voisin, C.: On a conjecture of Clemens on rational curves on hypersurfaces. J. Differential Geometry **44** 200–214 (1996)

Normality and Non Normality Of Certain Semigroups and Orbit Closures

Corrado De Concini

Dipartimento di Matematica "G. Castelnuovo", Università di Roma "La Sapienza" Piazzale A. Moro, 00100, Roma, Italia. deconcin@mat.uniroma1.it

Abstract. Given a representation $\rho : G \to \mathrm{GL}(N)$ of a semisimple group G, we discuss the normality or non normality of the cone over $\rho(G)$ using the wonderful compactification of the adjoint quotient of G and its projective normality [K]. These methods are then used to discuss the normality or non normality of certain other orbit closures including determinantal varieties.

1 Introduction

Given a finite dimensional representation $\rho : G \to \mathrm{GL}(V)$, where we assume that G is an algebraic group defined over an algebraically closed field of characteristic 0 and that ρ is rational, a natural object to consider is the cone Z_V over $\rho(G)$, i.e., the closure of the linear transformations in $\mathrm{End}(V)$ which are multiples of some $\rho(g)$ for $g \in G$. It is easy to see that Z_V is a semigroup and one may ask about geometric properties of this semigroup. In this paper we are going to analyze the normality of Z_V under the assumption that G is semisimple and that V is an highest weight module. In particular, we are going to perform this discussion in the case in which V is irreducible.

It turns out that in this case, Z_V is almost never normal. Indeed, we prove that it is normal if and only if V is a minusculrepresentation (for the definition see Section 3). On the other hand we are able in all cases to exhibit the normalization of Z_V, which turns out to be of the form Z_W for "the largest" G-module having the same highest weight as V, that is W is the sum of all irreducible modules whose highest weight is less than or equal to that of V in the dominant ordering. After this is achieved, we make the observation that our methods can be used to analyze a much larger class of orbit closures which include, among others, various varieties of matrices satisfying rank conditions, in particular determinantal varieties. Thus we slightly generalize our methods to treat this case as well.

We now briefly explain how our results are obtained. The main observation is that the coordinate ring of Z_V is also the homogeneous coordinate ring of the associated projective variety. This projective variety turns out to be the image under a suitable morphism of a certain completion of a homogeneous space (in the case of the semigroups over G, this is the wonderful compactification of G introduced and studied in [DCP]). Thus results about such compactifications can be used. In particular we crucially use their projective normality recently proved by Kannan in [K], not only directly, but also as the main idea behind our arguments.

To finish I would like to stress that many of our results are contained or implicit in the work of others. Affine semigroups for semisimple or reductive groups have been considered by various authors (see for example [Re1], [Re2], [RP], [Ri], [V]). Some of the questions considered below have been addressed in [F] and results closely related to those of this paper have been obtained in [T]. In particular a great deal of the content of our Theorem 3.1 can be obtained using [V] together with [K]. Under the normality assumption, the property of having rational singularities has been proved in similar terms in [Ri]. Nonetheless, we have decided to give complete proofs, including those of such known facts as the projective normality result of Kannan.

Our paper is divided as follows. After a section giving our notations, we discuss the normality of semigroups in Section 3. In Section 4 the generalization to other orbit closures is explained. In order to obtain this result, we need to slightly generalize the theory of wonderful completions of an adjoint group. Since the proofs of many assertions are essentially identical to those of the corresponding assertions in the case of group compactifications, they are sometimes omitted. Finally in the last section we discuss various examples.

2 Notation

Let k denote an algebraically closed field of characteristic 0. Let G be a semisimple simply connected algebraic group over k. Choose a maximal torus T and a Borel subgroup $B \supset T$ in G, and let $W = N(T)/T$ be the Weyl group of (G, T).

Consider the character group $P := X(T)$ of T, and let $\Delta \subset P$ denote the roots of (G, T) with Δ_+ the positive roots relative to B. Similarly consider the dual lattice $\check{P}$, and the set of coroots $\check{\Delta}$ with positive coroots $\check{\Delta}_+$. Set P_+ equal to the semigroup of dominant weights, the weights λ such that $\langle \lambda, \check{\alpha} \rangle \geq 0$ for each positive coroot $\check{\alpha}$. Here $\langle\ ,\ \rangle$ is the canonical W-invariant inner product on $P \otimes \mathbb{R}$. P_+ is a fundamental domain for the action of the Weyl group W on P.

Let us order P by setting $\lambda \geq \mu$ if $\lambda - \mu$ is a positive linear combination of positive roots. This order is called the dominant order.

Recall that the set of isomorphism classes of finite dimensional irreducible representations of G is in bijection with P_+, the bijection being defined as follows. Any such representation contains a unique line preserved by B called the highest weight line. Given a non zero vector v in such a line, then T acts on

v by multiplication by a character $\lambda \in P_+$. The "unique" (up to isomorphism) irreducible module corresponding to λ will be denoted by V_λ.

P can be identified with the Picard group of G/B. We fix this identification in such a way that, if $\lambda \in P_+$, the line bundle $\mathcal{L}_\lambda$ corresponding to λ is such that

$$H^0(G/B, \mathcal{L}_\lambda) \equiv V_\lambda.$$

Let us recall a few facts about the weight structure of V_λ. We say that a weight $\mu \in P$ appears in a representation V if there is a non zero T eigenvector $v \in V$ of weight μ. One has:

(1) μ appears in V if and only if $w\mu$ does for every $w \in W$.
(2) If μ appears in V_λ, then $\mu \leq \lambda$.
(3) μ appears in V_λ if and only if the unique dominant weight of the form $\nu = w\mu$ satisfies $\nu \leq \lambda$.

We shall say that a finite dimensional G-module U has highest weight $\lambda \in P_+$, if it has a non zero vector of weight λ and each weight μ which appears in U satisfies $\mu \leq \lambda$.

Recall that a dominant weight λ is called minuscule if it satisfies one of the following

(1) $\langle \lambda, \check{\alpha} \rangle \leq 1$ for all positive coroots $\check{\alpha}$.
(2) If μ is dominant and $\mu \leq \lambda$, then $\mu = \lambda$.
(3) A weight appears in V_λ if and only if it is in the W-orbit of λ.

Given now $\lambda \in P_+$, we define its saturation as the set

$$\Sigma(\lambda) = \{\mu \in P_+ \mid \mu \leq \lambda\}.$$

It is well known that λ is minuscule if and only if $\Sigma(\lambda) = \{\lambda\}$.

3 Some Semigroups

Given a G-module V, denote by $I \in \mathrm{End}(V)$ the identity map. Consider the morphism $\gamma : G \times G_m \to \mathrm{End}(V)$ defined by $\gamma((g, z)) = gzI = zgI$. The image of this morphism is the set of G translates of the homotheties.

Notice that clearly the image of γ lies in $\mathrm{GL}(V)$ and as a morphism

$$\gamma : G \times G_m \to \mathrm{GL}(V),$$

γ is a group homomorphism. Furthermore $\gamma(G \times G_m)$ is stable under the action of $G \times G$ by left and right multiplication.

We set Z_V equal to the closure of $\gamma(G \times G_m)$ in $\mathrm{End}(V)$. Note that, by continuity, Z_V is closed under composition, i.e., it is a semigroup, and it is stable under the action of $G \times G$ by left and right multiplication. The main goal of this paper will be to discuss the normality of these semigroups for certain representations V of G. First some notation. Fix a dominant weight $\lambda \in P_+$ and let $\Sigma(\lambda)$ be its saturation. Define $W_\lambda = \oplus_{\mu \in \Sigma_\lambda} V_\mu$.

We set

$$Z_\lambda := Z_{V_\lambda} \quad \text{and} \quad \mathcal{Z}_\lambda := Z_{W_\lambda}.$$

We can now state:

Theorem 3.1. 1) *$\mathcal{Z}_\lambda$ is a normal variety with rational singularities.*

2) *If V is a G-module of highest weight λ, then $\mathcal{Z}_\lambda$ is the normalization of Z_V, and it is equal to Z_V if and only if W_λ is a subrepresentation of V. In particular $\mathcal{Z}_\lambda$ is the normalization of Z_λ and it is equal to Z_λ if and only if λ is minuscule.*

The proof is an application of some of the results in [DCP], so before giving it, let us recall these facts.

In [DCP] one studies the following variety. Consider $\omega_1, \ldots, \omega_n$, the fundamental weights for (G, T). Let $p_i \in \mathbb{P}(\mathrm{End}(V_{\omega_i}))$ be the point representing the line spanned by the identity. Define X to be the closure in $\mathbb{P}(\mathrm{End}(V_{\omega_1})) \times \cdots \times \mathbb{P}(\mathrm{End}(V_{\omega_n}))$ of the orbit $G(p_1, \ldots, p_n)$. X is called the wonderful compactification of the adjoint group $\overline{G} = G/Z(G)$, and it has a number of very nice properties. Here we shall need some of them.

First of all X is a smooth $G \times G$-variety with open orbit $G(p_1, \ldots, p_n)$, the complement of which is a divisor with normal crossings and smooth irreducible components $D_1, \ldots, D_n$. Given a subset $I \subset \{1, \ldots, n\}$, the smooth subvariety $D_I := \cap_{i \in I} D_i$ is the closure of the $G \times G$-orbit $\mathcal{O}_I = D_I - \cup_{J \supsetneq I} D_J$ ($D_\emptyset = X$) and each $G \times G$-orbit equals one of the $\mathcal{O}_I$. In particular X contains a unique closed $G \times G$-orbit which can be seen to be isomorphic to $G/B \times G/B$.

One knows that the Picard group of $G/B \times G/B$ can be identified with $P \times P$, and one has that the homomorphism $\mathrm{Pic}(X) \to \mathrm{Pic}(G/B \times G/B)$ is an injection whose image is the lattice consisting of pairs of the form $(\lambda, -w_0(\lambda))$, w_0 being the longest element in W. Thus $\mathrm{Pic}(X)$ can be identified with P. Under this identification, the classes of the $\mathcal{O}(D_i)$ correspond to the simple roots α_i.

Furthermore take $\lambda \in P_+$ and U a G-module of highest weight λ.

Consider the point $p \in \mathbb{P}(\mathrm{End}(U))$ representing the line spanned by the identity. Set $X(U) = \overline{Gp}$. Then the obvious map $G(p_1, \ldots, p_n) \to X(U)$ given by $g(p_1, \ldots, p_n) \to gp$ extends to a morphism

$$\phi : X \to X(U) \to \mathbb{P}(\mathrm{End}(U))$$

(in fact if λ is regular ϕ gives an isomorphism of X onto $X(U)$).

Furthermore under the identification of $\mathrm{Pic}(X)$ with P, λ corresponds to the class of $\phi^*(\mathcal{O}(1))$, $\mathcal{O}(1)$ being the tautological line bundle on $\mathbb{P}(\mathrm{End}(U))$.

The above discussion obviously applies to both V_λ and W_λ. Furthermore it is clear that Z_λ is nothing else than the affine cone over $X(V_\lambda)$, while $\mathcal{Z}_\lambda$ is the affine cone over $X(W_\lambda)$. Thus the coordinate rings of Z_λ and $\mathcal{Z}_\lambda$ can be identified with graded $G \times G$ stable subrings of the ring

$$A := \oplus_n H^0(X, L_{n\lambda}),$$

L_μ being the line bundle on X corresponding to μ.

Let us take now for each simple root α_i the unique (up to a scalar) $G \times G$ invariant section of $H^0(X, L_{\alpha_i})$, whose set of zeroes is the divisor D_i. We can filter the ring

$$R := \oplus_{\mu \in P} H^0(X, L_\mu)$$

by the order of vanishing on the $D_i's$. We get a filtration $R = R_0 \supset R_1 \supset \cdots \supset R_m \supset \cdots$, where

$$R_i = \sum_{h_1+\cdots+h_n=i} s_1^{h_1} \cdots s_n^{h_n} R.$$

One then easily gets from [DCP]

Proposition 3.2. *The associated graded ring $GrR = \oplus_i R_i/R_{i+1}$, is isomorphic to the polynomial ring $C[x_1, \ldots, x_n]$, where x_i is the class of s_i in R_1/R_2 and*

$$C = \oplus_{\lambda \in P} H^0(G/B \times G/B, \mathcal{L}_{(\lambda, -w_0(\lambda))}),$$

$\mathcal{L}_{(\lambda,-w_0(\lambda))}$ being the line bundle on $G/B \times G/B$ corresponding to $(\lambda, -w_0(\lambda)) \in P \times P$.

Proof. Fix $\lambda \in P$. We have already seen that if we restrict the line bundle L_λ to $G/B \times G/B$, we get the line bundle $L_{(\lambda,-w_0(\lambda))}$. Also if we consider the restriction map

$$H^0(X, L_\lambda) \to H^0(G/B \times G/B, \mathcal{L}_{(\lambda,-w_0(\lambda))}),$$

we have that this is surjective with kernel $R_1(\lambda) = R_1 \cap H^0(X, L_\lambda)$.

Given two sequences $\underline{h} = \{h_1, \ldots, h_n\}$ and $\underline{k} = \{k_1, \ldots, k_n\}$, we shall say that $\underline{k} \geq \underline{h}$ if $k_i \geq h_i$ for each $i = 1, \ldots, n$. If we now fix such a sequence $\underline{h}$, we set $R_{\underline{h}}(\lambda)$ equal to the image of the map

$$H^0(X, L_{\lambda - \sum h_i\alpha_i}) \to H^0(X, L_\lambda)$$

given by multiplication by $s_1^{h_1} \cdots s_n^{h_n}$. Then by [DCP], we know that this map induces an isomorphism of $G \times G$-modules,

$$\psi_{\underline{h}}(\lambda) : R_{\underline{h}}(\lambda)/\sum_{\underline{k}>\underline{h}} R_{\underline{k}}(\lambda) \to H^0(X, L_{\lambda-\sum h_i\alpha_i})/R_1(\lambda - \sum h_i\alpha_i) \cong$$

$$\cong H^0(G/B \times G/B, \mathcal{L}_{(\lambda-\sum h_i\alpha_i, -w_0(\lambda-\sum h_i\alpha_i))}).$$

Since clearly

$$\mathrm{Gr}R = \oplus_{\lambda \in P, \underline{h}} R_{\underline{h}}(\lambda)/\sum_{\underline{k}>\underline{h}} R_{\underline{k}}(\lambda),$$

we get the required isomorphism

$$\psi : \mathrm{Gr}R \to C[x_1, \ldots, x_n]$$

by setting $\psi(a) = \psi_{\underline{h}}(\lambda)(a)x_1^{h_1} \cdots x_n^{h_n}$ for $a \in R_{\underline{h}}(\lambda)/\sum_{\underline{k}>\underline{h}} R_{\underline{k}}(\lambda)$. $\square$

As an application of this proposition, we have

Proposition 3.3. *Let $\lambda \in P_+$. Then the ring*

$$A := \oplus_n H^0(X, L_{n\lambda})$$

is normal with rational singularities.

Proof. By a result of Kempf and Ramanathan [KR, Theorem 2], the ring

$$\oplus_{(\lambda,\mu)\in P\times P} H^0(G/B \times G/B, \mathcal{L}_{(\lambda,\mu)})$$

is normal with rational singularities. Now if we let $T \times T$ act on this ring by $(t_1, t_2)s = \lambda(t_1)\mu(t_2)s$, for $s \in H^0(G/B \times G/B, \mathcal{L}_{(\lambda,\mu)})$, then the ring

$$C = \oplus_{\lambda\in P} H^0(G/B \times G/B, \mathcal{L}_{(\lambda,-w_0(\lambda))}),$$

is the subring of invariants under the action of the subgroup $\Gamma \subset T\times T$ defined as the intersection of the kernels of the characters $(\lambda, -w_0(\lambda))$ as λ varies in P. Thus by [B], it is also normal with rational singularities. It follows that also the polynomial ring $\mathrm{Gr}R = C[x_1, \ldots, x_n]$ is normal with rational singularities. Using a result of Elkik [E, Theorem 4], we then deduce that R itself is normal with rational singularities.

At this point, let us act on R with T by $ts = \lambda(t)s$ if $s \in H^0(X, L_\lambda)$. Then A is the subring of invariants under the action of the subgroup $\mathrm{Ker}(\lambda)$. Thus again by [B], it is normal and with rational singularities. □

Let us now observe the following easy Lemma. Consider the coordinate ring $k[G]$ of G. One knows that as a $G \times G$-module, $k[G] = \oplus_{\lambda\in P_+} \mathrm{End}(V_\lambda)$. We have

Lemma 3.4. *Given $\lambda, \mu \in P_+$, we set $C_{\lambda,\mu} = \{\nu \in P_+ \mid (V_\lambda \otimes V_\mu \otimes V_\nu^*)^G \neq 0\}$. Then, in $k[G]$,*

$$\mathrm{End}(V_\lambda)\mathrm{End}(V_\mu) = \oplus_{\nu\in C_{\lambda,\mu}} \mathrm{End}(V_\nu).$$

Proof. If we consider $k[G]$ as a G-module with respect to the G action induced by left multiplication, then $\mathrm{End}(V_\lambda)$ is the isotypic component of the irreducible module V_λ. It follows that if we consider $\mathrm{End}(V_\lambda)\mathrm{End}(V_\mu)$ as a G-module, each of its irreducible components has to be isomorphic to an irreducible module V_ν with $\nu \in C_{\lambda,\mu}$.

The decomposition $k[G] = \oplus_{\lambda\in P_+} \mathrm{End}(V_\lambda)$ is a decomposition into distinct pairwise non isomorphic irreducible $G \times G$-modules and $\mathrm{End}(V_\lambda)\mathrm{End}(V_\mu)$ is stable under the action of $G \times G$. From this we deduce that there is a subset $C'_{\lambda,\mu}$ of $C_{\lambda,\mu}$ such that

$$\mathrm{End}(V_\lambda)\mathrm{End}(V_\mu) = \oplus_{\nu\in C'_{\lambda,\mu}} \mathrm{End}(V_\nu).$$

In particular

$$\mathrm{End}(V_\lambda)\mathrm{End}(V_\mu) \subseteq \oplus_{\nu\in C_{\lambda,\mu}}\mathrm{End}(V_\nu).$$

From these facts we see that in order to prove equality, it suffices to see that, as a G-module, $\mathrm{End}(V_\lambda)\mathrm{End}(V_\mu)$ contains a copy of V_ν for every $\nu \in C_{\lambda,\mu}$.

To see this, consider the open G-orbit in $G/B \times G/B$, i.e., the orbit of (B, B^-), B^- being the opposite Borel subgroup to B with respect to T. Let $\pi : G \to G/B \times G/B$ be the map defined by $\pi(g) = g(B, B^-) = (gBg^{-1}, gB^-g^{-1})$. Notice that π can be written as the composition of the diagonal embedding $G \to G\times G$ and of the $G\times G$ action. Take the line bundle $L_{(\lambda,\mu)}$ on $G/B\times G/B$ with the property that $H^0(G/B\times G/B, L_{(\lambda,\mu)}) = V_\lambda\otimes V_\mu$ as a G-module. Since G is simply connected, its Picard group is zero and we get an embedding π^* of $H^0(G/B \times G/B, L_{(\lambda,\mu)})$ into $k[G]$. Since π is G-equivariant, and factors through the diagonal $G \to G \times G$, we easily deduce that the image of π^* is a copy of $V_\lambda \otimes V_\mu$ contained in $\mathrm{End}(V_\lambda)\mathrm{End}(V_\mu)$. In particular, it follows from the definition of $C_{\lambda,\mu}$, that we can find a copy of V_ν in $\mathrm{End}(V_\lambda)\mathrm{End}(V_\mu)$ for every $\nu \in C_{\lambda,\mu}$, as desired. □

We now recall the PRV conjecture proved by Kumar [Ku] and Mathieu [Ma].

Theorem 3.5. *Given two dominant weights λ, μ, if a dominant weight ν is of the form $w\lambda + w'\mu$, for some $w, w' \in W$, then $\nu \in C_{\lambda,\mu}$.*

From this we deduce

Theorem 3.6. [K] *Let $\lambda, \mu \in P$. Then the multiplication map*

$$H^0(X, L_\lambda) \otimes H^0(X, L_\mu) \to H^0(X, L_{\lambda+\mu})$$

is surjective.

Proof. As in the above Lemma 3.4, we can restrict to the open G-orbit and embed $H^0(X, L_\lambda)$ into $k[G]$ as the $G \times G$-submodule $\oplus_{\gamma\in\Sigma(\lambda)}\mathrm{End}(V_\gamma)$. Thus, following all our identifications and Lemma 3.4, we are reduced to prove that, given $\nu \in P_+$ with $\nu \leq \lambda + \mu$ there exist $\lambda' \leq \lambda$ and $\mu' \leq \mu$ both dominant, such that $\nu \in C_{\lambda',\mu'}$.

To see this, recall that the natural map $V_\lambda \otimes V_\mu \to V_{\lambda+\mu}$ is surjective (since it is non zero, G-equivariant and $V_{\lambda+\mu}$ is irreducible). So each weight appearing in $V_{\lambda+\mu}$ can be written as a sum of a weight appearing in V_λ and of a weight appearing in V_μ. Now notice that since $\nu \leq \lambda+\mu$ and ν is dominant, for every $w \in W$, $w\nu \leq \nu \leq \lambda+\mu$. We deduce that ν appears in $V_{\lambda+\mu}$. By the above remark write

$$\nu = \lambda'' + \mu''$$

with λ'' appearing in V_λ and μ'' in V_μ. We now know that there exist w, w' such that $\lambda' = w\lambda''$ and $\mu' = w'\mu''$ are dominant and appear in V_λ and in V_μ respectively. The PRV conjecture then implies that $\nu \in C_{\lambda',\mu'}$ and we are done. □

Remark 3.7. Although the variety X is defined in arbitrary characteristic and both the computation of the Picard group of X and Proposition 3.2 hold (see [S1]), Theorem 3.6 is false in positive characteristic, contrary to the claim contained in the Appendix of [K].

The easiest example is the following. Let $\mathbb{F}$ be a field of characteristic 2. Let $G = \mathrm{SL}(4)$ and let $\lambda = \omega_2$ be the second fundamental weight.

Consider a matrix $Y = (y_{i,j})$, $1 \leq i, j \leq 4$, of indeterminates. Given two sequences $1 \leq i_1 < \cdots < i_h \leq 4$ and $1 \leq j_1 < \cdots < j_h \leq 4$ (of course $h \leq 4$), we denote by $[i_1, \ldots, i_h | j_1, \ldots, j_h]$ the determinant of the minor of Y formed by the rows $i_1, \ldots, i_h$ and the columns $j_1, \ldots, j_h$. We can then identify $H^0(X, L_{\omega_2})$ with the subspace of $\mathbb{F}[z_{i,j}]$ spanned by 2×2 minors of Y and $H^0(X, L_{2\omega_2})$ with the span of the polynomials $[s,t|s',t'][u,v|u',v']$ and $[s,t,u|s',t',u'][v|v']$ with $1 \leq s,t,u,v,s',t',u',v' \leq 4$. The image of $H^0(X, L_{\omega_2}) \otimes H^0(X, L_{\omega_2})$ in $H^0(X, L_{2\omega_2})$ is the the span of the polynomials $[s,t|s',t'][u,v|u'v']$. A direct computation (see [Br] section 4 and 5), shows that the polynomial $[2,3,4|2,3,4][1|1]$ does not belong to this image.

Notice that if we set $V = \bigwedge^2 \mathbb{F}^4$, then the coordinate ring of Z_{ω_2} is the subring of $\mathbb{F}[y_{i,j}]$ generated by $H^0(X, L_{\omega_2})$. In [Br] it is also shown that this ring is neither normal nor Cohen–Macaulay. So we get that also Theorem 3.1 does not hold in this case.

Using Theorem 3.6 we can now prove the first part of our Theorem 3.1, namely that $\mathcal{Z}_\lambda$ is a normal variety with rational singularities.

To see this let us see that $k[\mathcal{Z}_\lambda] = A$. Since both rings are graded and generated in degree 1, ($k[\mathcal{Z}_\lambda]$ by definition and A by the Theorem 3.6) and since $k[\mathcal{Z}_\lambda] \subseteq A$, it suffices to see that the two rings coincide in degree 1. Now $A_1 = H^0(X, L_\lambda) = \oplus_{\mu \in \Sigma(\lambda)} \mathrm{End}(V_\mu)$. On the other hand, the degree one part $k[\mathcal{Z}_\lambda]_1$ of $k[\mathcal{Z}_\lambda]$, is the $G \times G$-module spanned by the identity in $\mathrm{End}(W_\lambda)$. Since $W_\lambda = \oplus_{\mu \in \Sigma(\lambda)} V_\mu$, it is immediate that $k[\mathcal{Z}_\lambda]_1 = \oplus_{\mu \in \Sigma(\lambda)} \mathrm{End}(V_\mu)$, proving our claim.

It remains to prove the second part of Theorem 3.1. By our assumption on V, we have that there is a subset $\Omega \subset \Sigma(\lambda)$ containing λ and positive integers n_μ for $\mu \in \Omega$, such that

$$V \simeq \oplus_{\mu \in \Omega} V_\mu^{\oplus n_\mu}.$$

If we take the identity map $I \in \mathrm{End}(V)$, we then deduce that the $G \times G$ span of I is isomorphic to $\oplus_{\mu \in \Omega} \mathrm{End}(V_\mu)$. It follows that we can assume, without loss of generality, $n_\mu = 1$ for all $\mu \in \Omega$. Also we deduce that $Z_V \subset \oplus_{\mu \in \Omega} \mathrm{End}(V_\mu)$. In particular, as we have already seen above, we get that $\mathcal{Z}_\lambda \subset \oplus_{\mu \in \Sigma(\lambda)} \mathrm{End}(V_\mu)$. We thus have that the obvious $G \times G$-equivariant projections

$$\oplus_{\mu \in \Sigma(\lambda)} \mathrm{End}(V_\mu) \to \oplus_{\mu \in \Omega} \mathrm{End}(V_\mu) \to \mathrm{End}(V_\lambda)$$

restrict to dominant morphisms

$$\mathcal{Z}_\lambda \to Z_V \to Z_\lambda.$$

Thus we get inclusions $k[Z_\lambda] \subset k[Z_V] \subset k[\mathcal{Z}_\lambda]$.

Notice that $k[Z_V] = k[\mathcal{Z}_\lambda]$ if and only if $\Omega = \Sigma(\lambda)$, since only in this case do the two rings coincide in degree 1. Also using the above inclusions, it is clear that in order to show our claims, it suffices to see that $k[\mathcal{Z}_\lambda]$ and $k[Z_\lambda]$ have the same quotient field and that $k[\mathcal{Z}_\lambda]$ is integral over $k[Z_\lambda]$.

Let us see that $k[\mathcal{Z}_\lambda]$ and $k[Z_\lambda]$ have the same quotient field. Indeed both the representation of $G \times G_m$ in $Gl(V_\lambda)$ and of $G \times G_m$ in $Gl(W_\lambda)$, have as kernel the subgroup $S \subset Z(G) \times G_m$, $Z(G)$ being the center of G, consisting of those pairs $(z,t) \in Z(G) \times G_m$ for which $\lambda(z)t = 1$. This clearly implies that $\mathcal{Z}_\lambda$ and Z_λ contain the dense open set $G \times G_m/S$, hence they are birational.

It remains to see that $k[\mathcal{Z}_\lambda]$ is integral on $k[Z_\lambda]$. For this, it is clearly sufficient to show that the degree one part of $k[\mathcal{Z}_\lambda]$ is integral over $k[Z_\lambda]$.

Restricting to the open $G \times G$-orbit, we can identify, for each m, the degree m part $k[Z_\lambda]_m$ of $k[Z_\lambda]$ with the subspace of $k[G]$ spanned by the products $f_1 f_2 \cdots f_m$, with $f_i \in \mathrm{End}(V_\lambda)$. Using Lemma 3.4 we deduce that as a $G \times G$-module,

$$k[Z_\lambda]_m = \bigoplus_{\{\nu \mid \mathrm{Hom}_G(V_\nu, V_\lambda^{\otimes m}) \neq 0\}} \mathrm{End}(V_\nu).$$

From this we see that our claim will immediately follow from the following statement.

Proposition 3.8. (Cf. also [T] Lemma 1) *Let μ, λ be dominant weights. Assume that $\mu \leq \lambda$. Then there exists a positive integer H such that $V_{H\mu}$ is an irreducible component of $V_\lambda^{\otimes H}$.*

Proof. We know that μ is a convex linear combination of the weights $w\lambda$, as w varies in the Weyl group W. So let us write

$$\mu = a_1 w_1 \lambda + \cdots + a_m w_m \lambda$$

with $a_1 + a_2 + \cdots + a_m = 1$, $a_i > 0$, a_i rational. Let us make induction on m. If $m = 1$ there is nothing to prove, since the fact that λ is the unique dominant weight in its W-orbit clearly implies that $\mu = \lambda$. Assume $m \geq 2$. Write

$$\mu = a_1 w_1 \lambda + (1 - a_1)\Big(\frac{a_2}{1 - a_1} w_2 \lambda + \cdots + \frac{a_m}{1 - a_1} w_m \lambda\Big).$$

Now notice that there is a positive integer N such that

$$\begin{aligned}\tilde{\mu} &= N\Big(\frac{a_2}{1 - a_1} w_2 \lambda + \cdots + \frac{a_m}{1 - a_1} w_m \lambda\Big) \\ &= \frac{a_2}{1 - a_1} w_2 N\lambda + \cdots + \frac{a_m}{1 - a_1} w_m N\lambda \in P,\end{aligned}$$

is a convex combination of $m-1$ W-translates of $N\lambda$. Thus also every W-translate of $\tilde{\mu}$ has the same properties and, by the inductive hypothesis, if we denote by $\overline{\mu}$ the unique W-translate of $\tilde{\mu}$ which lies in P_+, we obtain that there exists a M such that $V_{M\overline{\mu}}$ is an irreducible component of $V_{N\lambda}^{\otimes M}$. But we know that $V_{N\lambda}^{\otimes M}$ is a summand of $V_\lambda^{\otimes MN}$, so that $V_{M\overline{\mu}}$ is an irreducible component of $V_\lambda^{\otimes MN}$. Let us now consider

$$NM\mu = a_1 w_1 NM\lambda + (1 - a_1)M\tilde{\mu}.$$

Since $0 < a_1 < 1$, we can find two positive integers s and r with $r < s$ and $a_1 = r/s$. Clearing denominators, we obtain that

$$sNM\mu = w_1 rNM\lambda + (s - r)M\tilde{\mu}.$$

Now $(s-r)M\tilde{\mu}$ is a W-translate of $(s-r)M\overline{\mu}$, so that applying the PRV conjecture, we deduce that $V_{sNM\mu}$ is an irreducible component of $V_{rNM\lambda} \otimes V_{(s-r)M\overline{\mu}}$. On the other hand $V_{rNM\lambda}$ is an irreducible component of $V_\lambda^{\otimes rNM}$ while $V_{(s-r)M\overline{\mu}}$ is an irreducible component of $V_{M\overline{\mu}}^{\otimes s-r}$ which in turn is a direct summand of $V_\lambda^{\otimes(s-r)NM}$. Thus $V_{sNM\mu}$ is an irreducible component of

$$V_\lambda^{\otimes rNM} \otimes V_\lambda^{\otimes(s-r)NM} = V_\lambda^{\otimes sNM}.$$

Setting $H = sNM$ we obtain our claim. □

Remark 3.9. In [DCP, section 4], it is shown that whenever λ is regular, then both $X(V_\lambda)$ and $X(W_\lambda)$ are isomorphic to X. This can be used to deduce directly that $k[\mathcal{Z}_\lambda]$ is integral on $k[Z_\lambda]$. Also we have that Z_λ contains the origin as its unique singular point and the natural projection

$$f : \mathcal{Z}_\lambda \to Z_\lambda$$

is bijective. But as we have proved in Theorem 3.1, it is an isomorphism only for $G = \mathrm{SL}(2)^n$ and $\lambda = (\omega_1, \dots, \omega_1)$, in which case Z_λ is the cone over the Segre embedding of $(\mathbb{P}^3)^n$.

4 A Generalization

As in the previous section, G will denote a semisimple simply connected algebraic group over k, $T \subset G$ a maximal torus and $B \subset G$ a Borel subgroup containing T. We shall continue to denote by $\overline{G}$ the adjoint quotient of G.

We now take another semisimple simply connected group $\mathcal{G}$, a maximal torus $\mathcal{T}$ in $\mathcal{G}$ and a Borel subgroup $\mathcal{B} \supset \mathcal{T}$ in $\mathcal{G}$. We shall set $\tilde{P} = X(\mathcal{T})$, the character group of $\mathcal{T}$. Let $\tilde{\Delta} \subset \tilde{P}$ denote the set of roots, and let $\tilde{\Delta}_+ \subset \Delta$ be the positive roots relative to $\mathcal{B}$. Finally, let $\tilde{P}_+$ be the semigroup of dominant weights.

We now assume that $\mathcal{G}$ contains a parabolic subgroup $\mathcal{P} \supset \mathcal{B}$ having the following property. If $S \subset \mathcal{P}$ denotes the solvable radical of $\mathcal{P}$, we have a surjective homomorphism $\pi : \mathcal{P}/S \to \overline{G} \times \overline{G}$ with finitekernel. Equivalently we assume that the semisimple Levi factor of $\mathcal{P}$ is isogenous to $G \times G$. Composing

π with the quotient homomorphism $\mathcal{P} \to \mathcal{P}/S$, we get a surjection $\pi' : \mathcal{P} \to \overline{G} \times \overline{G}$.

We set K equal to the preimage under π' of the diagonal subgroup in $\overline{G} \times \overline{G}$.

Notice that we get an action of $G \times G$ on S and a surjective homomorphism $\gamma : S \rtimes (G \times G) \to \mathcal{P}$ with finite kernel. Using γ, we can consider any $\mathcal{P}$-module and hence any $\mathcal{G}$-module as a $S \rtimes (G \times G)$-module .

Let us consider now the wonderful compactification X of $\overline{G}$ and define

$$\mathcal{Y} = \mathcal{G} \times_{\mathcal{P}} X.$$

We want to make a study of some of the properties of $\mathcal{Y}$. This study is in fact essentially identical to that of X, so that we shall only sketch the proofs of the various assertions. First of all notice that, since we have an obvious $\mathcal{G}$-equivariant fibration

$$p : \mathcal{Y} \to \mathcal{G}/\mathcal{P},$$

with fiber X, we immediately deduce that all $\mathcal{G}$-orbits in $\mathcal{Y}$ are of the form $\mathcal{G} \times_{\mathcal{P}} \mathcal{O}$, $\mathcal{O}$ being a $G \times G$-orbit in X. This gives a codimension preserving bijection between $\mathcal{G}$-orbits in $\mathcal{Y}$ and $G \times G$-orbits in X with the property that since $\mathcal{G}/\mathcal{P}$ is projective, if $\mathcal{O}$ is any $G \times G$-orbit in X, then $\overline{\mathcal{G} \times_{\mathcal{P}} \mathcal{O}} = \mathcal{G} \times_{\mathcal{P}} \overline{\mathcal{O}}$. In particular each orbit closure in $\mathcal{Y}$ is smooth.

Recall that the complement of the open orbit, which is isomorphic to $\mathcal{G}/K$, is a divisor $\mathcal{D}$ with normal crossings and smooth irreducible components $\mathcal{D}_i$, $i = 1, \dots, n$, each of which is the closure of a $\mathcal{G}$-orbit. Furthermore, each orbit closure in $\mathcal{Y}$ is the transversal intersection of those among the $\mathcal{D}_i$'s which contain it. Finally $\mathcal{Y}$ contains a unique closed orbit $\cap_{i=1}^n \mathcal{D}_i$ which is isomorphic to $\mathcal{G} \times_{\mathcal{P}} (G/B \times G/B) \simeq \mathcal{G}/\mathcal{B}$.

We are now going to determine the Picard group of $\mathcal{Y}$. Recall that in the previous section we have seen that the homomorphism $i^* : \mathrm{Pic}(X) \to \mathrm{Pic}(G/B \times G/B)$ induced by the inclusion $i : G/B \times G/B \to X$ as the closed orbit is injective and has as image the lattice consisting of pairs of the form $(\lambda, -w_0(\lambda))$, w_0 being the longest element in the Weyl group W. Consider now the inclusions $j : \mathcal{G}/\mathcal{B} \to \mathcal{Y}$ as the closed orbit and the inclusion $h : G/B \times G/B \to \mathcal{G}/\mathcal{B}$ as the fiber over $[\mathcal{P}]$ of the fibration $\mathcal{G}/\mathcal{B} \to \mathcal{G}/\mathcal{P}$. We have

Proposition 4.1. *The homomorphism $j^* : \mathrm{Pic}(\mathcal{Y}) \to \mathrm{Pic}(\mathcal{G}/\mathcal{B})$ is injective and has as image the lattice $(h^*)^{-1} i^*(\mathrm{Pic}(X))$.*

Proof. We have a commutative diagram of inclusions

$$\begin{array}{ccc} G/B \times G/B & \xrightarrow{\ h\ } & \mathcal{G}/\mathcal{B} \\ \big\downarrow{\scriptstyle i} & & \big\downarrow{\scriptstyle j} \\ X & \xrightarrow{\ \tilde{h}\ } & \mathcal{Y} \end{array}$$

where $\tilde{h}$ is the inclusion of X as the fiber over $[\mathcal{P}]$ of the fibration $\mathcal{Y} \to \mathcal{G}/\mathcal{P}$.

Now observe that since $\mathcal{Y}$ has a finite number of orbits, it contains a finite number of fixed points under the action of $\mathcal{T}$. Thus applying [BB], we deduce that $\mathcal{Y}$ has a paving by affine spaces. The same is true both for $\mathcal{G}/\mathcal{P}$ and for X. Using this fact and the previous diagram, we get a diagram

$$\begin{array}{ccccccccc} 0 & \longrightarrow & \mathrm{Pic}(\mathcal{G}/\mathcal{P}) & \xrightarrow{(pj)^*} & \mathrm{Pic}(\mathcal{G}/\mathcal{B}) & \xrightarrow{h^*} & \mathrm{Pic}(G/B \times G/B) & \longrightarrow & 0 \\ & & \uparrow{\scriptstyle \mathrm{id}} & & \uparrow{\scriptstyle j^*} & & \uparrow{\scriptstyle i^*} & & \\ 0 & \longrightarrow & \mathrm{Pic}(\mathcal{G}/\mathcal{P}) & \xrightarrow{p^*} & \mathrm{Pic}(\mathcal{Y}) & \xrightarrow{\tilde{h}^*} & \mathrm{Pic}(X) & \longrightarrow & 0 \end{array}$$

whose horizontal sequences are exact. Now recall that i^* is injective. This clearly implies that also j^* is injective and has as image $(h^*)^{-1}i^*(\mathrm{Pic}(X))$ as desired. □

Now that we have computed the Picard group of $\mathcal{Y}$, we can proceed to analyze the space of sections of a line bundle on $\mathcal{Y}$. We can clearly assume that the homomorphism $\pi' : \mathcal{P} \to \overline{G} \times \overline{G}$ takes the Borel subgroup $\mathcal{B}$ to the Borel subgroup $\overline{B} \times \overline{B}$, which is the image in $\overline{G} \times \overline{G}$ of $B \times B$ and also takes the maximal torus $\mathcal{T}$ to the maximal torus $\overline{T} \times \overline{T}$, which is the image in $\overline{G} \times \overline{G}$ of $T \times T$. We can also assume that under the homomorphism $\gamma : S \rtimes (G \times G) \to \mathcal{P}$, $T \times T$ is mapped to $\mathcal{T}$. Set Q equal to the root lattice $X(\overline{T})$. Our various maps induce homomorphisms

$$Q \oplus Q \xrightarrow{\pi'^*} \tilde{P} \xrightarrow{\gamma^*} P \oplus P\,,$$

with the composition being the inclusion of the root lattice into the weight lattice for the maximal torus $T \times T$ in $G \times G$. Clearly we can identify $P \oplus P$ with $\mathrm{Pic}(G/B \times G/B)$, $\tilde{P}$ with $\mathrm{Pic}(\mathcal{G}/\mathcal{B})$ and γ^* with h^*. So $\mathrm{Pic}(\mathcal{Y})$ gets identified with the sublattice of $\tilde{P}$ consisting of those elements λ such that $\gamma^*(\lambda) = (\lambda', -w_0(\lambda'))$ for a suitable $\lambda' \in P$. In particular $\mathrm{Pic}(\mathcal{Y})$ contains a copy of the root lattice Q. This lattice consists of the elements $\tilde{\tau} = \pi'^*((\tau, -w_0\tau))$, $\tau \in Q$ (we want to stress that in all this discussion w_0 denotes the longest element in the Weyl group of G). Let $\{\alpha_1, \ldots, \alpha_n\} \subset Q$ denote the set of simple roots with respect to the Borel subgroup $B \subset G$ and $\{\tilde{\alpha}_1, \ldots, \tilde{\alpha}_n\}$ the corresponding subset of $\tilde{P}$.

Before we proceed, let us prove a well known and easy Lemma. Let $\mathcal{U} \subset \mathcal{P}$ denote the unipotent radical in $\mathcal{P}$. Given a $\mathcal{G}$-module M, set $M^{\mathcal{U}}$ equal to the subspace of vectors which are invariant under the action of $\mathcal{U}$. $M^{\mathcal{U}}$ is clearly a $G \times G$-module.

Lemma 4.2. *The $\mathcal{G}$-module M is irreducible if and only if the $G \times G$-module $M^{\mathcal{U}}$ is irreducible.*

Proof. M (resp. $M^{\mathcal{U}}$) is irreducible if and only if contains a unique line stable under $\mathcal{B}$ (resp. $B \times B$). But a line stable under $\mathcal{B}$ is automatically contained in $M^{\mathcal{U}}$ and a $B \times B$ stable line in $M^{\mathcal{U}}$ is automatically $\mathcal{B}$ stable, so that M is irreducible if and only if $M^{\mathcal{U}}$ is irreducible. □

We now have

Proposition 4.3. *Let* $\lambda \in \mathrm{Pic}(\mathcal{Y})$ *and let* L_λ *be the corresponding line bundle on* $\mathcal{Y}$, $\mathcal{L}_\lambda$ *its restriction to* $\mathcal{G}/\mathcal{B}$. *The the restriction map*

$$H^0(\mathcal{Y}, L_\lambda) \to H^0(\mathcal{G}/\mathcal{B}, \mathcal{L}_\lambda)$$

is surjective.

Proof. We can clearly assume that λ is dominant. Otherwise $H^0(\mathcal{G}/\mathcal{B}, \mathcal{L}_\lambda) = 0$ and there is nothing to prove. If λ is dominant, then $H^0(\mathcal{G}/\mathcal{B}, \mathcal{L}_\lambda) = M_\lambda^*$, M_λ being the irreducible module of highest weight λ. We get an associated morphism $\psi : \mathcal{G}/\mathcal{B} \to \mathbb{P}(M_\lambda)$. We are going to extend this morphism to $\mathcal{Y}$. This will clearly imply our claim.

As we have seen in the previous Lemma, the subspace $M_\lambda^{\mathcal{U}}$ is an irreducible $G \times G$-module whose highest weight is of the form $(\lambda', -w_0(\lambda'))$, since $\lambda \in \mathrm{Pic}(\mathcal{Y})$ for a suitable dominant weight $\lambda' \in P_+$. It follows that, as a $G \times G$-module, $M_\lambda^{\mathcal{U}}$ is isomorphic to $\mathrm{End}(V_{\lambda'})$.

Consider the identity map in $I \in \mathrm{End}(V_{\lambda'}) = M_\lambda^{\mathcal{U}} \subset M_\lambda$. The line spanned by I is clearly stable under the action of $\mathcal{P}$ and so is the corresponding point $[I] \in \mathbb{P}(M_\lambda^{\mathcal{U}})$. In the previous section, we have seen that we have a $G \times G$-equivariant morphism from X onto the closure of the $G \times G$-orbit of $[I]$. It is clear, from the above considerations, that this morphism is indeed $\mathcal{P}$-equivariant.

Using the $\mathcal{G}$ action and the inclusion $\mathbb{P}(M_\lambda^{\mathcal{U}}) \to \mathbb{P}(M_\lambda)$, we then obtain a $\mathcal{G}$-equivariant morphism

$$\tilde{\psi} : \mathcal{Y} = \mathcal{G} \times_{\mathcal{P}} X \to \mathbb{P}(M_\lambda)$$

which clearly extends ψ as desired. □

Once the above result has been established, most of the results which follow are proven exactly as in [DCP, section 8] or as in section 3, so we leave their proof to the reader or only sketch them. The first is the following.

Proposition 4.4. *We can order the divisors* $\mathcal{D}_i$, $i = 1, \ldots, n$ *in such a way that, under the above identifications, the class in* $\mathrm{Pic}(\mathcal{Y})$ *of* $\mathcal{O}(D_i)$ *is* $\tilde{\alpha}_i$.

Let us now choose for each $i = 1, \ldots, n$, a non zero section $t_i \in H^0(\mathcal{Y}, L_{\tilde{\alpha}_i})$ whose set of zeros is $\mathcal{D}_i$.

Consider the ring

$$\mathcal{R} = \bigoplus_{\lambda \in \mathrm{Pic}(\mathcal{Y})} H^0(\mathcal{Y}, L_\lambda).$$

As in section 3, given sequences $\underline{h} = \{h_1, \ldots, h_n\}$ and $\underline{k} = \{k_1, \ldots, k_n\}$ of non negative integers, we shall say that $\underline{k} \geq \underline{h}$ if $k_i \geq h_i$ for each $i = 1, \ldots, n$ and set $|\underline{h}| = h_1 + \ldots + h_n$. If we now fix such a sequence $\underline{h}$, we set $\mathcal{R}_{\underline{h}}(\lambda)$ equal to the image of the map

$$H^0(\mathcal{Y}, L_{\lambda - \sum h_i \tilde{\alpha}_i}) \to H^0(\mathcal{Y}, L_\lambda)$$

given by multiplication by $t_1^{h_1}\cdots t_n^{h_n}$. Clearly $\mathcal{R}_{\underline{k}}(\lambda)\subset\mathcal{R}_{\underline{h}}(\lambda)$ if and only if $\underline{k}\geq\underline{h}$ and $\oplus_{\lambda\in\mathrm{Pic}(\mathcal{Y})}\mathcal{R}_{\underline{h}}(\lambda)$ is the ideal generated by $t_1^{h_1}\cdots t_n^{h_n}$.

Theorem 4.5. (1) *For each* $\lambda\in\mathrm{Pic}(\mathcal{Y})$*,*

$$\mathcal{R}_{\underline{h}}(\lambda)/\sum_{\underline{k}>\underline{h}}\mathcal{R}_{\underline{k}}(\lambda)$$

is isomorphic as a $\mathcal{G}$*-module to* $H^0(\mathcal{G}/\mathcal{B},\mathcal{L}_{\lambda-\sum_i h_i\tilde{\alpha}_i})$. *In particular as a* $\mathcal{G}$*-module, we have an isomorphism*

$$H^0(\mathcal{Y},L_\lambda)\simeq\bigoplus_{(h_1,\dots,h_n)}H^0(\mathcal{G}/\mathcal{B},\mathcal{L}_{\lambda-\sum_i h_i\tilde{\alpha}_i}).$$

(2) *If we set*

$$\mathcal{C}=\bigoplus_{\lambda\in\mathrm{Pic}(\mathcal{Y})}H^0(\mathcal{G}/\mathcal{B},\mathcal{L}_\lambda),$$

and

$$\mathcal{R}_i=\bigoplus_{|\underline{h}|=i,\lambda\in\mathrm{Pic}(\mathcal{Y})}\mathcal{R}_{\overline{h}}(\lambda)=\sum_{|\underline{h}|=i}t_1^{h_1}\cdots t_n^{h_n}\mathcal{R},$$

then associated graded ring

$$\mathrm{Gr}\mathcal{R}=\oplus_{i\geq 0}\mathcal{R}_i/\mathcal{R}_{i+1}$$

is isomorphic to the polynomial ring $\mathcal{C}[x_1,\dots,x_n]$*, where for* $j=1,\dots,n$*,* x_j *is the image of the* t_j *in* $\mathcal{R}_1/\mathcal{R}_2$*.*

(3) *Let* $\lambda\in\mathrm{Pic}(\mathcal{Y})$ *be a dominant weight. Then the ring*

$$\mathcal{R}_\lambda=\bigoplus_{n\geq 0}H^0(\mathcal{Y},L_{n\lambda})$$

is normal with rational singularities.

Proof. The proof of (1) is identical to that given in [DCP] for the case of X. (2) then follows repeating the proof of Proposition 3.2 and (3) the one of Proposition 3.3. □

Before we proceed, let us remark that, by Lemma 4.2, the restriction of the map

$$h^*:M=H^0(\mathcal{G}/\mathcal{B},\mathcal{L}_\lambda)\to H^0(G/B\times G/B,\mathcal{L}_{\lambda|G/B\times G/B})$$

to $M^{\mathcal{U}}$ is an isomorphism for all dominant $\lambda\in\tilde{P}_+$. Using this we obtain

Proposition 4.6. *Let* $\lambda,\mu\in\mathrm{Pic}\ (\mathcal{Y})\cap\tilde{P}_+$*, then the multiplication map*

$$m:H^0(\mathcal{Y},L_\lambda)\otimes H^0(\mathcal{Y},L_\mu)\to H^0(\mathcal{Y},L_{\lambda+\mu})$$

is surjective.

Proof. Since the map m is $\mathcal{G}$-equivariant and $H^0(\mathcal{Y}, L_{\lambda+\mu})^{\mathcal{U}}$ generates $H^0(\mathcal{Y}, L_{\lambda+\mu})$ as a $\mathcal{G}$-module, it is sufficient to show that the restriction of m to the $\mathcal{U}$-invariants is surjective.

Using the notations of Section 3, notice that, by the definition of the divisors $\mathcal{D}_i$, the restriction of the section t_i to X is, up to scalar, the section s_i vanishing on the boundary divisor D_i of X. This implies that we have, for each $\underline{h} = (h_1, \ldots, h_n)$ and for any $\lambda \in \mathrm{Pic}(\mathcal{Y})$, a commutative diagram

$$\begin{array}{ccc} H^0(\mathcal{Y}, L_{\lambda-\sum h_i\tilde{\alpha}_i}) & \longrightarrow & H^0(\mathcal{Y}, L_\lambda) \\ \downarrow{\scriptstyle \tilde{h}^*} & & \downarrow{\scriptstyle \tilde{h}^*} \\ H^0(X, L_{\lambda-\sum h_i\tilde{\alpha}_i|X}) & \longrightarrow & H^0(X, L_{\lambda|X}) \end{array} .$$

Setting $h^*(\lambda) = (\lambda', -w_0(\lambda'))$ and, by abuse of notation, still denoting by $\tilde{h}^*$ the map induced by $\tilde{h}^*$ on a subquotient, we get a commutative diagram

$$\begin{array}{ccc} \mathcal{R}_{\underline{h}}(\lambda)/\sum_{\underline{k}>\underline{h}} \mathcal{R}_{\underline{k}}(\lambda) & \longrightarrow & H^0(\mathcal{G}/\mathcal{B}, \mathcal{L}_{\lambda-\sum_i h_i\tilde{\alpha}_i}) \\ \downarrow{\scriptstyle \tilde{h}^*} & & \downarrow{\scriptstyle h^*} \\ R_{\underline{h}}(\lambda')/\sum_{\underline{k}>\underline{h}} R_{\underline{k}}(\lambda') & \longrightarrow & H^0(G/B \times G/B, \mathcal{L}_{(\lambda'-\sum_i h_i\alpha_i, -w_0(\lambda'-\sum_i h_i\alpha_i))}) \end{array}$$

where the horizontal arrows are isomorphisms. Now assume that λ is dominant. Then it is easy to see, $\lambda' - \sum_i h_i\alpha_i$ is dominant if and only if $\lambda - \sum_i h_i\tilde{\alpha}_i$ is dominant. This implies, by the remark made before our Proposition, that for all $\underline{h}$,

$$h^* : H^0(\mathcal{G}/\mathcal{B}, \mathcal{L}_{\lambda-\sum_i h_i\tilde{\alpha}_i})^{\mathcal{U}} \to H^0(G/B \times G/B, \mathcal{L}_{(\lambda'-\sum_i h_i\alpha_i, -w_0(\lambda'-\sum_i h_i\alpha_i))})$$

is an isomorphism, so that also the restriction of $\tilde{h}^*$ to the space of $\mathcal{U}$ invariants is an isomorphism.

Take $\lambda, \mu \in \tilde{P}_+$. We get a diagram

$$\begin{array}{ccc} H^0(\mathcal{Y}, L_\lambda)^{\mathcal{U}} \otimes H^0(\mathcal{Y}, L_\mu)^{\mathcal{U}} & \xrightarrow{m} & H^0(\mathcal{Y}, L_{\lambda+\mu})^{\mathcal{U}} \\ \downarrow{\scriptstyle \tilde{h}^*\otimes\tilde{h}*} & & \downarrow{\scriptstyle \tilde{h}^*} \\ H^0(X, L_{\lambda|X}) \otimes H^0(X, L_{\mu|X}) & \xrightarrow{m} & H^0(X, L_{\lambda+\mu|X}) \end{array}$$

where the vertical arrows are isomorphisms.

By Theorem 3.6, we have that the map

$$H^0(X, L_{\lambda|X}) \otimes H^0(X, L_{\mu|X}) \xrightarrow{m} H^0(X, L_{\lambda+\mu|X})$$

is surjective and this, together with our previous considerations, implies our claim. □

We are now going to use the properties of $\mathcal{Y}$ to study certain orbit closures.

Let us take a representation M of $\mathcal{G}$ and a non zero vector $v \in M$ which, as we can suppose without loss of generality, spans M as a $\mathcal{G}$-module. The assumptions we are going to make on v are

Assumptions 4.7 1) There is a character $\chi : K \to k^*$ such that $kv = \chi(k)v$, for all $k \in K$.

2) Let $W \subset M$ be the $G \times G$-module spanned by v. Then W is a highest weight module.

Let us make some considerations. By assumption 1) the diagonal subgroup in $G \times G$ fixes v, so that the orbit map $G \times G \to (G \times G)v$ factors through the map $f : G \times G \to G$ given by $f((g_1, g_2)) = g_1 g_2^{-1}$, for all $g_1, g_2 \in G$. Thus, we get an $G \times G$-equivariant inclusion of the vector space W into the coordinate ring $k[G]$. In particular, using assumption 2), we deduce that there is a dominant $\lambda' \in P_+$ and a subset $\Omega' \subset \Sigma(\lambda')$ containing λ', such that, as a $G \times G$-module,

$$W \simeq \oplus_{\mu' \in \Omega'} \mathrm{End}(V_{\mu'}).$$

Also, by assumption 1), we have that $\mathcal{P}$ preserves the line spanned by v, so that W is stable under the action of $\mathcal{P}$ and $W \subset M^{\mathcal{U}}$. Since v spans M as a $\mathcal{G}$-module, we deduce that indeed $W = M^{\mathcal{U}}$. Using Lemma 4.2 and our description of $\mathrm{Pic}(\mathcal{Y})$, we then deduce that there is a subset $\Omega \subset \mathrm{Pic}(\mathcal{Y})$ mapped bijectively onto Ω' by $\tilde{h}^* : \mathrm{Pic}(\mathcal{Y}) \to \mathrm{Pic}(X) = P$ such that, as a $\mathcal{G}$-module,

$$M \simeq \oplus_{\mu \in \Omega} M_\mu.$$

Definition 4.8. The variety $Y(v, \Omega)$ is the cone over the orbit $\mathcal{G}v$, i.e., if we let G_m act on M by homotheties,

$$Y(v, \Omega) = \overline{(\mathcal{G} \times G_m)v}.$$

In order to simplify notations we denote by U the G-module $\oplus_{\mu' \in \Omega'}(V_{\mu'})$. We have

Lemma 4.9. *Set* $Z = \overline{(\mathcal{P} \times G_m)v} \subset W$. *Then*

(1) *Z is isomorphic as a $G \times G$ variety to Z_U.*

(2) $Y(v, \Omega) = \mathcal{G}Z$.

Proof. (1) Notice that up to rescaling, we can assume that the isomorphism

$$\psi : W \to \oplus_{\mu' \in \Sigma} \mathrm{End}(V_{\mu'})$$

has the property that $\psi(v) = \sum_{\mu \in \Sigma} \mathrm{Id}_{V_\mu}$. On the other hand, we have a $G \times G$-equivariant inclusion of $\oplus_{\mu \in \Sigma} \mathrm{End}(V_\mu)$ into $\mathrm{End}(U)$ taking $\sum_{\mu \in \Sigma} \mathrm{Id}_{V_\mu}$ to the identity. Composing, we get a $G \times G$-equivariant inclusion $\tilde{\psi} : W \to \mathrm{End}(U)$ with $\tilde{\psi}(v) = \mathrm{Id}_U$. Clearly the restriction of $\tilde{\psi}$ to Z gives the required isomorphism with Z_U.

(2) Since both $Y(v,\Omega)$ and Z are cones, it suffices to show that $\mathcal{G}\tilde{Z} = \tilde{Y}(v,\Omega)$ for the projective varieties $\tilde{Z} = (Z - \{0\})/G_m$ and $\tilde{Y}(v,\Omega) = (Y(v,\Omega) - \{0\})/G_m$. Notice that $\tilde{Z}$ is $\mathcal{P}$-stable, so that the morphism $\mathcal{G} \times \tilde{Z} \to \tilde{Y}(v,\Omega)$ factors through $\mathcal{G} \times_{\mathcal{P}} \tilde{Z}$. Since $\mathcal{G}/\mathcal{P}$ and $\tilde{Z}$ are complete, also $\mathcal{G} \times_{\mathcal{P}} \tilde{Z}$ and hence its image $\mathcal{G}\tilde{Z}$ are complete, proving that $\mathcal{G}\tilde{Z}$ coincides with $\tilde{Y}(v,\Omega)$. □

Notice that if $\mathcal{G} = \mathcal{P} = G \times G$, then the variety $Y(v,\Omega)$ coincides with the variety Z_U considered in Section 3.

The following Lemma strongly restricts our choice of Ω. Let $\lambda \in \Omega$ be the unique element such that $\tilde{h}^*(\lambda) = \lambda'$. Given $\mu' = \lambda' - \sum h_i \alpha_i \in \Sigma(\lambda')$, we set $\rho_\lambda(\mu') = \lambda - \sum h_i \tilde{\alpha}_i$. Notice that the set $\Omega(\lambda) := \rho_\lambda(\Sigma(\lambda'))$ coincides with the set of highest weights of irreducible components of the $\mathcal{G}$-module $H^0(\mathcal{Y}, L_\lambda)$.

Lemma 4.10. $\Omega = \rho_\lambda(\Omega')$. *In particular* $\Omega \subset \Omega(\lambda)$.

Proof. By what we have seen in Section 3, we have a $G \times G$-equivariant morphism

$$\psi : X \to \tilde{Z}.$$

So we deduce from part 2) of the previous Lemma that we have a surjective $\mathcal{G}$-equivariant morphism

$$\phi : \mathcal{Y} \to \tilde{Y}(v,\Omega)$$

whose restriction to X equals ψ. The line bundle on $\mathcal{Y}$ which is the pull back of $\mathcal{O}(1)$ on $\tilde{Y}(v,\Omega)$ is equal to L_λ for some $\lambda \in$ Pic $(\mathcal{Y})$ with the property that $\tilde{h}^*(\lambda) = \lambda'$. It follows that the $\mathcal{G}$-module M is a direct summand in $H^0(\mathcal{Y}, L_\lambda)^*$. Thus Proposition 4.6 implies our claim. □

We are ready to show:

Theorem 4.11. 1) *The variety* $Y(v,\Omega)$ *is normal with rational singularities if and only if* $\Omega = \Omega(\lambda)$.

2) *For a general* $\Omega \subset \Omega(\lambda)$, $Y(v,\Omega(\lambda))$ *is the normalization of* $Y(v,\Omega)$. *In particular* $Y(v,\{\lambda\})$ *is normal if and only if* λ' *is minuscule.*

Proof. The argument given in the above lemma implies that we can identify the coordinate ring $k[Y(v,\Omega)]$ of $Y(v,\Omega)$ with the subring of the ring $\mathcal{R}_\lambda = \oplus_{n \geq 0} H^0(\mathcal{Y}, L_\lambda^n)$ generated by the $\mathcal{G}$ submodule $M^* \subset H^0(\mathcal{Y}, L_\lambda^n)$. In particular, notice that, up to isomorphism, $Y(v,\Omega)$ depends, as a $\mathcal{G}$-variety, only on Ω and not on the choice of a specific vector v (of course provided that the Assumptions 4.7 are satisfied).

We can now prove Theorem 4.11 exactly as we have shown Theorem 3.1. If $\Omega = \Omega(\lambda)$, we get that $k[Y(v,\Omega(\lambda))]$ and $\mathcal{R}_\lambda$ coincide in degree one by Theorem 4.5. Since by Proposition 4.6 they are both generated by their degree one components, we deduce that $k[Y(v,\Omega(\lambda))] = \mathcal{R}_\lambda$. In particular $k[Y(v,\Omega(\lambda))]$ is normal with rational singularities.

It remains to see that, if $\Omega \subsetneq \Omega(\lambda)$, then $k[Y(v,\Omega)]$ is not normal. This follows in a way completely analogous to the corresponding statement for Z_V, which has been seen in Section 3, so we leave the details to the reader. □

Theorem 4.11 can be extended as follows. Suppose $G = G_1 \times \cdots \times G_s$. Let $M_1, \ldots, M_s$ be $\mathcal{G}$-modules and $v_1, \ldots, v_s$ be vectors with $v_i \in M_i$ each satisfying Assumptions 4.7. Assume furthermore that for each $i = 1, \ldots, s$, and $j \neq i$, $G_j \times G_j$ fixes v_i. By what we have already seen, for each i, M_i is a highest weight module of highest weight λ_i and we get a subset $\lambda_i \in \Omega_i \subset \mathrm{Pic}(\mathcal{Y})$, such that $M_i \simeq \oplus_{\mu \in \Omega_i} M_\mu$. Also, we have the subsets $\Omega'_i = \tilde{h}^*(\Omega_i) \subset P_+$. Denote by S the subspace in $M = M_1 \oplus \cdots \oplus M_s$ spanned by the vectors $v_1, \ldots, v_s$.

We then define $Y(v_1, \ldots, v_s; \Omega_1, \ldots, \Omega_s)$ as the closure of $\mathcal{G}S \subset M$. One obtains

Theorem 4.12. 1) *The variety $Y(v_1, \ldots, v_s; \Omega_1, \ldots, \Omega_s)$ is normal with rational singularities if and only if $\Omega_i = \Omega(\lambda_i)$ for each $i = 1, \ldots, s$.*

2) *For a general sequence $\Omega_1, \ldots, \Omega_s$, with $\Omega_i \subset \Omega(\lambda_i)$, the normalization of $Y(v_1, \ldots, v_s; \Omega_1, \ldots, \Omega_s)$ is given by $Y(v_1, \ldots, v_s, \Omega_1(\lambda_1), \ldots, \Omega_s(\lambda_s))$. In particular, $Y(v_1, \ldots, v_s; \{\lambda_1\}, \ldots, \{\lambda_s\})$ is normal if and only if λ'_i is minuscule for each $i = 1, \ldots, s$.*

Proof. Let $\Gamma \subset \mathcal{T}$ be the intersection of the kernels of the characters λ_i. It is easy to see that the definition of the λ_i's implies that the coordinate ring of $Y(v_1, \ldots, v_s, \Omega_1(\lambda_1), \ldots, \Omega_s(\lambda_s))$ can be identified with the ring of Γ invariants in

$$\mathcal{R} = \bigoplus_{\lambda \in \mathrm{Pic}(\mathcal{Y})} H^0(\mathcal{Y}, L_\lambda),$$

the $\mathcal{T}$ action being given by $ts = \lambda(t)s$, if $s \in H^0(\mathcal{Y}, L_\lambda)$, $t \in \mathcal{T}$.

This immediately implies that $Y(v_1, \ldots, v_s, \Omega_1(\lambda_1), \ldots, \Omega_s(\lambda_s))$ is normal with rational singularities. We leave the rest of the proof to the reader. □

5 Some Examples

In this section, using the notations of the previous sections, we are going to make a number of examples of varieties of the form $Y(v_1, \ldots, v_s; \Omega_1, \ldots, \Omega_s)$.

Example 5.1. Assume $\mathcal{G}$ is arbitrary and $\mathcal{P} = \mathcal{B}$ is a Borel subgroup. Then necessarily $G = \{e\}$ the trivial group. We can then take any dominant weight $\lambda \in \tilde{P}$ and consider the irreducible $\mathcal{G}$-module M_λ. Then necessarily v is a highest weight vector and the variety $Y(v, \{\lambda\})$ is just the affine cone over the unique closed orbit in $\mathbb{P}(M_\lambda)$. Notice that in this case, $Y(v, \{\lambda\})$ is normal with rational singularities in accordance with our result (every representation of the trivial group is minuscule). However, we remark that we have used this fact to prove our result.

Example 5.2. Let us start with the case of one of the semigroups Z_V. Take $G = \mathrm{SL}(n)$ and let V be its fundamental representation $\bigwedge^h k^n$. Then it is easy to see, as we have already remarked in Section 3 in a special case, that the coordinate ring $k[Z_V]$ is nothing else than the subring of the polynomial ring $k[x_{i,j}]$, $i, j = 1, \dots, n$, generated by the determinants of the $h \times h$ minors of the matrix $(x_{i,j})$. More generally, assume $\mathcal{G} = \mathrm{SL}(n) \times \mathrm{SL}(m)$ and $G = \mathrm{SL}(h)$ with $h \leq \min(m, n)$. Take $V = \mathrm{Hom}(\bigwedge^r k^n, \bigwedge^r k^m)$ for some $r \leq h$. Consider k^h as a subspace of both $k^m$and k^n in the obvious way, so that we have an inclusion of $\mathrm{End}(\bigwedge^r k^h)$ into $\mathrm{Hom}(\bigwedge^r k^n, \bigwedge^r k^m)$. Take v to be the identity map in $\mathrm{End}(\bigwedge^r k^h)$. Then, since $\bigwedge^r k^h$ is a minuscule $\mathrm{SL}(h)$-module , we deduce that the corresponding variety $Y(v, \Omega)$ is normal with rational singularities. It is clear from our description, that the coordinate ring of $Y(v, \Omega)$ can be described as follows. Consider the ring $R = k[x_{i,j}]/I_h$ with $i = 1, \dots, n$, $j = 1, \dots, m$ and I_h equal to the ideal generated by determinants of $h + 1 \times h + 1$ minors of the matrix $(x_{i,j})$. Then $k[Y(v, \Omega)]$ is the subring of R generated as a k algebra, by the determinants of $r \times r$ minors of $(x_{i,j})$ (in particular if $r = 1$, $Y(v, \Omega)$ is the determinantal variety of $n \times m$ matrices of rank less than or equal than h). The fact that this ring is normal with rational singularities, at least when $h = \min(n, m)$, has been originally shown in [Br], see also [BrC].

Example 5.3. Suppose now $\mathcal{G} = \mathrm{Sl}(n_0) \times \mathrm{Sl}(n_1) \times \cdots \times \mathrm{Sl}(n_r)$, fix a sequence of non negative integers $(h_1, h_2, \dots, h_r)$ with $h_1 \leq \min(n_0, n_1)$, $h_i \leq \min(n_{i-1} - h_{i-1}, n_i)$, for $2 \leq i \leq r$. Set $v = (v_1, v_2, \dots, v_r) \in \mathrm{Hom}(k^{n_0}, k^{n_1}) \oplus \mathrm{Hom}(k^{n_1}, k^{n_2}) \oplus \cdots \oplus \mathrm{Hom}(k^{n_{r-1}}, k^{n_r})$ where, letting I_h denotes the identity $h \times h$ matrix, we have

$$v_i = \begin{pmatrix} I_{h_i} & 0 \\ 0 & 0 \end{pmatrix}$$

if i is odd,

$$v_i = \begin{pmatrix} 0 & 0 \\ 0 & I_{h_i} \end{pmatrix}$$

if i is even.

Notice that $v_{i+1} v_i = 0$ for each $i = 1, \dots, r - 1$. $G = \mathrm{Sl}(h_1) \times \cdots \times \mathrm{Sl}(h_r)$ and the $G \times G$-module spanned by v is just $\mathrm{End}(k^{h_1}) \oplus \cdots \oplus \mathrm{End}(k^{h_r})$ so that the conditions of Theorem 4.12 are satisfied. Furthermore for each i, $\Omega_i = \{\omega_1\}$, a minuscule weight. We deduce that the variety

$$Y(v_1, v_2, \dots, v_r; \{\omega_1\}, \dots, \{\omega_1\})$$

is normal with rational singularities. The fact $v_{i+1} v_i = 0$ for each $i = 1, \dots, r - 1$ clearly implies that $Y(v_1, v_2, \dots, v_r; \{\omega_1\}, \dots, \{\omega_1\})$ is nothing else than the variety of complexes with rank conditions $(h_1, h_2, \dots, h_r)$, i.e., it is the variety of sequences $(\psi_1, \dots, \psi_r)$ in $\mathrm{Hom}(k^{n_0}, k^{n_1}) \oplus \cdots \oplus \mathrm{Hom}(k^{n_{r-1}}, k^{n_r})$ such that $\psi_{i+1} \psi_i = 0$ for $i = 1, \dots, r - 1$ and $\mathrm{rk}\, \psi_i \leq h_i$ for $i = 1, \dots, r$. The fact that varieties of complexes have rational singularities is well known (see [Ke], [DS] or, for a more recent reference [MT]).

Example 5.4. Suppose now $\mathcal{G} = \mathrm{SL}(n)\times\mathrm{SL}(m)$, and fix a pair of non negative integers (h, s) with $h + s \leq \min(n, m)$. Define $v = (v_1, v_2) \in \mathrm{Hom}(k^n, k^m) \oplus \mathrm{Hom}(k^m, k^n)$ by

$$v_1 = \begin{pmatrix} I_h & 0 \\ 0 & 0 \end{pmatrix}, \quad v_2 = \begin{pmatrix} 0 & 0 \\ 0 & I_s \end{pmatrix}.$$

(As before, I_h and I_s are identity matrices of size h and k respectively). $G = \mathrm{SL}(h)\times\mathrm{SL}(s)$ and the $G\times G$-module spanned by v is just $\mathrm{End}(k^h)\oplus\mathrm{End}(k^s)$, so $\Omega_i = \{\omega_1\}$ for $i = 1, 2$. Reasoning as in the previous example, we deduce that $Y(v_1, v_2; \{\omega_1\}, \{\omega_1\})$ is normal with rational singularities. Now notice that $v_1v_2 = 0$ and $v_2v_1 = 0$. From this, it is immediate that $Y(v_1, v_2; \{\omega_1\}, \{\omega_1\})$ is the variety of circular complexes with rank conditions (h, s), i.e., it is the variety of pairs (ψ_1, ψ_2) in $\mathrm{Hom}(k^n, k^m) \oplus \mathrm{Hom}(k^m, k^n)$ such that $\psi_1\psi_2 = 0$, $\psi_2\psi_1 = 0$ and $\mathrm{rk}\,\psi_1 \leq h$, $\mathrm{rk}\,\psi_2 \leq s$. These varieties have been studied in [S] and in [MT1].

Example 5.5. As a final example, let V be an n-dimensional vector space with a non degenerate symmetric or antisymmetric bilinear form (in this case $\dim V$ is even). We let G be the group of isometries with respect to the form: i.e., $G = \mathrm{SO}(V)$ if our form is symmetric, $G = \mathrm{Sp}(V)$ if it is antisymmetric. Given a linear tranformation $A \in \mathrm{End}\,(V)$, we denote by ${}^tA$ its adjoint with respect to the form. Then the variety Z_V is the variety of linear tranformations $A \in \mathrm{End}\,(V)$ such that ${}^tAA = A{}^tA = tI$, for some $t \in k$ (I is the identity) and, if $G = \mathrm{SO}(V)$ and $\dim V$ is even, $\det A = t^n$. Since V is a minuscule representation if and only if $\dim V$ is even, we deduce that in this case Z_V is normal with rational singularities. If on the other hand $\dim V$ is odd, the normalization of Z_V is given by Z_W, with $W = V \oplus k$.

References

[BB] Białynicki–Birula, A.: Some theorems on actions of algebraic groups. Ann. of Math. (2) **98** 480–497 (1973)

[B] Boutot, J.–F.: Singularitès rationnelles et quotients par les groupes rèductifs. Invent. Math. **88**(1) 65–68 (1987)

[Br] Bruns, W.: Algebras defined by powers of determinantal ideals. J. Algebra **142**(1) 150–163 (1991)

[BrC] Bruns, W., Conca, A.: Algebras of minors. J. Algebra **246**(1) 311–330 (2001)

[DS] De Concini, C., Strickland, E.: On the variety of complexes. Adv. in Math. **41**(1) 57–77 (1981)

[DCP] De Concini, C., Procesi, C.: *Complete symmetric varieties.* Invariant theory (Montecatini, 1982), Lecture Notes in Math., vol. 996 (Springer, Berlin 1983) pp. 1–44

[E] Elkik, R.: Singularitès rationnelles et dèformations. Invent. Math. **47**(2) 139–147 (1978)

[F] Faltings, G.: Explicit resolution of local singularities of moduli-spaces. J. Reine Angew. Math. **483** 183–196 (1997)

[K] Kannan, S.S.: Projective normality of wonderful compactification of semisimple adjoint groups. Math. Z. **239**(4) 673–682 (2002)

[Ke] Kempf, G.R.: Images of homogeneous vector bundles and varieties of complexes. Bull. Amer. Math. Soc. **81**(5) 900–901 (1975)

[KR] Kempf, G.R., Ramanathan, A.: Multicones over Schubert varieties. Invent. Math. **87**(2) 353–363 (1987)

[Ku] Kumar, S.: Proof of the Parthasarathy–Ranga Rao–Varadarajan conjecture. Invent. Math. **93**(1) 117–130 (1988)

[Ma] Mathieu, O.: Construction d'un groupe de Kac–Moody et applications. Compositio Math. **69**(1) 37–60 (1989)

[MT] Mehta, V.B., Trivedi, V.: Variety of complexes and F-splitting. J. Algebra **215**(1) 352–365 (1999)

[MT1] Mehta, V.B., Trivedi, V.: The variety of circular complexes and F-splitting. Invent. Math. **137**(2) 449–460 (1999)

[Re1] Renner, L.E.: Classification of semisimple algebraic monoids. Trans. Amer. Math. Soc. **292**(1) 193–223 (1985)

[Re2] Renner, L.E.: Regular algebraic monoids. Semigroup Forum **63**(2) 107–113 (2001)

[RP] Renner, L., Putcha, M.: *Linear algebraic monoids*. Positivity in Lie theory: open problems, de Gruyter Exp. Math., vol. 26 (de Gruyter, Berlin 1998) pp. 271–281

[Ri] Rittatore, A.: Algebraic monoids and group embeddings. Transform. Groups **3**(4) 375–396 (1998)

[S] Strickland, E.: On the exactness of the circular complex. J. Algebra **85**(2) 382–389 (1983)

[S1] Strickland, E.: A vanishing theorem for group compactifications. Math. Ann. **277**(1) 165–171 (1987)

[T] Timashev, D.A.: Equivariant compactifications of reductive groups. arXiv:math. AG/0207034.

[V] Vinberg, E.B.: On reductive algebraic semigroups. Lie groups and Lie algebras: E. B. Dynkin's Seminar, Amer. Math. Soc. Transl. Ser. 2 **169** (Amer. Math. Soc., Providence, RI 1995) pp. 145–182

Deformation Rigidity Of the 20-dimensional F_4-homogeneous Space Associated To a Short Root

Jun–Muk Hwang[1,⋆] and Ngaiming Mok[2,⋆⋆]

[1] Korea Institute for Advanced Study, 207-43 Cheongryangri-dong Seoul 130-012, Korea. jmhwang@ns.kias.re.kr
[2] Department of Mathematics, The University of Hong Kong Pokfulam Road, Hong Kong. nmok@hkucc.hku.hk

In continuation of [HM1], [HM4] and [Hw], we work on the following conjecture.

Conjecture. *Let $\pi : \mathcal{X} \to \Delta := \{t \in \mathbb{C}; |t| < 1\}$ be a smooth projective morphism from a complex manifold $\mathcal{X}$ to the unit disc. Suppose the fiber $X_t := \pi^{-1}(t)$ is biholomorphic to a rational homogeneous space of Picard number 1 for each $t \neq 0$. Then X_0 is also biholomorphic to the rational homogeneous space.*

Recall that a rational homogeneous space of Picard number 1 is the homogeneous space of a complex simple Lie group modulo a maximal parabolic subgroup. We say that the rational homogeneous space of Picard number 1 is of type $(\mathfrak{g}, \alpha)$ when $\mathfrak{g}$ is the complex simple Lie algebra and α is the simple root determining the maximal parabolic subgroup. By the results of [HM1], [Hw] and [HM4], the conjecture was proved for all rational homogeneous spaces of Picard number 1 excepting the following three cases.

(1) type (C_l, α) where α is a short root,
(2) type (F_4, α_1),
(3) type (F_4, α_2).

Here the numbering of the simple roots $\{\alpha_1, \alpha_2, \alpha_3, \alpha_4\}$ of F_4 is chosen such that α_1 and α_2 are short and the highest long root is of the form $2\alpha_1 + 4\alpha_2 + 3\alpha_3 + 2\alpha_4$. The proof of [HM1], [Hw], [HM4] uses the natural distribution of smallest rank on the homogeneous space. In these settled cases, this distribution agrees with the distribution defined by the variety of minimal rational tangents (cf. definition in Section 3 below), which enables

⋆ Supported by Grant No. 98-0701-01-5-L from the KOSEF.
⋆⋆ Supported by a CERG of the Research Grants Council of Hong Kong.

one to recover the structure of the homogeneous space from the data of rational curves. In the remaining three cases, these two distributions do not agree. However in case (3), the distribution defined by the variety of minimal rational tangents agree with a natural distribution on the homogeneous space albeit not the minimal one. This suggests that the proof of [HM4] can be modified to attack the case (3). This is precisely what we will do in this article. Our main result is the following.

Main Theorem. *Let G be a complex simple Lie group of type F_4 and let P be the maximal parabolic subgroup associated to the root α_2. Let $\pi : \mathcal{X} \to \Delta := \{t \in \mathbb{C} \mid |t| < 1\}$ be a smooth projective morphism from a complex manifold $\mathcal{X}$ to the unit disc. Suppose each fiber $X_t := \pi^{-1}(t)$ is biholomorphic to G/P for each $t \neq 0$. Then X_0 is also biholomorphic to G/P.*

From now on, G/P denotes the homogenous space of type (F_4, α_2). As in [HM4], the essential step of the proof of the main theorem lies in establishing that the symbol algebra of the differential system on the rational homogenous is determined by the projective geometry of the variety of minimal rational tangents. In [HM4], Proposition 5, this fact was proved for the homogeneous spaces associated with long roots by realizing that the condition on the symbol algebra arising from the tangent lines to the variety of minimal rational tangents is equivalent to the finiteness condition in the Serre presentation of the simple Lie algebra. The main difference of the current paper from [HM4] is that the symbol algebra structure cannot be recovered easily by such algebraic results as the Serre presentation. In the current situation, one has to recover the structure more directly by studying the projective geometry of the variety of minimal rational tangents. For this reason, a large part, three out of four sections, of the paper is devoted to a geometric study of the variety of minimal rational tangents for G/P, which is essentially independent from the deformation rigidity problem.

The projective variety corresponding to the variety of minimal rational tangents for G/P will be denoted by Z. In the first section, we will study Z as a complex manifold. Here the key result is a certain rigidity for Z. In the second section, the projective geometry of Z under a natural embedding in the projective space will be discussed. The third section is to show that this Z is indeed isomorphic to the variety of minimal rational tangents of G/P. The Main Theorem will be proved in Section 4.

For the remaining cases (1) and (2), the distribution spanned by the variety of minimal rational tangents is the full tangent space. In this sense, the approach of studying the symbol algebra of the distribution defined by the variety of minimal rational tangents is of no help. In another paper [HM5], we have settled cases (1) and (2) in the affirmative by a new method involving the study of holomorphic vector fields on uniruled projective manifolds.

1 Z As a Complex Manifold

Our variety Z as a complex manifold is quite simple. Let $\mathcal{E}$ be a vector bundle of rank 4 on $\mathbb{P}_1$ with a subbundle $\mathcal{V}$ of rank 3 such that $\mathcal{V} \cong \mathcal{O}(1)^3$ and $\mathcal{E}/\mathcal{V} \cong \mathcal{O}$. Note that the subbundle $\mathcal{V}$ is uniquely determined by this property. We define

$$Z := \mathbb{G}(2; \mathcal{E}^*),$$

the Grassmannian bundle of 2-planes in the dual bundle $\mathcal{E}^*$. We denote by $\eta : Z \to \mathbb{P}_1$ the natural projection to $\mathbb{P}_1$. Let ξ be the Plücker bundle on Z satisfying

$$\eta_* \xi = \bigwedge^2 \mathcal{E}.$$

Let $\mathcal{A}^* \subset \mathcal{E}^*$ be the line subbundle annihilating the subbundle $\mathcal{V}$ so that the quotient bundle $\mathcal{E}^*/\mathcal{A}^*$ is canonically isomorphic to $\mathcal{V}^*$. Let $S \subset Z = \mathbb{G}(2, \mathcal{E}^*)$ be the Schubert subbundle corresponding to the 2-planes in $\mathcal{E}^*$ containing the line bundle $\mathcal{A}^*$. Then there is a canonical identification $S = \mathbb{P}\mathcal{V}^*$. The projection $\eta|_S : S \to \mathbb{P}_1$ will be denoted by χ. Under the identification $S = \mathbb{P}\mathcal{V}^*$, the restriction of ξ to S, which will be denoted by the same letter ξ, agrees with the hyperplane line bundle of the projectivization so that

$$\chi_* \xi = \mathcal{V}.$$

A rational curve on Z will be called a line if it has degree 1 with respect to ξ and a conic if the degree is 2. It is easy to see that a line is contained in a fiber of η or in S. In fact, any rational curve which is contained neither in a fiber of η nor in S can be deformed under the action of the automorphism group of Z as a fiber bundle to a reducible curve with components both in fibers of η and in S. Thus

Lemma 1.1. *For a line C in Z, $-K_Z \cdot C \leq 4$.*

The rest of the section is devoted to the proof of the following rigidity.

Proposition 1.2. *Let $\psi : \mathcal{Z} \to \Delta$ be a smooth projective morphism and ζ be a line bundle on $\mathcal{Z}$. Suppose for each $t \neq 0$, the fiber $Z_t := \psi^{-1}(t)$ is biholomorphic to Z and $\zeta|_{Z_t}$ is isomorphic to the line bundle ξ on Z. Further assume that for any flat family of curves $\{C_t \subset Z_t \mid t \in \Delta\}$,*

(a) *if C_t has degree 1 with respect to ζ, the limit curve C_0 is irreducible, and*

(b) *if C_t has degree 2 with respect to ζ, then the limit curve C_0 is either irreducible or a union of two irreducible curves of degree 1 with respect to ζ.*

Then Z_0 is also biholomorphic to Z.

We will say that a point of a uniruled complex manifold is **very general** if any rational curve through that point is **free** in the sense that the pull-back of the tangent bundle of the complex manifold to the normalization of the rational curve is a semi-positive bundle on $\mathbb{P}_1$. It is well-known that the complement of the set of very general points is a countable union of proper subvarieties. The following stability result will be useful.

Lemma 1.3. *Let $\cup_{t\in\Delta} M_t$ be a smooth family of uniruled complex manifolds. Then for any rational curve C passing through a very general point of M_0, there exists a flat family of rational curves $C_t \subset M_t$ with $C_0 = C$.*

Proof. This is a direct consequence of Kodaira's stability ([Ko]). Note that the obstruction to deforming a holomorphic map $f : \mathbb{P}_1 \to M_0$ out of M_0 is $H^1(\mathbb{P}_1, f^*T(M_0))$ which vanishes when $C = f(\mathbb{P}_1)$ is free on M_0. □

The following lemma is a direct consequence of the basic deformation theory of rational curves.

Lemma 1.4. *Let $C \subset M$ be a free rational curve on a complex manifold. Suppose there exists a point $P \in C$ and an m-dimensional family of deformations of C fixing P such that the members of the family are all distinct rational curves. Then $-K_M \cdot C \geq 2 + m$.*

Now we start the proof of Proposition 1.2. For $t \neq 0$, let $\mathcal{Q}_t$ be the foliation on Z_t given by the fibers of the Grassmannian bundle $\eta : Z \to \mathbb{P}_1$. The leaves are hypersurfaces in Z_t biholomorphic to $\mathbb{G}(2;\mathbb{C}^4)$, which is biholomorphic to the 4-dimensional smooth hyperquadric $\mathbb{Q}_4$. Let $\mathcal{Q}_0$ be the foliation on a Zariski open subset of Z_0 defined by taking limits of $\mathcal{Q}_t$. Leaves of $\mathcal{Q}_0$ can be compactified to hypersurfaces in Z_0.

Lemma 1.5. *Let $y \in Z_0$ be a generic point and $\theta : \Delta \to \mathcal{Z}$ be a section of ψ with $\theta(0) = y$. Let Q_t be the $\mathcal{Q}_t$-leaf through $\theta(t)$ on $Z_t, t \neq 0$. Then the limit Q_0 is an irreducible hypersurface in Z_0.*

Proof. Without loss of generality, we can assume that y is a very general point of Z_0. If Q_0 is reducible, choose families of points $\{A_t \in Q_t \mid t \in \Delta\}$ and $\{B_t \in Q_t \mid t \in \Delta\}$ so that A_0 and B_0 are generic points of distinct components of Q_0. We may assume that A_0 is a very general point. On the 4-dimensional quadric $Q_t, t \neq 0$, we can find an irreducible conic $c_t \subset Q_t$ containing both A_t and B_t. The limit $c_0 \subset Q_0$ cannot be irreducible because it contains A_0 and B_0 lying on different components of Q_0. Thus c_0 must be the union of two irreducible curves of degree 1 with respect to ζ by the assumption (b). Fixing $\{A_t \mid t \in \Delta\}$ and varying $\{B_t \mid t \in \Delta\}$, we get irreducible rational curves C of degree 1 through A_0 which cover a component of Q_0. Thus $-K_{Z_0} \cdot C \geq 5$ by Lemma 1.4. Since A_0 is very general, Lemma 1.3 says that C is the limit of a family of curves of degree 1 on Z_t, a contradiction to Lemma 1.1. □

Lemma 1.6. *The hypersurfaces Q, Q' which are closures of two generic leaves of $\mathcal{Q}_0$ do not intersect.*

Proof. Let Q_0 be as in Lemma 1.5. Suppose $B_0 \in Q_0$ is contained in the closure of any other leaf. Pick a generic point $A_0 \in Q_0$ and local holomorphic sections of $\mathcal{Z}$ given by $A_t, B_t \in Q_t$, extending A_0 and B_0. The set of irreducible conics through A_t and B_t cover Q_t for $t \neq 0$. We claim the same is true for $t = 0$. Otherwise Q_0 is covered by a family of curves of degree 1 through A_0 or by a family of curves of degree 1 through B_0 by the assumption (b). A generic member C of this family passes through a very general point. Thus C is free and its deformation fixing one point cover Q_0, giving $-K_{Z_0} \cdot C \geq 5$ by Lemma 1.4., which leads to a contradiction as in the proof of Lemma 1.5. It follows that we get an irreducible conic c through B_0 on Q_0 which is not contained in a generic $Q' \neq Q_0$, while it intersects Q' at B_0. Thus $Q_0 \cdot c = Q' \cdot c > 0$. But c is the limit of conics lying on Q_t which must have zero intersection with Q_t, a contradiction. □

Since the closures of two general leaves of $\mathcal{Q}_0$ are linearly equivalent divisors on Z_0, Lemma 1.6 implies that the linear system defined by these divisors is base-point free. In other words, the leaves of $\mathcal{Q}_0$ give a base-point free pencil on Z_0, defining a morphism $Z_0 \to \mathbb{P}_1$. A generic leaf Q_0 must be projective and smooth. It is a limit of the quadrics Q_t, thus Q_0 itself is a 4-dimensional quadric. So we have a morphism $\nu : \mathcal{Z} \to \Delta \times \mathbb{P}_1$ whose fibers are $\mathbb{Q}_4$ except possibly at finitely many points in $\{0\} \times \mathbb{P}_1$.

Lemma 1.7. *Every fiber of $Z_0 \to \mathbb{P}_1$ has a generically reduced component.*

Proof. We can choose a family of curves $C_t \subset Z_t$ of degree 1 with respect to ξ such that it is a section of $Z_t \to \mathbb{P}_1$. The limit C_0 must be irreducible of degree 1 by the assumption (a) and a section of $Z_0 \to \mathbb{P}_1$. Since $Q_t \cdot C_t = 1$ for $t \neq 0$, C_0 has also intersection number 1 with any fiber of $Z_0 \to \mathbb{P}_1$. Since C_0 is not contained in any fiber, this implies that any fiber must be reduced at the point of intersection with C_0. □

We conclude that every fiber of $\nu : \mathcal{Z} \to \Delta \times \mathbb{P}_1$ is $\mathbb{Q}_4$ by the following.

Lemma 1.8. *Let $\nu : Y \to \mathbb{B}$ be a flat morphism from a complex manifold Y over a 2-dimensional complex ball $\mathbb{B}$ with center $0 \in \mathbb{B}$. Suppose that every fiber $\nu^{-1}(b)$ over $b \in \mathbb{B}$ is biholomorphic to a smooth quadric $\mathbb{Q}$. If the fiber $\nu^{-1}(0)$ has a generically reduced component, then it is also biholomorphic to $\mathbb{Q}$.*

Proof. Since the statement is local at $0 \in \mathbb{B}$, we will shrink the 2-dimensional ball whenever necessary without mentioning it explicitly. Take a local section of ν slicing the fiber $\nu^{-1}(0)$ at a generic point of the generically reduced component. Choose a tubular neighborhood Ω of this section so that $\nu|_\Omega : \Omega \to \mathbb{B}$ is of maximal rank everywhere. Let Ω_b be the fiber over $b \in B$ which is a complex manifold. By assumption, $\Omega_b, b \neq 0$, is an open subset in $\mathbb{Q}$ and is equipped with a flat holomorphic conformal structure. This conformal structure can be extended to a flat conformal structure on Ω_0 by Hartogs (cf. [HM1], Proposition 1). Thus we have a holomorphic immersion $\varphi : \Omega \to \mathbb{Q} \times \mathbb{B}$

which preserves the conformal structure. Either by [HM1], Proposition 3 or by [HM3], Theorem 1.2, we can extend $\varphi|_{\Omega - \Omega_0}$ to a biholomorphic map

$$\Phi : Y - \nu^{-1}(0) \to \mathbb{Q} \times (\mathbb{B} - \{0\}).$$

Choose any non-vanishing vector field ω on $\mathbb{B}$ and lift it to $\tilde{\omega}$ on $\mathbb{Q} \times \mathbb{B}$. The pull-back $\Phi^* \tilde{\omega}$ defines a vector field on $Y - \nu^{-1}(0)$ lifting ω. By Hartogs, we have an extension $\hat{\omega}$ of $\Phi^* \tilde{\omega}$ to Y. Then $\hat{\omega}$ is a lifting of ω to Y. Thus any vector field on $\mathbb{B}$ can be lifted to a vector field on Y, which implies that ν is a locally trivial family. □

It follows that $\nu : \mathcal{Z} \to \mathbb{P}_1 \times \Delta$ is the Grassmannian bundle $\mathbb{G}(2, \mathcal{F})$ of a vector bundle $\mathcal{F}$ of rank 4 on $\mathbb{P}_1 \times \Delta$. For $t \neq 0$, we know that $\mathcal{F}^t := \mathcal{F}|_{\mathbb{P}_1 \times \{t\}} \cong \mathcal{E}^* \cong \mathcal{O} \oplus \mathcal{O}(-1)^3$. To complete the proof of Proposition 1.2, it suffices to show that $\mathcal{F}^0 \cong \mathcal{O} \oplus \mathcal{O}(-1)^3$.

Let $pr : \mathbb{P}_1 \times \Delta \to \Delta$ be the projection to the second factor. Let σ be a non-vanishing section of the invertible sheaf $pr_* \mathcal{F}$ and let σ^t be the corresponding section of $\mathcal{F}^t$ on $\mathbb{P}_1$. For $t \neq 0$, the section σ^t of $\mathcal{F}^t$ is nowhere vanishing on $\mathbb{P}_1$. If σ^0 has a zero, we can easily get a contradiction to the assumption (a). Thus σ^0 is nowhere vanishing and σ defines a trivial line subbundle $\mathcal{O}$ of $\mathcal{F}$ with locally free quotient $\mathcal{G} = \mathcal{F}/\mathcal{O}$. Let $\mathcal{G}^t$ be the restriction of $\mathcal{G}$ on $\mathbb{P}_1 \times \{t\}$. Then $\mathcal{G}^t \cong \mathcal{O}(-1)^3$ for all $t \neq 0$. Applying the same argument using the assumption (a) to $pr_*(\mathcal{G} \otimes \mathcal{O}(1))$, we see that $\mathcal{G}^0 \cong \mathcal{O}(-1)^3$. It follows that $\mathcal{F}^0 \cong \mathcal{O} \oplus \mathcal{O}(-1)^3$. This completes the proof of Proposition 1.2.

Remark. Lemma 1.8 works when we replace the hyperquadric $\mathbb{Q}$ by any irreducible Hermitian symmetric space. The same proof works if the holomorphic conformal structure is replaced by the holomorphic structure modelled on an irreducible Hermitian symmetric space of rank ≥ 2 (see, e.g., [HM1] Section 1 for definition).

2 *Z* As a Projective Variety

The Plücker line bundle ξ on Z is very ample, defining an embedding of Z into $\mathbb{P}H^0(Z, \xi)^*$. Note that

$$H^0(Z, \xi) = H^0(\mathbb{P}_1, \eta_* \xi) = H^0(\mathbb{P}_1, \bigwedge^2 \mathcal{E}) \cong H^0(\mathbb{P}_1, \mathcal{O}(1)^3 \oplus \mathcal{O}(2)^3)$$

is 15-dimensional. Thus $Z \subset \mathbb{P}_{14}$. The restriction of this embedding to $S \subset Z$ is equivalent to the Segre embedding of $S \cong \mathbb{P}_2 \times \mathbb{P}_1$. In fact, the restriction map $H^0(Z, \xi) \to H^0(S, \xi)$ is surjective and

$$H^0(S, \xi) = H^0(\mathbb{P}_1, \chi_* \xi) = H^0(\mathbb{P}_1, \mathcal{V}) \cong H^0(\mathbb{P}_1, \mathcal{O}(1)^3)$$

is 6-dimensional. From now on, we will regard Z as a projective subvariety of $\mathbb{P}_{14}$ under this embedding. In this section, we will study some geometric properties of tangent lines of Z.

Let us fix a 2-dimensional vector space W and a 3-dimensional vector space V. Let us identify our base curve $\mathbb{P}_1$ with $\mathbb{P}W$ and the hyperplane line bundle $\mathcal{O}(1)$ with the dual tautological line bundle h on $\mathbb{P}W$ such that $H^0(\mathbb{P}W, h) = W^*$. Also make an identification of S with $\mathbb{P}\mathcal{V}^*$ where $\mathcal{V} = V^* \otimes h$, so that the subspace of $H^0(Z, \xi)^*$ spanned by $S \subset \mathbb{P}H^0(Z, \xi)^*$ is canonically identified with $V \otimes W$. Under these identifications,

$$H^0(S, \xi) = H^0(\mathbb{P}V \otimes h^*, \xi) = H^0(\mathbb{P}W, V^* \otimes h) = V^* \otimes W^*$$

in a canonical manner. The kernel of the restriction map $H^0(Z, \xi) \to H^0(S, \xi)$ can be naturally identified with the subspace of

$$H^0(Z, \xi) = H^0(\mathbb{P}W, \bigwedge^2 \mathcal{E})$$

consisting of sections of the subbundle $\bigwedge^2 \mathcal{V} \subset \bigwedge^2 \mathcal{E}$ on $\mathbb{P}W$. So the kernel is

$$H^0(\mathbb{P}W, \bigwedge^2 \mathcal{V}) = H^0(\mathbb{P}W, \bigwedge^2 (V^* \otimes h)) = H^0(\mathbb{P}W, \bigwedge^2 V^* \otimes h^2) = \bigwedge^2 V^* \otimes S^2 W^*.$$

This gives the natural exact sequence

$$0 \longrightarrow V \otimes W \longrightarrow H^0(Z, \xi)^* \longrightarrow \bigwedge^2 V \otimes S^2 W \longrightarrow 0.$$

Summarizing,

Proposition 2.1. *Once we make an identification of $\mathbb{P}_1$ with $\mathbb{P}W$ and S with $\mathbb{P}V \times \mathbb{P}W$ then the subspace of $H^0(Z, \xi)^*$ spanned by $S \subset \mathbb{P}H^0(Z, \xi)^*$ is canonically identified with $V \otimes W$ and the quotient space is canonically identified with $\bigwedge^2 V \otimes S^2 W$.*

To study the projective geometry of $Z \subset \mathbb{P}H^0(Z, \xi)^*$, it is convenient to fix a splitting of

$$0 \longrightarrow \mathcal{V} \longrightarrow \mathcal{E} \longrightarrow \mathcal{O} \longrightarrow 0$$

given by a trivial line subbundle $\mathcal{A} \subset \mathcal{E}$ complementary to $\mathcal{V}$. Then

$$\bigwedge^2 \mathcal{E} = (\bigwedge^2 \mathcal{V}) \oplus (\mathcal{V} \otimes \mathcal{A}).$$

Thus $H^0(Z, \xi)$ has splitting

$$H^0(Z, \xi) = H^0(\mathbb{P}W, \bigwedge^2 \mathcal{V}) \oplus H^0(\mathbb{P}W, \mathcal{V} \otimes \mathcal{A}) = (\bigwedge^2 V^* \otimes S^2 W^*) \oplus (V^* \otimes W^*).$$

To simplify the notation, set

$$U = H^0(Z,\xi)^*,\ U_1 := V \otimes W,\ U_2 := \bigwedge^2 V \otimes S^2W$$

and identify $U := U_1 \oplus U_2$ by the above splitting. We denote by $\widetilde{Z} \subset U$ the affine cone defining Z. Then

$$\widetilde{Z} = \{(v \otimes w, (v_1 \wedge v_2) \otimes w^2) \in U \mid w \in W \text{ and } v, v_1, v_2 \in V \text{ with } v \wedge v_1 \wedge v_2 = 0\}.$$

Now we use the symmetry of Z. The vector spaces U, U_1 and U_2 have natural $\mathrm{SL}(V) \times \mathrm{SL}(W)$-module structure. As an $\mathrm{SL}(V)$-module, $\bigwedge^2 V$ is isomorphic to V^*. To simplify the notation, we will make the identification

$$U_2 = V^* \otimes S^2W.$$

This identification is uniquely defined up to a multiplicative constant. We will fix a choice of such an identification. Then the affine cones of S and Z can be written as

$$\widetilde{S} = \{v \otimes w \in U_1 \mid v \in V, w \in W\}$$

$$\widetilde{Z} = \{(v \otimes w, v^* \otimes w^2) \mid w \in W, v \in V, v^* \in V^*, < v^*, v >= 0\}$$

where $< v^*, v >$ denotes the evaluation of v^* at v. Let

$$\widetilde{R} := \{v^* \otimes w^2 \in U_2 \mid v \in V^*, w \in W\}$$

and $R \subset \mathbb{P}U_2$ be the corresponding projective subvariety. The varieties $S \subset \mathbb{P}U_1, R \subset \mathbb{P}U_2$ and $Z \subset \mathbb{P}U$ are invariant under the $\mathrm{SL}(V) \times \mathrm{SL}(W)$-action on $\mathbb{P}U$.

Remark. Here and in what follows, whenever we identify an irreducible $\mathrm{SL}(V)$-module or an irreducible $\mathrm{SL}(W)$-module with another, the identification is unique up to a choice of multiplicative constant by Schur lemma. We will implicitly make a fixed choice of the identification.

The projective geometry of $Z \subset \mathbb{P}U$ we are interested in is about the linear span of the variety of tangent lines of Z in $\bigwedge^2 U$.

Lemma 2.2. *The tangent lines to S, regarded as points of $\mathbb{P}(\bigwedge^2 U_1)$ by Plücker map, span $\bigwedge^2 U_1$.*

Proof. The irreducible K-module decomposition of $\bigwedge^2 U_1$ is

$$\bigwedge^2 (V \otimes W) = (\bigwedge^2 V \otimes S^2W) \oplus (S^2V \otimes \bigwedge^2 W).$$

Consider a plane in $\widetilde{S}$ of the form

$$\mathbb{C}(v_1 \otimes w) + \mathbb{C}(v_2 \otimes w)$$

for some $v_1 \neq v_2 \in V$ and $w \in W$. This corresponds to the element $(v_1 \wedge v_2) \otimes (w \odot w)$ in $\bigwedge^2 V \otimes S^2W$. On the other hand, a plane in $\widetilde{S}$ of the form $(v \otimes w_1) \wedge (v \otimes w_2)$ for some $v \in V$ and $w_1 \neq w_2 \in W$ corresponds to the element $(v \odot v) \otimes (w_1 \wedge w_2)$ in $S^2V \otimes \bigwedge^2 W$. Thus there are points corresponding to tangent lines to S in both irreducible $\mathrm{SL}(V) \times \mathrm{SL}(W)$-components of $\bigwedge^2 U_1$. Since the linear span of the tangent lines to S in $\bigwedge^2 U_1$ is an $\mathrm{SL}(V) \times \mathrm{SL}(W)$-invariant subspace, it must be the whole of $\bigwedge^2 U_1$. □

Lemma 2.3. *The linear span of the tangent lines to R in $\bigwedge^2 U_2$ contains the unique subspace of codimension 3 in $\bigwedge^2 U_2$ with $\mathrm{SL}(V) \times \mathrm{SL}(W)$-module structure $(\bigwedge^2 V^* \otimes S^4W) \oplus (S^2V^* \otimes \bigwedge^2(S^2W))$.*

Proof. The irreducible $\mathrm{SL}(V) \times \mathrm{SL}(W)$-module decomposition of $\bigwedge^2 U_2$ is

$$\begin{aligned}\bigwedge^2(V^* \otimes S^2W) &= \left[\bigwedge^2 V^* \otimes S^2(S^2W)\right] \oplus \left[S^2V^* \otimes \bigwedge^2(S^2W)\right] \\ &= \bigwedge^2 V^* \oplus \left(\bigwedge^2 V^* \otimes S^4W\right) \oplus \left(S^2V^* \otimes \bigwedge^2(S^2W)\right)\end{aligned}$$

using the $\mathrm{SL}(W)$-module decomposition $S^2(S^2W) = \mathbb{C} \oplus S^4W$. A plane in $\widetilde{R}$ of the form $\mathbb{C}(v_1^* \otimes w^2) + \mathbb{C}(v_2^* \otimes w^2)$ for some $v_1^* \neq v_2^* \in V^*$ and $w \in W$ corresponds to the element $(v_1^* \wedge v_2^*) \otimes (w^2 \odot w^2)$ in $\bigwedge^2 V^* \otimes S^4W$. Consider the family of secant lines of R through the point $[v^* \otimes w_1^2]$ and the family of points $[v^* \otimes (w_1 + tw_2)^2]$, where $0 \neq t \in \mathbb{C}$ and $w_1 \neq w_2 \in W$. They correspond to the elements

$$(v^* \otimes w_1^2) \wedge (v^* \otimes (w_1 + tw_2)^2) \in \bigwedge^2(V^* \otimes S^2W).$$

As t approaches to 0, this family of secant lines converges to the tangent line

$$\lim_{t \to 0} \frac{(v^* \otimes w_1^2) \wedge (v^* \otimes (w_1 + tw_2)^2)}{t} = 2(v^* \otimes v^*) \otimes (w_1^2 \wedge w_2^2)$$

which is contained in $S^2V^* \otimes \bigwedge^2(S^2W)$. Since the linear span of the tangent lines to R is $\mathrm{SL}(V) \times \mathrm{SL}(W)$-invariant in $\bigwedge^2 U_2$, the proof of Lemma 2.3 is complete. □

Let us recall the $\mathrm{SL}(W)$-module decomposition

$$W \otimes S^2W = W \oplus S^3W$$

given by the symmetrization $W \otimes S^2W \to S^3W$ and its two-dimensional kernel. Also recall the $\mathrm{SL}(V)$-module decomposition

$$V \otimes V^* = \mathbf{ad}_V \oplus \mathbb{C}$$

where $\mathbf{ad}_V$ corresponds to the traceless endomorphisms and $\mathbb{C}$ corresponds to the homothety endomorphisms, under the identification $V \otimes V^* = \mathrm{End}(V)$. So we have the equivalence of $\mathrm{SL}(V) \times \mathrm{SL}(W)$-modules

$$\begin{aligned}(V \otimes W) \wedge (V^* \otimes S^2W) &= (V \otimes W) \otimes (V^* \otimes S^2W) \\ &= (V \otimes V^*) \otimes (W \otimes S^2W) \\ &= (\mathbf{ad}_V \otimes W) \oplus (\mathbf{ad}_V \otimes S^3W) \oplus W \oplus S^3W.\end{aligned}$$

Proposition 2.4. *Let $\Sigma \subset \bigwedge^2 U$ be the linear span of the tangent lines to Z. Then in the $\mathrm{SL}(V) \times \mathrm{SL}(W)$-module decomposition of $\bigwedge^2 U$*

$$\begin{aligned}\bigwedge^2(U_1 \oplus U_2) &= \bigwedge^2 U_1 \oplus \bigwedge^2 U_2 \oplus (U_1 \otimes U_2) \\ &= \bigwedge^2 U_1 \oplus \bigwedge^2(V^* \otimes S^2W) \oplus [(V \otimes W) \wedge (V^* \otimes S^2W)] \\ &= \bigwedge^2 U_1 \oplus \Big[\bigwedge^2 V^* \oplus (\bigwedge^2 V^* \otimes S^4W) \oplus (S^2V^* \otimes \bigwedge^2(S^2W))\Big] \\ &\oplus [(\mathbf{ad}_V \otimes W) \oplus (\mathbf{ad}_V \otimes S^3W) \oplus W \oplus S^3W]\end{aligned}$$

Σ contains

$$\bigwedge^2 U_1 \oplus \Big[(\bigwedge^2 V^* \otimes S^4W) \oplus (S^2V^* \otimes \bigwedge^2(S^2W))\Big] \oplus [(\mathbf{ad}_V \otimes W) \oplus (\mathbf{ad}_V \otimes S^3W) \oplus S^3W].$$

Proof. We have already seen in Lemma 2.2. and Lemma 2.3. that $\bigwedge^2 U_1, \bigwedge^2 V^* \otimes S^4W, S^2V^* \otimes \bigwedge^2(S^2W)$ are contained in Σ.

Choose $v_1, v_2 \in V$ and $v_1^*, v_2^* \in V^*$ such that

$$< v_1^*, v_1 >= 0, \ < v_1^*, v_2 > + < v_2^*, v_1 >= 0, \ < v_1^*, v_2 > - < v_2^*, v_1 > \neq 0.$$

Then for any non-zero element $w \in W$,

$$(v_1 \otimes w + v_1^* \otimes w^2) \in \widetilde{Z} = \{(v \otimes w, v^* \otimes w^2); w \in W, v \in V, v^* \in V^*, < v^*, v >= 0\}.$$

The plane

$$\mathbb{C}(v_1 \otimes w + v_1^* \otimes w^2) + \mathbb{C}(v_2 \otimes w + v_2 \otimes w^2) \subset U$$

is tangent to $\tilde{Z}$. In fact, the line

$$\begin{aligned}&\{(v_1 \otimes w + v_1^* \otimes w^2) + \lambda(v_2 \otimes w + v_2^* \otimes w^2); \lambda \in \mathbb{C}\} \\ &= \{(v_1 + \lambda v_2) + (v_1^* + \lambda v_2^*) \otimes w^2); \lambda \in \mathbb{C}\}\end{aligned}$$

is tangent to $\widetilde{Z}$ because

$$< (v_1^* + \lambda v_2^*, v_1 + \lambda v_2 >= \lambda^2 < v_2^*, v_2 >$$

contains no linear term in λ. In $\bigwedge^2 U$, this plane is represented by the element

$$(v_1 \otimes w + v_1^* \otimes w^2) \wedge (v_2 \otimes w + v_2^* \otimes w^2)$$

$$= (v_1 \wedge v_2) \otimes w^2 + (v_1^* \wedge v_2^*) \otimes w^4 + (v_1 \otimes w) \wedge (v_2^* \otimes w^2) - (v_2 \otimes w) \wedge (v_1^* \otimes w^2).$$

The first two terms on the right hand side are in $\bigwedge^2 V \otimes S^2 W$ and $\bigwedge^2 V^* \otimes S^4 W$. Upon identification of $U_1 \wedge U_2$ with $U_1 \otimes U_2$, the last two terms corresponds to

$$(v_1 \otimes v_2^*) \otimes w^3 - (v_2 \otimes v_1^*) \otimes w^3 = (v_1 \otimes v_2^* - v_2 \otimes v_1^*) \otimes w^3.$$

But this element of $(V^* \otimes V) \otimes S^3 W = (\mathbf{ad}_V + \mathbb{C}\,\mathrm{Id}_V) \otimes S^3 W$ does not lie in $\mathbf{ad}_V \otimes S^3 W$ because $< v_1^*, v_2 > - < v_2^*, v_1 > \neq 0$. Nor does it lie in $\mathbb{C}\,\mathrm{Id}_V \otimes S^3 W$ because $v_1 \otimes v_2^* - v_2 \otimes v_1^*$ has rank 2 as an endomorphism of V. Since this element is contained in Σ, we see that both $\mathbf{ad}_V \otimes S^3 W$ and $S^3 W$ are contained in Σ from the $\mathrm{SL}(V) \times \mathrm{SL}(W)$-invariance.

Now fix $v \in V$ and $v^* \in V^*$ with $< v^*, v >= 0$ and choose two independent vectors $w_1, w_2 \in W$. Consider the point $v \otimes w_1 + v^* \otimes w_1^2$ and a family of points

$$\{[v \otimes (w_1 + tw_2) + v^* \otimes (w_1 + tw_2)^2]; t \in \mathbb{C}\}$$

in Z. The secant lines

$$[v \otimes w_1 + v^* \otimes w_1^2] \wedge [v \otimes (w_1 + tw_2) + v^* \otimes (w_1 + tw_2)^2]$$

converges to the tangent line

$$\begin{aligned} &(v \otimes w_1 + v^* \otimes w_1^2) \wedge (v \otimes w_2 + 2v^* \otimes w_1 w_2) \\ &\quad = 2(v \otimes w_1) \wedge (v^* \otimes w_1 w_2) - (v \otimes w_2) \wedge (v^* \otimes w_1^2) \\ &\qquad + v^2 \otimes (w_1 \wedge w_2) + 2(v^*)^2 \otimes (w_1^2 \wedge w_1 w_2). \end{aligned}$$

The last two terms on the right hand side lie in $S^2 V \otimes \bigwedge^2 W \subset \bigwedge^2 U_1$ and $S^2 V^* \otimes \bigwedge^2 (S^2 W)$ respectively. Upon identifying $U_1 \wedge U_2$ with $U_1 \otimes U_2$, the first two terms correspond to

$$2(v \otimes v^*) \otimes (w_1 \otimes w_1 w_2) - (v \otimes v^*) \otimes (w_2 \otimes w_1^2) = (v \otimes v^*) \otimes (2w_1 \otimes w_1 w_2 - w_2 \otimes w_1^2).$$

This element of Σ lies in $\mathbf{ad}_V \otimes (W \otimes S^2 W) = \mathbf{ad}_V \otimes (W \oplus S^3 W)$ because $< v^*, v >= 0$. Since $2w_1 \otimes w_1 w_2 - w_2 \otimes w_1^2$ is not contained in $S^4 W$, it must have a component in $\mathbf{ad}_V \otimes W$. It follows that $\mathbf{ad}_V \otimes W$ is contained in Σ. □

3 Z As the Variety Of Minimal Rational Tangents For G/P

Let us recall some general facts before we look at our homogeneous variety G/P (cf. [HM1], Section 2, [HM3], Section 1). Let X be a Fano manifold of Picard number 1 and $x \in X$ be a generic point. Let $\mathcal{K}_x$ be the union of all components of the normalized Chow space of rational curves of minimal degree through x. It has finitely many components and each of them is a smooth projective variety. If the anti-canonical degree of members of $\mathcal{K}_x$ is $p+2$, then $\mathcal{K}_x$ is of pure dimension p, and for a generic member C of any component of $\mathcal{K}_x$,

$$T(X)|_C = \mathcal{O}(2) \oplus [\mathcal{O}(1)]^p \oplus \mathcal{O}^{n-1-p}.$$

Define the tangent map $\tau_x : \mathcal{K}_x \to \mathbb{P}T_x(X)$ by assigning the tangent vector at x to each member of $\mathcal{K}_x$ which is smooth at x. By [Ke], this can be extended to a morphism which is finite over its image. The image is denoted by $\mathcal{C}_x$ and called the **variety of minimal rational tangents** at x. Suppose X is embedded in some projective space $\mathbb{P}_N$ and a minimal rational curve through a generic point x is a line in $\mathbb{P}_N$. Since lines through x in $\mathbb{P}_N$ are determined by their tangent vectors at x, τ_x is an embedding.

We need to recall some definitions in the theory of differential systems ([Ya]). Given a distribution D on a complex manifold X, define the weak derived system $D^k, k \geq 1$, inductively by

$$D^1 = D,$$
$$D^k = D^{k-1} + [D, D^{k-1}].$$

For a generic point $x \in X$ in a neighborhood of which D^k's are subbundles of $T(X)$, we define the **symbol algebra** of D at x as the graded nilpotent Lie algebra $D^1_x + D^2_x/D^1_x + \cdots + D^r_x/D^{r-1}_x$ where r is chosen so that $D^{r+1} = D^r$. This integer r is called the depth of the system.

When X is a Fano manifold of Picard number 1, let $\mathcal{W}_x \subset T_x(X)$ be the linear span of $\mathcal{C}_x$ at a generic point $x \in X$ and $\mathcal{W}$ be the distribution defined by $\mathcal{W}_x$'s on a Zariski open subset of X. The distribution $\mathcal{W}$ has the following two properties.

Proposition 3.1. ([HM1], Proposition 10) *For any plane $\Lambda \subset \mathcal{W}_x$ corresponding to a tangent line to the smooth locus of $\mathcal{C}_x$, there exists a local analytic integral surface of the distribution $\mathcal{W}$ which contains x as a smooth point and has Λ as its tangent space at x. In particular, the Frobenius bracket tensor $[\ ,\ ]_x : \wedge^2\mathcal{W}_x \to T_x(X)/\mathcal{W}_x$ at a generic point $x \in X$ annihilates any element corresponding to tangent lines to the smooth locus of $\mathcal{C}_x$.*

Proposition 3.2. ([HM4], Proposition 4) *At a generic point $x \in X$, the symbol algebra of $\mathcal{W}$ has dimension $n = \dim(X)$.*

Now we introduce our homogeneous variety G/P. Let us start with the Lie algebra part. This is a special case of [HM2] Section 2. Let $\mathfrak{g}$ be the complex simple Lie algebras of type F_4. Let $\mathfrak{h}$ be a Cartan subalgebra of $\mathfrak{g}$ and Φ be the corresponding system of roots. We choose a system of simple roots

$\{\alpha_1, \alpha_2, \alpha_3, \alpha_4\}$. Here the numeration of simple roots are made such that α_1 and α_2 are short roots and the maximal root is

$$2\alpha_1 + 4\alpha_2 + 3\alpha_3 + 2\alpha_4.$$

We will be interested in the maximal parabolic subalgebra $\mathfrak{p}$ of $\mathfrak{g}$ corresponding to the simple root α_2. Let us recall the definition of $\mathfrak{p}$. For $-4 \leq k \leq k$, we define Φ_k as the set of all roots $\sum_{i=1}^{4} m_i\alpha_i$ with $m_2 = k$. For $\alpha \in \Phi$, let $\mathfrak{g}_\alpha$ be the corresponding root space. Define

$$\mathfrak{g}_0 = \mathfrak{h} \oplus \bigoplus_{\alpha \in \Phi_0} \mathfrak{g}_\alpha,$$

$$\mathfrak{g}_k = \bigoplus_{\alpha \in \Phi_k} \mathfrak{g}_\alpha, \quad k \neq 0.$$

Then the parabolic subalgebra $\mathfrak{p}$ can be defined as

$$\mathfrak{p} = \mathfrak{g}_0 \oplus \mathfrak{g}_{-1} \oplus \mathfrak{g}_{-2} \oplus \mathfrak{g}_{-3} \oplus \mathfrak{g}_{-4}.$$

Define

$$\mathfrak{k} = \mathfrak{g}_0,$$
$$\mathfrak{u} = \mathfrak{g}_{-1} \oplus \mathfrak{g}_{-2} \oplus \mathfrak{g}_{-3} \oplus \mathfrak{g}_{-4}.$$

Then $\mathfrak{p} = \mathfrak{u} + \mathfrak{k}$ is a Levi decomposition of $\mathfrak{p}$. It is easy to check that the semi-simple part of $\mathfrak{k}$ is isomorphic to $\mathfrak{sl}_2 \oplus \mathfrak{sl}_3$.

Now let G be a simply connected complex Lie group with Lie algebra $\mathfrak{g}$ and $P \subset G$ be the subgroup with Lie algebra $\mathfrak{p}$. The homogeneous space G/P is a projective variety of dimension 20 with $\mathrm{Pic}(G/P) = \mathbb{Z}L$ for an ample generator L and $c_1(G/P) = 7L$. Let $o \in G/P$ be the base point. The tangent space $T_o(G/P)$ is canonically isomorphic to $\mathfrak{g}/\mathfrak{p}$. The isotropy representation of P on $T_o(G/P)$ is equivalent to the adjoint representation of P on $\mathfrak{g}/\mathfrak{p}$. The choice of a Levi factor $\mathfrak{k}$ as above gives a reductive subgroup $K \subset P$ with Lie algebra $\mathfrak{k}$. As a complex Lie group, K is isogenous to $\mathbb{C}^* \times \mathrm{SL}(V) \times \mathrm{SL}(W)$ where V and W are as in Section 2. After such a choice of Levi factor, we have the identification

$$T_o(G/P) = \mathfrak{g}/\mathfrak{p} = \mathfrak{g}_1 + \mathfrak{g}_2 + \mathfrak{g}_3 + \mathfrak{g}_4$$

which is the decomposition into irreducible K-modules. In fact, this is the weight space decomposition of the action of the subgroup of K corresponding to $\mathbb{C}^*$: $\lambda \in \mathbb{C}^*$ acts as the scalar multiplication by λ^i on $\mathfrak{g}_i$. This decomposition depends on the choice of the Levi factor. But the filtration on $T_o(G/P)$ given by

$$D_{i,o} := \mathfrak{g}_1 + \cdots + \mathfrak{g}_i,$$
$$D_{1,o} \subset D_{2,o} \subset D_{3,o} \subset D_{4,o} = T_o(G/P)$$

is naturally defined. This filtration induces natural subbundles of the tangent bundle

$$D_1 \subset D_2 \subset D_3 \subset D_4 = T(G/P).$$

In fact, this is the weak derived system of the distribution D_1. Let $H_i \subset \mathbb{P}\mathfrak{g}_i$ be the highest weight variety with respect to the K-action. Then the following is well-known.

Proposition 3.3. ([HM2], Proposition 5) *For any $a \in \mathbb{P}T_o(G/P)$, the closure of the P-orbit of a under the isotropy action contains H_1. For any $a \in \mathbb{P}T_o(G/P) - \mathbb{P}\mathfrak{g}_1$, the closure of the P-orbit of a contains H_2.*

The line bundle L is very ample, giving an embedding $G/P \subset \mathbb{P}_{271}$. Moreover G/P is covered by lines of $\mathbb{P}_{271}$. Let $\mathcal{K}_x$ be the subvariety consisting of all lines passing through $x \in G/P$. The tangent morphism $\tau_x : \mathcal{K}_x \to \mathcal{C}_x \subset \mathbb{P}T_x(G/P)$ is an embedding. From $c_1(X) = 7L$, $\mathcal{C}_x$ is 5-dimensional. By the homogeneity, the structure of $\mathcal{C}_x$ as a subvariety of $\mathbb{P}T_x(G/P)$ does not depend on x. At the base point $o \in G/P$, $\mathcal{C}_o$ is invariant under the isotropy action of P on $\mathbb{P}T_o(G/P)$. In particular, it is invariant under the K-action for any choice of the Levi factor $K \subset P$. The structure of $\mathfrak{g}_i$ as an irreducible K-module can be easily read off from the basic theory of roots:

$$\begin{aligned} \mathfrak{g}_1 &= V \otimes W, \\ \mathfrak{g}_2 &= \bigwedge^2 V \otimes S^2 W, \\ \mathfrak{g}_3 &= W, \\ \mathfrak{g}_4 &= V. \end{aligned}$$

Thus as K-modules, $D_{1,o} = U_1$ and $D_{2,o} = U$. Under this identification, we aim to show that $\mathcal{C}_x$ is equivalent to Z. The following characterization of Z is useful.

Proposition 3.4. *The variety Z is the only K-invariant projective subvariety of dimension ≤ 5 in $\mathbb{P}U$ which intersects both S and R.*

Proof. Viewing elements of U_1 as a linear map from V to W, we can list all the K-orbits in $\mathbb{P}U_1$:

{Elements of rank 1} of dimension 3,
{Elements of rank 2} of dimension 5.

Similarly, viewing elements of U_2 as linear maps from V^* to S^2W, we can list all K-orbits $\mathbb{P}U_2$. Let $Q := \{[w^2] \in \mathbb{P}S^2W; w \in W\}$ be the conic curve in the plan $\mathbb{P}S^2W$. Then there are exactly five K-orbits in $\mathbb{P}U_2$:

{Elements of rank 1 with images contained in Q} of dimension 3,
{Elements of rank 1 with images not contained in Q} of dimension 4,
{Elements of rank 2 with images tangent to Q} of dimension 6,

{Elements of rank 2 with images not tangent to Q} of dimension 7,
{Elements of rank 3} of dimension 8.

Let Z' be an irreducible K-invariant projective subvariety of $\mathbb{P}U$ which intersects both S and R. Then Z' is contained in neither $\mathbb{P}U_1$ nor $\mathbb{P}U_2$. Thus Z' contains a point $[(u_1, u_2)] \in \mathbb{P}(U_1 + U_2)$ with $u_1 \neq 0$ and $u_2 \neq 0$. Recall that $\lambda \in \mathbb{C}^* \subset K$ acts as the multiplication by λ on u_1 and as the multiplication by λ^2 on u_2. Thus the closure of the $\mathbb{C}^*$ orbit of the point $[(u_1, u_2)]$ contains both $[u_1] \in \mathbb{P}U_1$ and $[u_2] \in \mathbb{P}U_2$. Z' contains the K-orbits of $[u_1]$ and $[u_2]$. This implies that the K-orbits of $[u_1]$ and $[u_2]$ must have dimension ≤ 4. The above dimension counting implies that u_1 is in $\widetilde{S}$ and u_2 lies in the affine cone of the closure of the 4-dimensional orbit in $\mathbb{P}U_2$. In other words, $u_1 = v \otimes w$ and $u_2 = v^* \otimes s$ for some $v \in V, v^* \in V^*, w \in W$ and $s \in S^2W$. For fixed w and s, the SL(V)-orbit of $[(u_1, u_2)]$ has dimension 5 unless $< v^*, v >= 0$. For fixed v^* and v, the SL(W)-orbit of $[(u_1, u_2)]$ has dimension ≥ 2 unless s is proportional to w^2. Thus the K-orbit of $[(u_1, u_2)]$ has dimension > 5 unless $[(u_1, u_2)]$ lies on Z. We conclude that $Z' = Z$. □

Proposition 3.5. *The variety of minimal rational tangents $\mathcal{C}_o \subset \mathbb{P}T_o(G/P)$ is irreducible and contained in $\mathbb{P}D_{2,o}$. Under the identification of $D_{2,o}$ with U as K-modules, $\mathcal{C}_o$ corresponds to Z. In particular, the normalized Chow space $\mathcal{K}_o$ is biholomorphic to Z.*

Proof. Note that the irreducibility is not entirely obvious, because a priori there may exist more than one family of lines on G/P. We know that $\mathcal{C}_o$ is smooth and of pure dimension 5. We claim that $\mathcal{C}_o$ does not contain a component inside $\mathbb{P}D_{1,o}$. Suppose there exists a component inside $\mathbb{P}D_{1,o}$. Then the component must be equal to $\mathbb{P}D_{1,o}$ by considering dimensions. This implies that there is no other component from the smoothness of $\mathcal{C}_o$ and Proposition 3.3. Then D_1 agrees with the distribution $\mathcal{W}$ spanned by the variety of minimal rational tangents. By Proposition 3.1, D_1 must be integrable, a contradiction to Proposition 3.2. Thus no component of $\mathcal{C}_o$ is contained in $\mathbb{P}D_{1,o}$ and each component $\mathcal{C}'_o$ of $\mathcal{C}_o$ must contain both $H_1 = S$ and $H_2 = R$ by Proposition 3.3. Then $\mathcal{C}'_o \cap \mathbb{P}D_{2,o}$ is a K-invariant projective subvariety in $\mathbb{P}U$ of dimension ≤ 5 which intersects both S and R. It follows that $\mathcal{C}_o$ is contained in $\mathbb{P}D_{2,o}$ and is isomorphic to $Z \subset \mathbb{P}U$ as a projective variety, by Proposition 3.4. The last statement of Proposition 3.5. follows from the fact that the tangent morphism τ_x is an embedding. □

Proposition 3.6. *In the statement of Proposition 2.4, Σ is exactly the subspace of $\bigwedge^2 U$ corresponding to the factors isomorphic to*

$$\bigwedge^2 U_1 \oplus \Big[(\bigwedge^2 V^* \otimes S^4 W) \oplus (S^2 V^* \otimes \bigwedge^2 (S^2 W)) \Big] \oplus [(\mathbf{ad}_V \otimes W) \oplus (\mathbf{ad}_V \otimes S^3 W) \oplus S^3 W].$$

In particular, $\bigwedge^2 U / \Sigma \cong W \oplus V$.

Proof. When we make an identification of $D_{2,o}$ with U, $\Sigma \subset \bigwedge^2 D_{2,o}$ is annihilated by the Lie bracket of the symbol algebra of D_2 at o by Proposition 3.1. If Σ is larger, $(D_2)^2/D_2$ must have dimension $< 5 = \dim(W \oplus V)$, where $(D_2)^2$ denotes the weak derived system of the distribution D_2. But the Lie bracket for the symbol algebra must be the one inherited from the Lie bracket of the Lie algebra $\mathfrak{g}$, thus $(D_2)^2/D_2 = \mathfrak{g}_3 + \mathfrak{g}_4$ is of dimension 5. □

4 Proof Of Main Theorem

Let $\pi : \mathcal{X} \to \Delta$ be as in Main Theorem. It is easy to see $\mathrm{Pic}(\mathcal{X}) = \mathrm{Pic}(G/P) = \mathrm{Pic}(X_0)$. Let $\mathcal{L}$ be an ample generator of $\mathrm{Pic}(\mathcal{X})$. Then $c_1(X_0) = 7\mathcal{L}$. X_0 is a Fano manifold of Picard number 1.

Proposition 4.1. *Let $x \in X_0$ be a generic point and $\mathcal{K}_x$ be the normalized Chow variety of rational curves of degree 1 with respect to $\mathcal{L}$. Then $\mathcal{K}_x \cong Z$.*

Proof. Let $\{x_t \in X_t\}$ be a section of π with $x_0 = x$. Let

$$\psi : \mathcal{Z} = \cup_{t \in \Delta} \mathcal{K}_{x_t} \to \Delta$$

be the family of normalized Chow spaces of minimal rational curves through x_t. Then ψ is a smooth projective morphism by the same proof as Proposition 4 in [HM1]. We know $\psi^{-1}(t) \cong Z, t \neq 0$, by Proposition 3.5. The tangent morphisms $\tau_{x_t} : \mathcal{K}_{x_t} \to \mathcal{C}_{x_t} \subset \mathbb{P}T_{x_t}(X_t)$ form a flat family of finite morphisms. By pulling back the hyperplane line bundle of $\mathbb{P}T_{x_t}(X_t)$, we get a line bundle ζ on $\mathcal{Z}$ whose restriction to $\psi^{-1}(t)$ is isomorphic to the Plücker line bundle ξ on Z. It follows that $\psi^{-1}(0) \cong Z$ by Proposition 1.2. and the next lemma whose proof is the same as those of Lemma 1 and Lemma 2 in [HM4]. Instead of repeating the proof, let us just mention that it is a direct consequence of the fact that the curves C_t corresponds to surfaces in X_t of degree 1 for (a) and of degree 2 for (b) with respect to the ample line bundle $\mathcal{L}$. □

Lemma 4.2. *Let $\psi : \mathcal{Z} \to \Delta$ be the family of normalized Chow spaces of minimal rational curves through generic points x_t of X_t and ζ be the line bundle on $\mathcal{Z}$ as defined above. Then for any flat family of curves $\{C_t \subset Z_t \mid t \in \Delta\}$,*

(a) *if C_t has degree 1 with respect to ζ, the limit curve C_0 is irreducible, and*

(b) *if C_t has degree 2 with respect to ζ, then the limit curve C_0 is either irreducible or a union of two irreducible curves of degree 1 with respect to ζ.*

Let $D_i^t, 1 \leq i \leq 4, t \neq 0$, be the distributions on X_t induced by the distributions D_i on G/P under the biholomorphism $X_t \cong G/P$. Let D_i^0 be the limit distribution which is well-defined on a Zariski open subset of X_0. On a Zariski open subset of X_0, we have another distribution $\mathcal{W}$ defined as the linear span of the variety of minimal rational tangents $\mathcal{C}_x$ as x varies over generic points of X. Recall that on $G/P \cong X_t, t \neq 0$, the distribution defined by the linear span of the variety of minimal rational tangents is D_2 by Proposition 3.5. The key ingredient in the proof of Main Theorem is the following.

Proposition 4.3. *On a Zariski open subset of X_0, the two distributions $\mathcal{W}$ and D_2^0 agree and the tangent morphism $\tau_x : Z \cong \mathcal{K}_x \to \mathbb{P}\mathcal{W}_x$ at a generic $x \in X_0$ is the embedding defined by the complete linear system $|\xi|$ on Z. In particular, the variety of minimal rational tangents $\mathcal{C}_x \subset \mathbb{P}\mathcal{W}_x$ is isomorphic to $Z \subset \mathbb{P}U$ as projective varieties.*

For the proof of Proposition 4.3., we need the following algebraic result, which is an analogue of [HM4] Proposition 6 in our setting.

Lemma 4.4. *Let $\mathfrak{n} = \sum_{i=1}^{\infty} \mathfrak{n}_i$ be a graded Lie algebra generated by $\mathfrak{n}_1 = U$ so that the kernel of the Lie bracket $[\ ,\ ] : \bigwedge^2 \mathfrak{n}_1 \to \mathfrak{n}_2$ is exactly $\Sigma \subset \bigwedge^2 U$ defined in Proposition 3.6. Then $\mathfrak{n}_i = 0$ for $i \geq 3$.*

Proof. We will check this explicitly by brute force. The roots $\Phi_i, 1 \leq i \leq 4$ defining $\mathfrak{g}_i$ can be explicitly written as follows ([HM2] p. 226), where we denote a root $a\alpha_1 + b\alpha_2 + c\alpha_3 + d\alpha_4$ by *abcd*:

$\Phi_1 = \{0100, 0110, 0111, 1100, 1110, 1111\}$,
$\Phi_2 = \{0210, 0211, 0221, 1210, 1211, 1221, 2210, 2211, 2221\}$,
$\Phi_3 = \{1321, 2321\}$,
$\Phi_4 = \{2421, 2431, 2432\}$.

By the assumption, we can identify $\mathfrak{n}_1 = U$ with $\mathfrak{g}_1 + \mathfrak{g}_2$ and $\mathfrak{n}_2 = \mathfrak{g}_3 + \mathfrak{g}_4$ as vector spaces. We choose a basis of $\mathfrak{n}_1$ corresponding to root vectors in $\mathfrak{g}_1 + \mathfrak{g}_2$ and a basis of $\mathfrak{n}_2$ corresponding to root vectors in $\mathfrak{g}_3 + \mathfrak{g}_4$. By abuse of notation, we will denote a root vector corresponding to the root $a\alpha_1 + b\alpha_2 + c\alpha_3 + d\alpha_4$ by *abcd*. In other words, the basis of U can be written as

$$\{abcd \in \Phi_1 \cup \Phi_2\}$$

and the basis of $\mathfrak{n}_2$ can be written as

$$\{abcd \in \Phi_3 \cup \Phi_4\}.$$

In terms of these bases, the Lie bracket $[,] : \bigwedge^2 \mathfrak{n}_1 \to \mathfrak{n}_2$ behaves like the Lie bracket for root vectors in the Lie algebra $\mathfrak{g}$ up to non-zero multiplicative constants, except that $[\mathfrak{g}_1, \mathfrak{g}_1] = 0$ in $\mathfrak{n}$. In other words, we have the following identities in $\mathfrak{n}$ up to non-zero multiplicative constants:

$[abcd, a'b'c'd'] = 0$ for $abcd, a'b'c'd' \in \Phi_1 \cup \Phi_2$ if $(a+a')(b+b')(c+c')(d+d')$ do not appear in $\Phi_3 \cup \Phi_4$,

$1321 = [0100, 1221] = [0110, 1211] = [0111, 1210] = [1100, 0221] = [1110, 0211] = [1111, 0210]$,

$2321 = [0100, 2221] = [0110, 2211] = [0111, 2210] = [1100, 1221] = [1110, 1211] = [1111, 1210]$,
$2421 = [0210, 2211] = [0211, 2210] = [1210, 1211]$,
$2431 = [0210, 2221] = [0221, 2210] = [1210, 1221]$,
$2432 = [0211, 2221] = [0221, 2211] = [1211, 1221]$.

To prove Lemma 4.4, it suffices to show that the following brackets vanish in the Lie algebra $\mathfrak{n}$ when we make the identification of vector spaces $\mathfrak{n}_1 = \mathfrak{g}_1 + \mathfrak{g}_2$ and $\mathfrak{n}_2 = \mathfrak{g}_3 + \mathfrak{g}_4$:

$$[\mathfrak{g}_1, \mathfrak{g}_4], [\mathfrak{g}_2, \mathfrak{g}_3], [\mathfrak{g}_2, \mathfrak{g}_4], [\mathfrak{g}_3, \mathfrak{g}_3], [\mathfrak{g}_4, \mathfrak{g}_4], [\mathfrak{g}_3, \mathfrak{g}_4], [\mathfrak{g}_1, \mathfrak{g}_3].$$

Among these the vanishing of the first 6 brackets follows from the above formulae. Let us check explicitly.

$[\mathfrak{g}_1, \mathfrak{g}_4] = 0$
$[0100, 2421] = [0100, [0210, 2211]] = 0$,
$[0100, 2431] = [0100, [0221, 2210]] = 0$,
$[0100, 2432] = [0100, [0221, 2211]] = 0$,
$[0110, 2421] = [0110, [0211, 2210]] = 0$,
$[0110, 2431] = [0110, [0221, 2210]] = 0$,
$[0110, 2432] = [0110, [0211, 2221]] = 0$,
$[0111, 2421] = [0111, [0210, 2211]] = 0$,
$[0111, 2431] = [0111, [0210, 2221]] = 0$,
$[0111, 2432] = [0111, [0211, 2221]] = 0$,
$[1100, 2421] = [1100, [0210, 2211]] = 0$
$[1100, 2431] = [1100, [0210, 2221]] = 0$,
$[1100, 2432] = [1100, [0211, 2221]] = 0$,
$[1110, 2421] = [1110, [0210, 2211]] = 0$,
$[1110, 2431] = [1110, [0210, 2221]] = 0$,
$[1110, 2432] = [1110, [0221, 2211]] = 0$,
$[1111, 2421] = [1111, [0211, 2210]] = 0$,
$[1111, 2431] = [1111, [0221, 2210]] = 0$,
$[1111, 2432] = [1111, [0221, 2211]] = 0$.

$[\mathfrak{g}_2, \mathfrak{g}_3] = 0$
$[0210, 1321] = [0210, [1100, 0221]] = 0$,
$[0210, 2321] = [0210, [1100, 1221]] = 0$,
$[0211, 1321] = [0211, [1100, 0221]] = 0$,
$[0211, 2321] = [0211, [1100, 1221]] = 0$,
$[0221, 1321] = [0221, [1111, 0210]] = 0$,
$[0221, 2321] = [0221, [1111, 1210]] = 0$,
$[1210, 1321] = [1210, [1100, 0221]] = 0$,
$[1210, 2321] = [1210, [0100, 2221]] = 0$,
$[1211, 1321] = [1211, [1100, 0221]] = 0$,
$[1211, 2321] = [1211, [0100, 2221]] = 0$,
$[1221, 1321] = [1221, [1111, 0210]] = 0$,
$[1221, 2321] = [1221, [0110, 2211]] = 0$,

$$[2210, 1321] = [2210, [0100, 1221]] = 0,$$
$$[2210, 2321] = [2210, [0100, 2221]] = 0,$$
$$[2211, 1321] = [2211, [0100, 1221]] = 0,$$
$$[2211, 2321] = [2211, [0100, 2221]] = 0,$$
$$[2221, 1321] = [2221, [0111, 1210]] = 0,$$
$$[2221, 2321] = [2221, [0111, 2210]] = 0.$$

$[\mathfrak{g}_2, \mathfrak{g}_4] = 0$

$$[0210, 2421] = [0210, [1210, 1211]] = 0,$$
$$[0210, 2431] = [0210, [1210, 1221]] = 0,$$
$$[0210, 2432] = [0210, [1211, 1221]] = 0,$$
$$[0211, 2421] = [0211, [1210, 1211]] = 0,$$
$$[0211, 2431] = [0211, [1210, 1221]] = 0,$$
$$[0211, 2432] = [0211, [1211, 1221]] = 0,$$
$$[0221, 2421] = [0221, [1210, 1211]] = 0,$$
$$[0221, 2431] = [0221, [1210, 1221]] = 0,$$
$$[0221, 2432] = [0221, [1211, 1221]] = 0,$$
$$[1210, 2421] = [1210, [0210, 2211]] = 0,$$
$$[1210, 2431] - [1210, [0210, 2221]] = 0,$$
$$[1210, 2432] = [1210, [0211, 2221]] = 0,$$
$$[1211, 2421] = [1211, [0210, 2211]] = 0,$$
$$[1211, 2431] = [1211, [0210, 2221]] = 0,$$
$$[1211, 2432] = [1211, [0211, 2221]] = 0,$$
$$[1221, 2421] = [1221, [0210, 2211]] = 0,$$
$$[1221, 2431] = [1221, [0210, 2221]] = 0,$$
$$[1221, 2432] = [1221, [0211, 2221]] = 0,$$
$$[2210, 2421] = [2210, [1210, 1211]] = 0,$$
$$[2210, 2431] = [2210, [1210, 1221]] = 0,$$
$$[2210, 2432] = [2210, [1211, 1221]] = 0,$$
$$[2211, 2421] = [2211, [1210, 1211]] = 0,$$
$$[2211, 2431] = [2211, [1210, 1221]] = 0,$$
$$[2211, 2432] = [2211, [1211, 1221]] = 0,$$
$$[2221, 2421] = [2221, [1210, 1211]] = 0,$$
$$[2221, 2431] = [2221, [1210, 1221]] = 0,$$
$$[2221, 2432] = [2221, [1211, 1221]] = 0.$$

$[\mathfrak{g}_3, \mathfrak{g}_4] = 0$

$$[1321, 2421] = [[0100, 1221], [0210, 2211]] = 0,$$
$$[1321, 2431] = [[1100, 0221], [0210, 2221]] = 0,$$
$$[1321, 2432] = [[1111, 0210], [1211, 1221]] = 0,$$
$$[2321, 2421] = [[0100, 2221], [1210, 1211]] = 0,$$
$$[2321, 2431] = [[0110, 2211], [1210, 1221]] = 0,$$
$$[2321, 2432] = [[0111, 2210], [1211, 1221]] = 0.$$

$[\mathfrak{g}_3, \mathfrak{g}_3] = 0$

$$[1321, 2321] = [[1111, 0210], [0111, 2210]] = 0.$$

$[\mathfrak{g}_4, \mathfrak{g}_4] = 0$

$$[2421, 2431] = [[0210, 2211], [1210, 1221]] = 0,$$

$[2421, 2432] = [[0210, 2211], [1211, 1221]] = 0,$
$[2431, 2432] = [[0210, 2221], [1211, 1221]] = 0.$

It remains to check that $[\mathfrak{g}_1, \mathfrak{g}_3] = 0$. This does not follow directly from the above identities between the special bases, because it is not true in the Lie algebra $\mathfrak{g}$. Recall that under the identification $(\bigwedge^2 U)/\Sigma = \mathfrak{g}_3 + \mathfrak{g}_4$, the subspace $\mathfrak{g}_3$ corresponds to the W-component of the K-module decomposition of $\bigwedge^2 U$ given in Proposition 2.4. Thus we can see $[\mathfrak{g}_1, \mathfrak{g}_3] = 0$ from Jacobi identity, once we establish the following claim :

(Claim) *The part corresponding to the factor $U_1 \wedge W$ in $U_1 \wedge (\bigwedge^2 U)$ is generated by the vectors of the form $u \wedge (u' \wedge u'')$ with $u \wedge u' \in \Sigma$ and $u \wedge u'' \in \Sigma$.*

Let us prove the claim. For general choices of non-zero vectors $v \in V, v^* \in V^*$ and $w_1 \neq w_2 \in W$, the element $(v \otimes w_1) \wedge (v^* \otimes w_2^2)$ of $(V \otimes W) \wedge (V^* \otimes S^2 W)$ has a component in W-component of the decomposition

$$(V \otimes W) \wedge (V^* \otimes S^2 W) = (\mathbf{ad}_V \otimes W) \oplus (\mathbf{ad}_V \otimes S^3 W) \oplus W \oplus S^3 W.$$

Choose $v_1 \in V$ satisfying $< v^*, v_1 >= 0$ and a general $w \in W$. Then the element

$$(v_1 \otimes w) \wedge [(v \otimes w_1) \wedge (v^* \otimes w_2^2)]$$

of $U \wedge (\bigwedge^2 U)$ is of the form we want because

$$(v_1 \otimes w) \wedge (v \otimes w_1) \in \bigwedge^2 U_1 \subset \Sigma \quad \text{and}$$
$$(v_1 \otimes w_2) \wedge (v^* \otimes w_2^2) \in \mathbf{ad}_V \otimes (W \otimes S^2 W) \subset \Sigma.$$

On the other hand, since w, w_1, w_2 are chosen in a general manner, it has non-zero components in both irreducible K-factors of $(V \otimes W) \wedge W \cong (V \otimes \bigwedge^2 W) \oplus (V \otimes S^2 W)$. This proves the claim by K-invariance and the proof of Lemma 4.4. is complete. □

Now we are ready to prove Proposition 4.3.

Proof (of Proposition 4.3.). Let us continue to use the notation in the proof of Proposition 4.1. We can view the tangent morphisms $\tau_{x_t} : \mathcal{K}_{x_t} \to \mathbb{P}T_{x_t}(X_t)$ as a family of finite morphisms $\tau_{x_t} : Z \to \mathbb{P}T_{x_t}(X_t)$.

By Proposition 3.5., $\tau_{x_t} : Z \to \mathbb{P}T_{x_t}(X_t)$ has image inside $\mathbb{P}D^t_{2,x_t}$ and is an embedding defined by the complete linear system $|\xi|$ for all $t \neq 0$. It follows that $\tau_{x_0} : Z \to \mathbb{P}T_{x_0}(X_0)$ is a morphism into $\mathbb{P}D^0_{2,x_0}$ defined by a subsystem of $|\xi|$. So $\mathcal{C}_x$ is isomorphic to the image of $Z \subset \mathbb{P}U$ under a projection $U \to U'$. This has the following implication on the symbol algebra of $\mathcal{W}$ at x. The symbol algebra of $\mathcal{W}$ at x is a nilpotent graded Lie algebra $\mathfrak{w} = \mathfrak{w}_1 + \mathfrak{w}_2 + \cdots$ generated by $\mathfrak{w}_1 = \mathcal{W}_x$. The Lie bracket $[\ ,\ ] : \bigwedge^2 \mathfrak{w}_1 \to \mathfrak{w}_2$ annihilates all tangent lines to $\mathcal{C}_x$. Since $\mathcal{C}_x$ is a projection of Z, $\mathfrak{w}$ is a quotient algebra of the nilpotent graded Lie algebra $\mathfrak{n} = \mathfrak{n}_1 + \cdots$ generated by U with the Lie bracket $[\ ,\ ] : \bigwedge^2 \mathfrak{n}_1 \to \mathfrak{n}_2$ determined by the subspace $\Sigma \subset \bigwedge^2 U$ spanned by tangent lines to Z. But by Lemma 4.4., $\mathfrak{n} = \mathfrak{n}_1 + \mathfrak{n}_2$ and $\dim(\mathfrak{n}_2) = 5$. Thus the symbol algebra $\mathfrak{w}$ is just

$$\mathfrak{w}_1 + \mathfrak{w}_2$$

with $\dim(\mathfrak{w}_2) \leq 5$. Using Proposition 3.2., we conclude that $\dim \mathcal{W}_x = 15 = U$. Thus $\mathcal{W} = D_2^0$ and $\tau_x : Z \to \mathbb{P}\mathcal{W}_x$ is the embedding given by the complete linear system $|\xi|$. □

Proposition 4.5. *The symbol algebra of D_1^0 at a generic point of X_0 is isomorphic to that of D_1 on G/P.*

Proof. Let us denote by $\mathcal{D}_i$ the distribution on $\mathcal{X}$ defined outside a set of codimension ≥ 2 which agrees with D_i^t on each $X_t, t \in \Delta$. By Proposition 4.3 there exists a subvariety $E \subset X_0$ such that $\mathcal{C}_x \subset \mathbb{P}\mathcal{D}_{2,x}$ is isomorphic to $Z \subset \mathbb{P}U$ for each $x \in \mathcal{X} - E$. Let $\mathcal{S}_x \subset \mathcal{C}_x$ be the subvariety corresponding to the Schubert subvariety $S \subset Z$. Let

$$\mathcal{S} := \bigcup_{x \in \mathcal{X} - E} \mathcal{S}_x.$$

Then $\mathcal{S} \to \mathcal{X} - E$ is a holomorphic fiber bundle with fiber biholomorphic to S. The distribution $\mathcal{D}_1$ is exactly the linear span of $\mathcal{S}$ in $T(\mathcal{X})$.

Choose a neighborhood $\mathcal{U} \subset \mathcal{X}$ of a generic point $x \in X^0$ such that there exists a trivialization

$$\mathcal{S}|_{\mathcal{U}} \cong \mathbb{P}(V \otimes h^*) \times \mathcal{U}.$$

By Proposition 2.1, the choice of such a trivialization naturally induces a trivialization of the vector bundles

$$\mathcal{D}_1|_{\mathcal{U}} = (V \otimes W) \otimes \mathcal{O}_{\mathcal{U}},$$

$$\mathcal{D}_2/\mathcal{D}_1|_{\mathcal{U}} = (\bigwedge^2 V \otimes S^2 W) \otimes \mathcal{O}_{\mathcal{U}}.$$

By Proposition 3.6, this induces a trivialization of $[\mathcal{D}_1, \mathcal{D}_2]/\mathcal{D}_2 = \mathcal{D}_3/\mathcal{D}_2$ and $[\mathcal{D}_2, \mathcal{D}_2]/\mathcal{D}_3 = \mathcal{D}_4/\mathcal{D}_3$ of the type

$$\mathcal{D}_3/\mathcal{D}_2|_{\mathcal{U}} = U \otimes \mathcal{O}_{\mathcal{U}},$$

$$\mathcal{D}_4/\mathcal{D}_3|_{\mathcal{U}} = V \otimes \mathcal{O}_{\mathcal{U}}.$$

These trivializations are not canonically determined by that of $\mathcal{S}|_{\mathcal{U}}$, but are determined once we have made uniform choices of the identifications of irreducible SL(V) and SL(W) modules involved (cf. Remark in Section 1). Once we have made these trivializations and fixed bases for V and W, the symbol algebra of the weak derived system of $\mathcal{D}_1$ is completely determined by the structure coefficients of the Lie algebra, which are holomorphic functions on $\mathcal{U}$. These holomorphic functions are constant on $\mathcal{U} - X^0$ because the structure coefficients are fixed constants on G/P. Thus they must be the same constant on $\mathcal{U} \cap X^0$. We conclude that the symbol algebra of the system generated by D_1^0 is isomorphic to that of the one generated by $D_1^t, t \neq 0$. □

Now we recall the following result of K. Yamaguchi.

Proposition 4.6. *Let X be a 20-dimensional complex manifold with a differential system D'_1 of rank 6 and depth 4 such that the symbol algebra of the weak derived system $D'_i, 1 \leq i \leq 4$, is isomorphic to that of the system D_1 on G/P. Then there exists an open subset $\mathcal{U} \subset X$ and an open immersion $\varphi : \mathcal{U} \to G/P$ such that under the differential $d\varphi : T(\mathcal{U}) \to T(\varphi(\mathcal{U}))$,*

$$d\varphi(D'_i|_{\mathcal{U}}) = D_i|_{\varphi(\mathcal{U})} \text{ for each } i = 1, 2, 3, 4.$$

K. Yamaguchi ([Ya], Proposition 5.5) showed that the obstructions to constructing such an open immersion φ always vanish identically.

Now we can finish the proof of Main Theorem as in the last paragraph of [HM4]. Let $\mathcal{U} \subset X_0$ and $\varphi : \mathcal{U} \to G/P$ be the ones defined on our X_0 by Proposition 4.5 and Proposition 4.6. From the upper-semi-continuity of $h^0(X_t, T(X_t))$, the Lie algebra $\mathrm{aut}(X_0)$ of infinitesimal automorphisms of X_0 has dimension $\geq \dim(\mathfrak{g})$. By Corollary 5.4 of [Ya], the Lie algebra of infinitesimal automorphisms of $\varphi(\mathcal{U})$ preserving D_1 is isomorphic to $\mathfrak{g}$. Thus the Lie algebra of infinitesimal automorphisms of $\mathcal{U}$ preserving D^0_1 is isomorphic to $\mathfrak{g}$. By considering dimensions, $\mathrm{aut}(X_0) \cong \mathfrak{g}$ and the isomorphism is induced by φ. In particular, G acts on X_0 with the isotropy subgroup at a generic point isomorphic to P, implying $X_0 \cong G/P$.

References

[HM1] Hwang, J.–M., Mok, N.: Rigidity of irreducible Hermitian symmetric spaces of the compact type under Kähler deformation. Invent. math. **131** 393–418 (1998)

[HM2] Hwang, J.–M., Mok, N.: Holomorphic maps from rational homogeneous spaces of Picard number 1 onto projective manifolds. Invent. math. **136** 209–231 (1999)

[HM3] Hwang, J.–M., Mok, N.: Cartan–Fubini type extension of holomorphic maps for Fano manifolds of Picard number 1. Journal Math. Pures Appl. **80** 563–575 (2001)

[HM4] Hwang, J.–M., Mok, N.: Deformation rigidity of the rational homogeneous space associated to a long simple root. Ann. scient. Ec. Norm. Sup. **35** 173–184 (2002)

[HM5] Hwang, J.–M., Mok, N.: Prolongations of infinitesimal linear automorphisms of projective varieties and rigidity of rational homogeneous spaces of Picard number 1 under Kähler deformation. Preprint

[Hw] Hwang, J.–M.: Rigidity of homogeneous contact manifolds under Fano deformation. J. reine angew. Math. **486** 153–163 (1997)

[Ke] Kebekus, S.: Families of singular rational curves. J. Alg. Geom. **11** 245–256 (2002)

[Ko] Kodaira, K.: On stability of compact submanifolds of complex manifolds. Amer. J. Math. **85** 79–94 (1963)

[Ya] Yamaguchi, K.: Differential systems associated with simple graded Lie algebras. In: *Progress in Differential Geometry.* Adv. Study Pure Math. **22** 413–494 (1993)

Severi–Brauer Varieties and Symmetric Powers

Daniel Krashen[1*] and David J. Saltman[2**]

[1] Department of Mathematics University of California, Los Angeles Los Angeles, CA 90095, USA. dkrashen@math.ucla.edu
[2] Department of Mathematics The University of Texas Austin, Texas 78712, USA. saltman@math.utexas.edu

Abstract. The general question behind this paper is the possibility of nonrational varieties V with rational symmetric powers. We extend known results and show Severi–Brauer varieties provide a counterexample. On the other hand, we show that unramified cohomology cannot be the means to find such an example.

Introduction

For the second author, this paper began with a question at the conference. The question was the following. Let V be a variety (over F). Suppose the symmetric power $S^n(V)$ is rational. Does this force V to be rational or even unirational? Recall that V is rational if the function field $F(V)$ is rational, i.e., purely transcendental over F, and unirational if $F(V) \subset L$ where L/F is rational.

In this generality, the answer was, in fact, known. Let V be a nonrational conic curve (so V has no rational points and is, therefore, not unirational). Since $V \subset \mathbb{P}^2(F)$, the symmetric square is clearly rational since it is (birationally) the variety of lines in $\mathbb{P}^2$. Of course, the map associates two points on the conic with the line between them.

Such an easy argument suggests that the above fact can be generalized. A conic can also be thought of as the Severi–Brauer variety of a quaternion algebra. In this paper we consider Severi–Brauer varieties V of central simple algebras A/F. We show that the symmetric power $S^n(V)$ is rational, if n is the degree of A. This variety was studied previously by Merkurjev [M]. We also give descriptions of the other symmetric powers of V.

Our result can be viewed as partly known, though our proof is completely new as far as we know. One can show (it takes a bit of work) that $S^n(V)$ as above is birational to the variety of maximal separable commutative subalgebras of A, and therefore also to the variety of maximal tori of A^*. This later

* The first author is grateful for support under NSF grant DMS–9983726.
** The second author is grateful for support under NSF grant DMS–9970213.

variety was shown to be rational in [C], [BS]. The first author, in a future paper, will explore the connection between other symmetric powers of V and subalgebras of A.

Of course Severi–Brauer varieties are always rational over F if F is algebraically closed. Thus these varieties cannot provide an example in the case that is most interesting, namely, a nonrational V over an algebraically closed field such that a symmetric power $S^n(V)$ is rational. In the second section we show that this may be a challenging problem. One of the most useful ways of showing that a field is nonrational is to use unramified cohomology. In this section we show unramified cohomology can never be used to show V nonrational when $S^n(V)$ is rational.

In this context it is useful to recall some weaker notions of rationality. If K/F is a field extension we say K is *stably rational* if there is a field $L \supset K$ such that L/K and L/F are rational. This is related to the notion that K/F and K'/F are *stably isomorphic* if and only if there is an $L \supset K$, $L \supset K'$ such that L/K and L/K' are rational. Finally, we say K/F is retract rational if the following holds. K is the field of fractions of an affine domain R (written $K = q(R)$) and there are F maps $R \to F[x_1, \ldots, x_n](1/s)$ and $F[x_1, \ldots, x_n](1/s) \to R$ with $R \to F[x_1, \ldots, x_n](1/s) \to R$ the identity. Here $F[x_1, \ldots, x_n](1/s)$ is the polynomial ring $F[x_1, \ldots, x_n]$ localized at a nonzero $s \in F[x_1, \ldots, x_n]$. Of course we say a variety V/F is rational, or stably rational, or retract rational, if $F(V)/F$ is. We say varieties V/F, V'/F are stably isomorphic is the same is true of $F(V)/F$ and $F(V')/F$. As a final bit of notation, if K is a field, let K_s denote the separable closure of K.

It is typical to consider the symmetric product as an operation on varieties, but for our purposes it is best to think about fields. Suppose K/F is a regular field extension. Then $K^n = K \otimes_F \ldots \otimes_F K$ is a domain with field of fractions $q(K^n)$. There is an obvious action of the symmetric group S_n on K^n and hence on $q(K^n)$. We define the *symmetric power* $S^n(K/F)$ to be equal to the invariant field $q(K^n)^{S_n}$. It is very easy to see that if K/F is rational, or stably rational, or retract rational, then $S^n(K/F)$ is also. When F is clear we will write this field as $S^n(K)$.

1 Severi–Brauer Varieties

We begin with a construction more general than the Severi–Brauer varieties, or at least their fields of fractions. We consider the following construction, which we recall from, e.g., [LN]. Let G be a finite group, and M a lattice over G. Let L/F be a G extension of fields, and $\alpha \in \mathrm{Ext}_G(M, L^*)$. Form the Laurent polynomial ring $L[M]$ which can be also thought of as the group algebra of M over L. It is easy to see one can define an action of G on $L[M]$ such that the extension $0 \to L^* \to L[M]^* \to M \to 0$ corresponds to α. With this so called twisted G action, we write $L[M]$ as $L_\alpha[M]$. We can consider the invariant ring $L_\alpha[M]^G$ and its field of fractions $L_\alpha(M)^G$. Note that the former is the affine ring of a principal homogeneous space over the torus with affine ring $L[M]^G$ (untwisted action; i.e., $\alpha = 0$).

We want to compute the symmetric power of this field, and so we proceed as follows. Following the definition, $q(L_\alpha(M) \otimes_F \ldots \otimes_F L_\alpha(M))$ (n times) is the field of fractions of the group ring

$$(L^n)[M \oplus \ldots \oplus M]$$

with a twisted action by $G \oplus \ldots \oplus G$. The action by S_n can be defined via the usual action on L^n and $M^n = M \oplus \ldots \oplus M$. The symmetric power $S^n(L_\alpha(M))$ is the field of fractions of the fixed ring of the above under the group $W = (G \oplus \ldots \oplus G) \rtimes S_n$ which is, of course, a wreath product.

In L^n there is a "multiplication" idempotent e such that $eL^n \cong L$ and in fact $e(a_1 \otimes \ldots \otimes a_n)$ maps to $a_1 \ldots a_n$. The stabilizer of e in W is $G \oplus S_n$ where $G \subset G \oplus \ldots \oplus G$ is the diagonal. The index of $G \oplus S_n$ in W is $|G|^{n-1}$ which is also the number of primitive idempotents in L^n. It follows that these primitive idempotents are all W conjugates of e, and as a W representation $L^n(M^n)$ is the induced representation $\mathrm{Ind}_{G \oplus S_n}^{W}((eL^n)(M^n))$. In particular, projection onto $(eL^n))M^n)$ induces an isomorphism $L^n(M^n)^W \cong (eL^n)(M^n)^{G \oplus S_n}$.

Define the α induced *diagonal* $\beta \in \mathrm{Ext}_{G \oplus S_n}(M^n, L^*)$ to be the element induced by inflation from $(\alpha \oplus \ldots \oplus \alpha) \in (\mathrm{Ext}_G(M, L^*))^n = \mathrm{Ext}_G(M^n, L^*)$. It is now clear that:

Lemma 1.1. *$S^n(L_\alpha(M)^G) = L_\beta(M^n)^{G \oplus S_n}$ where β is the diagonal element induced by α.*

We require a special case of these twisted actions as follows. Let G be a finite group, $H \subset G$ a subgroup and L/F a G Galois extension. Let $\mathbb{Z}[G/H]$ be the permutation G lattice with basis u_{gH} one can identify with the cosets of H in G. Let $I = I[G/H] \subset \mathbb{Z}[G/H]$ be the G sublattice generated by elements of the form $u_{gH} - u_H$. In [LN] was shown that there is a natural one to one correspondence between $\alpha \in H^2(G, L^*)$ which are split on H and elements of $\mathrm{Ext}_G(I, L^*)$. Let $L_\alpha(I)$ be the field with associated twisted action.

Note that from [S2] 0.4 the following universal property is easy. Suppose $K \supset L$ is an L algebra with G action extending that on L. Assume α splits in K^*. Then by [S2] 0.4, there is a G module map $f : (L_\alpha[I[G/H]])^* \to K^*$ which is the identity on L^*. f induces an L algebra morphism $\phi_f : L_\alpha[I[G/H]] \to K$. More explicitly, choose c a $G - H$ cocycle in α. Let $e_{gH} \in L_\alpha[I[G/H]]^*$ split c and map to $u_{gH} - u_H \in I[G/H]$. Choose $d_{gH} \in K^*$ which splits c. Then ϕ_f is defined by the condition $\phi_f(e_{gH}) = d_{gH}$.

The above universal property for $L_\alpha[I[G/H]]$ extends to one for $L_\beta[I^n]$. Consider $e_{gH} \in L_\beta[I^n]$ via the embedding of $I = I[G/H]$ in the first direct summand of I^n. Let $S_{n-1} \subset S_n$ be the stabilizer of this first direct summand.

Lemma 1.2. *Suppose $\bar{K}/K$ is S_n Galois and there are $d_{gH} \in (L \otimes \bar{K}^{S_{n-1}})^*$ splitting c. Then there is a $G \oplus S_n$ preserving L algebra morphism $\phi : L_\beta[I^n] \to L \otimes_F \bar{K}$ with $\phi(e_{gH}) = d_{gH}$.*

Proof. We define ϕ_f as above and then extend to $L_\beta[I^n]$ in the unique way that preserves the S_n action. □

Clearly an $\alpha \in H^2(G, L^*)$ split by H defines a Brauer group element in $\mathrm{Br}(F)$ which is split by L^H. Thus there is an algebra in this Brauer class with maximal subfield L^H. In [S3] this algebra is described a bit more and we recall that here. Since α is split on H we observe in [S3] that α contains a cocycle $c : G \times G \to L^*$ such that $c(G, H) = 1$. Form the crossed product $B = \sum_{g \in G} Lu_g$ where $u_g x = g(x)u_g$ and $u_g u_{g'} = c(g, g')u_{gg'}$ for all $x \in L$, $g, g' \in G$. As usual we can assume $u_1 = 1$. In B we can form the left B ideal, J, generated by all $u_h - 1$ for all $h \in H$. Then we define the $G - H$ crossed product $A = \Delta(L^H/F, G - H, c)$ by setting $A^\circ = \mathrm{End}_B(B/J)$. We show in [S3] that A is central simple, is in the Brauer class of B, and has L^H as maximal subfield. The reader could also easily verify these facts for themself. Note that $B \otimes_F A^\circ = \mathrm{End}_F(B/J)$ and we can view B/J as a left $B \otimes_F A^\circ$ module via this map.

By construction B contains $C = \oplus_{h \in H} Lu_h$ which can be identified with the trivial crossed product $\Delta(L/L^H, H)$. A maximal separable subring $K \subset A^\circ$ is called *good* if the map $l \otimes k \to (l \otimes k)(1 + J)$ is an isomorphism $\phi_K : L \otimes_F K \cong B/J$. Here $L \otimes_F K$ is being viewed as a subring of $B \otimes_F A^\circ$. Via its canonical action on L, $L \otimes_F K$ is a C module. $L \otimes_F K$ is also a K space via the multiplication action of K on K, and so $L \otimes_F K$ is a $C \otimes_F K$ module.

Lemma 1.3. *ϕ_K is a $C \otimes_F K$ module morphism.*

Proof. Let $l'u_h \otimes k' \in C \otimes_F K$ and $l \otimes k \in L \otimes_F K$. Then

$$\phi_K((l'u_h \otimes k')(l \otimes k)) = \phi_K(l'h(l) \otimes k'k) = (l'h(l) \otimes K'K)(1 + J).$$

As elements of $C \otimes_F K$, $l'h(l) \otimes k'k = (l' \otimes k')(u_h \otimes 1)(l \otimes k)(u_h \otimes 1)^{-1}$. Furthermore, $(u_h \otimes 1)(1 + J) = u_h + J = 1 + J$ so $(u_h \otimes 1)^{-1}(1 + J) = 1 + J$. Thus $(l' \otimes k')(u_h \otimes 1)(l \otimes k)(u_h \otimes 1)^{-1}(1 + J) = (l' \otimes k')(u_h \otimes 1)(l \otimes k)(1 + J) = (l'u_h \otimes k')\phi_K(l \otimes k)$ as needed. □

Assume K is good. Then the isomorphism ϕ_K induces

$$\psi_K : \mathrm{End}_K(L \otimes_F K) \cong \mathrm{End}_K(B/J) = B \otimes_F K$$

since K is self centralizing. Write $LK = L \otimes_F K$ and view this as a G Galois extension of K. Note $B \otimes_F K$ is the (Azumaya) crossed product algebra $\Delta(L/F, G, c) \otimes_F K = \Delta(LK/K, G, c)$. We have $\psi_K : \Delta(LK/K, G, c) \cong \mathrm{End}_K(LK)$. For any $g \in G$, $g \otimes 1 \in \mathrm{End}_K(LK) = \mathrm{End}_K(L \otimes K)$ and we consider $\psi_K^{-1}(g \otimes 1) = v_g$. Since ϕ_K is an LK morphism, $v_g(\eta) = g(\eta)v_g$ for all $\eta \in LK$. Thus $v_g = d_g^{-1}u_g$ for some $d_g \in (LK)^*$. Since ϕ_K is a C morphism, $v_h = u_h$ or $d_h = 1$ for all $h \in H$. Since $v_g v_{g'} = v_{gg'}$, the d_g's split the cocycle c. We call d_g the *G-H splitting* associated to a good K. Note that $\phi_K^{-1}(u_g + J) = u_g(1 \otimes 1) = d_g v_g(1 \otimes 1) = d_g$. It follows that the L span of the d_g is LK. If $d'_g \in (LK)^*$ are a G-H splitting of c where $LK = \sum Ld'_g$, we call the d'_g a *good* splitting. The above paragraph then gives a mapping from good $K \subset A^\circ$ to good G-H splittings in $(LK)^*$.

Conversely, suppose K/F is separable of degree $[G : H]$ and $B \otimes_F K$ is split. Let $LK = L \otimes_F K$, and let $d_g \in (LK)^*$ be a good G-H splitting of c. Note this means $d_h = 1$ all $h \in H$ and the d_g L span LK. Set $v_g = d_g^{-1} u_g \in B \otimes_F K = \Delta(LK/K, G, c)$ so using the v_g $B \otimes_F K = \Delta(LK/K, G) = \mathrm{End}_K(LK)$. Since $u_h(1 \otimes 1) = h(1) \otimes 1 = 1 \otimes 1$ (remember $v_h = u_h$), there is a unique B module morphism $\phi : B/J \to LK$. We have $\phi(u_g + J) = u_g(1 \otimes 1) = d_g \in LK$. Since the d_g are good, ϕ is an isomorphism inducing a $\psi : \mathrm{End}_F(B/J) \cong \mathrm{End}_F(LK)$ and ψ^{-1} defines an embedding of K into $A^\circ = \mathrm{End}_B(B/J)$. The image of K is a maximal separable subfield and d_g is obviously the associated good G-H splitting.

We can now state:

Theorem 1.4. *Let A/F be a central simple algebra of degree n and X its Severi Brauer variety. Then $S^n(X)$ is rational over F.*

Proof. We can write $A = \Delta(L^H/F, G - H, \alpha)$. Then X has function field $L_\alpha(I[G/H])^G$. We saw above that $S^n(X)$ has function field $L_\beta(I[G/H]^n)^{G \oplus S_n}$ where β is the diagonal element induced by α.

The idea of the proof is that this last field is isomorphic to the function field of the "variety of maximal separable subfields of A°". This was partially done in [M], and will be included in a future paper by the first author. For our current purposes we will instead use a projective space birational to this variety, had we defined it.

Let $\alpha_0 \in A^\circ$ be such that $F(\alpha_0)$ is good. Let $V' \subset A^\circ$ be an $n^2 - n$ dimensional F-subspace such that $F(\alpha_0) \cap V' = (0)$. Set $V = F\alpha_0 + V'$. Let $U' \subset V$ be the Zariski open subset such that α is a K point of U' if and only if the following holds. First, $K(\alpha) \subset A^\circ \otimes_F K$ is maximal separable and good. Second, that $K(\alpha) \cap (V \otimes_F K) = K\alpha$. Since α_0 is an F point of U', U' is nonempty. It is immediate the conditions defining U' are homogeneous, so there is an associated open subset $U \subset \mathbb{P}(V)$ with inverse image $U' \subset V$.

Let K be the function field of $P(V)$, and $\alpha \in U'_K \subset V \otimes_F K$ an element associated to the generic point of U. Set $K' = K(\alpha)$ which is clearly a good maximal separable subring of $A^\circ \otimes_F K$. Write $LK' = (L \otimes_F K)K'$ and $d_g \in (LK')^*$ the associated splitting. Let $\bar{K} \supset K' \supset K$ be S_n Galois over K such that $K' = \bar{K}^{S_{n-1}}$. Note that $\bar{K}$ always exists and we need not argue it is a field (though it is). Then $L\bar{K} = (L \otimes_F K) \otimes_K \bar{K}$ is Galois over K with group $G \oplus S_n$.

Since the d_g split $c \in \alpha$, there is a $\Phi : L_\beta[I^n] \to L\bar{K}$ as in 1.2. Let $\bar{S}$ be total quotient ring of the image of Φ, $S' = \bar{S}^{G \oplus S_{n-1}}$ and $S = \bar{S}^{G \oplus S_n}$. Note that $\bar{S}, S'$ are direct sums of fields and S is a field. Clearly $S' \subset K'$, so $S' \cap (V \otimes_F S) = S'\alpha'$, for some α'. Since $\alpha = k\alpha'$ for some $k \in K$, $S(\alpha')$ is n dimensional separable. More importantly, since α represents the generic point of $\mathbb{P}(V)$, $S = K$. Since K and $L_\beta[I^n]^{G \oplus S_n}$ both have transcendence degree $n(n-1)$ over F, Φ is an injection and hence induces an isomorphism $L_\beta(I^n)^{G \oplus S_n} \cong K$. But $K = F(\mathbb{P}(V))$ is rational over F. $\square$

This result may be applied to describe other symmetric powers $S^k(X)$ for a given Severi–Brauer variety X.

Theorem 1.5. *Let A/F be a central simple algebra of degree n and X its Severi–Brauer variety. Let $k < n$. Then $S^k(X)$ is birationally isomorphic to $V_k(A) \times \mathbb{P}^{k(k-1)}$, where $V_k(A)$ is the k-th generalized Severi–Brauer variety of kn-dimensional right ideals of A.*

Proof. Let $X_k = V_k(A)$. If $J \in X_k(A)$ is a point of X_k defined over some field extension of F, we follow the notation of [K] and define $V(J) = \{I \in X | I \subset J\}$. If J is defined over a field $L \supset F$, then it is easy to see that $V(J)$ is also defined over L as a closed subvariety of X. Define a rational map $f : \prod^k X \dashrightarrow X_k$ via $f(I_1, \ldots, I_k) = \sum I_i$. It is easy to check that this is a rational morphism which is surjective at the algebraic closure and hence dominant. In addition, one can check that the closure of the fiber $f^{-1}(J)$ over a kn-dimensional ideal J is simply $\overline{f^{-1}(J)} = \prod^k V(J)$. Since the map f is invariant under the action of the symmetric group S_k acting on $\prod^k V(J)$, we get an induced map $f' : S^k(X) \dashrightarrow X_k$. Chasing through the diagrams which define the relevant fiber squares, we find that $\overline{f'^{-1}(J)} = (\overline{f^{-1}(J)})/S_k = S^k(V(J))$.

Now, let $J \in X_k$ be the generic point, which we think of simply as a point of X_k defined over the function field $F(X_k)$. We will show that the fiber $f'^{-1}(J)$ is rational, which implies our result. Set $P = V(J)$, so P is a closed subvariety of X defined over $F(X_k)$. By [K] Lemma 2, $P = V(D)$ for some algebra D which is Brauer equivalent to $A \otimes F(X_k)$, and with $\deg(D) = k$. By the previous paragraph, it follows that $\overline{f'^{-1}(J)} = S^k(P) = S^k(V(D))$. But since D has degree k, this is rational by the previous theorem. Therefore as the generic fiber $f'^{-1}(J)$ is birationally isomorphic to $\overline{f'^{-1}(J)}$, it is also rational. □

For $k > n$, we state the situation up to stable isomorphism. For convenience of notation, if A is a central simple F-algebra of degree n, we set $F_k(A)$ to be the function field of $V_k(A)$, where if $k > n$, we define $V_k(A) = V_{k'}(A)$ where $k' < n$, and $n | k - k'$. We further define $V_0(A) = Spec(F)$. Set $F(A) = F_1(A)$.

Corollary 1.6. *Let k be arbitrary. Then $S^k(F(A))$ is stably isomorphic to $F_k(A)$. In particular, $S^{nl}(F(A))$ is stably rational.*

Proof. Suppose $k = nl + r$ where $r < n$. We want to show that $S^k(F(A))$ is stably isomorphic to $F_r(A)$. Set $B = M_{l+1}(A)$. Then $F(B)$ is rational over $F(A)$, so by lemma (the one at the end of the first section), $S^k(B)$ is rational over $S^k(A)$. By the previous theorem, $S^k(B)$ is rational over $F_k(B)$, and $F_k(B) = F_{n(l+1)-k}(B^\circ) = F_{n-r}(B^\circ)$ which is rational over $F_{n-r}(A^\circ) = F_r(A)$. □

2 Unramified Cohomology

Of course, one would like an example like 1.4 but over an algebraically closed field. That is, one would like a nonretract rational K/F such that $S^n(K)/F$ is retract rational for F algebraically closed. In this section we will prove a result that will indicate that this might be hard. That is, we will show that one cannot use unramified cohomology to construct a nonretract rational K/F such that $S^n(K/F)$ is retract rational.

We begin by recalling a bit about unramified cohomology. For convenience assume the ground field F contains all roots of one. If A is any torsion abelian group, we will write A' to mean the elements of A whose order is prime to the characteristic of F. Of course, if F has characteristic 0, $A' = A$. Let $\mu \subset F^*$ be the group of roots of one, and K_s the separable closure of the field K. Denote by $H^i(K, M)$ the Galois cohomology group $H^i(G_K, M)$ where G_K is the Galois group of K_s/K and M is a continuous G_K module (for example μ or K_s^*). Note that by standard arguments $H^i(K, \mu) \cong H^i(K, K_s^*)'$ for $i \geq 2$.

Suppose K/F is any field, and $R \subset K$ is discrete valuation domain containing F with field of fractions K. Then for any $i \geq 2$, R induces a residue map $\mathrm{ram}_R : H^i(K, \mu) \to H^{i-1}(\bar{R}, \mu)$ where $\bar{R}$ is the residue field of R. Note that the assumptions mean all discrete valuation rings have unmixed characteristic.

We recall the definition of the residue map, slightly generalizing [S1] p. 213 to cover the nonzero characteristic case. Since $\mu \subset F^*$, we can identify μ with $(\mathbb{Q}/\mathbb{Z})'$. Let $\mathbb{Q}'$ be the inverse image of $(\mathbb{Q}/\mathbb{Z})'$ and let $\hat{\mathbb{Z}}'$ be the prime to the characteristic part of the profinite completion $\hat{\mathbb{Z}}$, or equivalently $\mathrm{Hom}(\mu, \mathbb{Q}/\mathbb{Z})$. Define Π to be the abelian group $\mu \oplus \mathbb{Q}'$ with the $\hat{\mathbb{Z}}'$ action $\sigma(s + \mathbb{Z}, r) = (s + r + \mathbb{Z}, r)$ where σ is the canonical generator of $\hat{\mathbb{Z}}'$. If G is any profinite group, we view Π as a $G \oplus \hat{\mathbb{Z}}'$ module using the trivial G action. Note that since $\mathbb{Q}'$ is uniquely n divisible for n prime to the characteristic, $H^i(G \oplus \hat{\mathbb{Z}}', \mu) \cong H^i(G \oplus \hat{\mathbb{Z}}', \Pi)'$.

Arguing just as in [S1] (using the Hochschild–Serre spectral sequence) the inflation map $H^i(G, \mu \oplus \mathbb{Z})' \to H^i(G \oplus \hat{\mathbb{Z}}', \Pi)'$ is an isomorphism for $i \geq 2$. More easily, the coboundary maps defines an isomorphism $H^{i-1}(G, \mu) \cong H^i(G, \mathbb{Z})'$. Define $\mathrm{ram} : H^i(G \oplus \hat{\mathbb{Z}}', \mu) \to H^{i-1}(G, \mu)$ to be the composition

$$H^i(G \oplus \hat{\mathbb{Z}}', \mu) \to H^i(G \oplus \hat{\mathbb{Z}}', \Pi)' \cong H^i(G, \mu \oplus \mathbb{Z})' \to H^i(G, \mathbb{Z})' \cong H^{i-1}(G, \mu)$$

where the arrows are the obvious maps and the isomorphisms are described above.

We are ready to define the residue map as follows. Given K, R, form the completion $\hat{K}$ with respect to R. If $K' \supset \hat{K}$ is the maximal tamely ramified extension, then associated to K' there is an exact sequence

$$1 \to P \to G_{\hat{K}} \to G_{\bar{R}} \oplus \hat{\mathbb{Z}}' \to 1$$

where P is a profinite p group if $0 \neq p$ is the characteristic of K and $P = 1$ otherwise. In the nonzero characteristic case, $H^i(P, \mu) = 0$ for all $i > 0$ and so (by Hochschild-Serre again) inflation induces an isomorphism $H^i(G \oplus \hat{\mathbb{Z}}', \mu) \cong H^i(G_{\hat{K}}, \mu)$. The residue map ram_R is the composition $H^i(K, \mu) \to H^i(\hat{K}, \mu) \cong H^i(G_{\bar{R}} \oplus \hat{\mathbb{Z}}', \mu) \to H^{i-1}(G_{\bar{R}}, \mu)$ where the last arrow is the ram map above. We need the following basic properties of this residue map which are well known but a bit hard to find.

Proposition 2.1. *Let $K_1 \subset L_1$, be fields and R a discrete valuation ring with $q(R) = K_1$.*

a) *Suppose $S \subset L_1$ is a discrete valuation domain, $q(S) = L_1$, and S over R. Let* $\mathrm{Res} : H^i(K_1, \mu) \to H^i(L_1, \mu)$ *be the restriction map and set $e = e(S/R)$. Then for $i \geq 2$ the following diagram commutes:*

$$\begin{array}{ccc} H^i(L_1, \mu) & \longrightarrow & H^{i-1}(\bar{S}, \mu) \\ {\scriptstyle \mathrm{Res}} \uparrow & & \uparrow {\scriptstyle e\,\mathrm{Res}} \\ H^i(K_1, \mu) & \longrightarrow & H^{i-1}(\bar{R}, \mu) \end{array}$$

b) *Suppose L_1/K_1 is finite separable, and $S_1, \ldots, S_r$ are the discrete valuation domains with $q(S_j) = L_1$ and such that the S_j lie over R. Let*

$$\mathrm{Cor} : H^i(L_1, \mu) \to H^i(K_1, \mu)$$

be the corestriction map. The following diagram also commutes.

$$\begin{array}{ccc} H^i(L_1, \mu) & \longrightarrow & \oplus_j H^{i-1}(\bar{S}_j, \mu) \\ {\scriptstyle \mathrm{Cor}} \downarrow & & \downarrow {\scriptstyle \Sigma\,\mathrm{Cor}} \\ H^i(K_1, \mu) & \longrightarrow & H^{i-1}(\bar{R}, \mu) \end{array}$$

where all horizontal arrows are residue maps and $\Sigma\,\mathrm{Cor}$ *means the sum of the corestriction maps for $\bar{S}_j/\bar{R}$.*

Proof. a) Let $\hat{L}_1$, $\hat{K}_1$ be the completions of L_1 at S and K_1 at R respectively. Then we can view $\hat{K}_1 \subset \hat{L}_1$. This induces a map $G_{\hat{L}_1} \to G_{\hat{K}_1}$ and then a map $\iota : G_{\bar{S}} \oplus \hat{\mathbb{Z}}' \to G_{\bar{R}} \oplus \hat{\mathbb{Z}}'$. If π_R, π_S are primes of R,S, then $\pi_R = u\pi_S^e$ for $u \in S$ a unit. It follows that the diagram

$$\begin{array}{ccc} G_{\bar{S}} \oplus \hat{\mathbb{Z}}' & \longrightarrow & \hat{\mathbb{Z}}' \\ {\scriptstyle \iota} \downarrow & & \downarrow {\scriptstyle e} \\ G_{\bar{R}} \oplus \hat{\mathbb{Z}}' & \longrightarrow & \hat{\mathbb{Z}}' \end{array}$$

commutes. Let Π_R, Π_S be the $G_{\bar{R}} \oplus \hat{\mathbb{Z}}'$ respectively $G_{\bar{R}} \oplus \hat{\mathbb{Z}}'$ module induced from Π by the respective projections. Let Π' be the $G_{\bar{S}} \oplus \hat{\mathbb{Z}}'$ module induced from Π_R by ι. It is easy to see that $\Pi' \cong \Pi_S$ and in fact that there is a commutative diagram:

$$\begin{array}{ccccc} \mu & \longrightarrow & \Pi_S & \longrightarrow & \mathbb{Q}' \\ \| & & \|{\scriptstyle S} & & \downarrow{\scriptstyle e} \\ \mu & \longrightarrow & \Pi' & \longrightarrow & \mathbb{Q}' \end{array}$$

Part a) is now immediate.

As for b), let $\hat{K}$ be the completion of K with respect to R and let $\hat{L}_j$ be the corresponding object for L and S_j. Then $L \otimes_K \hat{K} \cong \oplus_j \hat{L}_j$ and the diagram

$$\begin{array}{ccc} H^i(L,\mu) & \longrightarrow & \oplus_j H^i(\hat{L}_j,\mu) \\ {\scriptstyle \mathrm{Cor}}\downarrow & & \downarrow{\scriptstyle \Sigma\,\mathrm{Cor}} \\ H^i(K,\mu) & \longrightarrow & H^i(\hat{K},\mu) \end{array}$$

commutes. This it suffices to prove b) in the case $K = \hat{K}$, and there is only one extension S. By composition it suffices to consider the two cases L/K is unramified and L/K is totally ramified.

If L/K is unramified, we can choose the prime π to be prime in both R and S, so the inclusion $G_{\bar{S}} \oplus \hat{\mathbb{Z}}' \subset G_{\bar{R}} \oplus \hat{\mathbb{Z}}'$ respects the direct sums and is the identity in $\hat{\mathbb{Z}}'$. This case of b) follows from the naturality of corestriction. On the other hand, if L/K is totally ramified then $G_{\bar{R}} = G_{\bar{S}}$ and any element of $H^i(G_{\bar{L}},\mu) \cong H^i(G_{\bar{S}} \oplus \hat{\mathbb{Z}}',\mu) \cong H^i(G_{\bar{S}} \oplus \hat{\mathbb{Z}}',\Pi)' \cong H^i(G_{\bar{S}},\mu \oplus \mathbb{Z})' = H^i(G_{\bar{R}},\mu \oplus \mathbb{Z})'$ is in the image of $H^i(\hat{K},\mu)$. Thus b) follows from a). □

We define the unramified cohomology $H^i(K,\mu)_u \subset H^i(K,\mu)$ to be the intersection of the kernels of ram_R for all R as above. The point about this cohomology is that it detects nonrationality. Let $\mathrm{Im}(H^i(F,\mu) \subset H^i(K,\mu)$ be the image under restriction. If K/F is retract rational, then $H^i(K,\mu)_u = \mathrm{Im}(H^i(F,\mu))$ for all $i \geq 2$. From the 2.1 above we immediately have:

Corollary 2.2. *Suppose $K_1 \subset L_1$ are fields containing F. Then the restriction map $H^i(K,\mu) \to H^i(L,\mu)$ takes $H^i(K,\mu)_u$ to $H^i(L,\mu)_u$. If L/K is finite separable, then the corestriction $H^i(L,\mu) \to H^i(K,\mu)$ also preserves unramified elements.*

We want to show that specialization preserves unramified elements. To this end, suppose $K_1 \supset F$ is a field. Assume R is a discrete valuation domain with $q(R) = K_1$ and $\alpha \in H^i(K,\mu)$ is in the kernel of ram_R. Then α maps to $\hat{\alpha} \in H^i(\hat{K},\mu) \cong H^i(G_{\bar{R}} \oplus \hat{\mathbb{Z}}',\mu) \cong H^i(G_{\bar{R}},\mu \oplus \mathbb{Z})'$. Since $\mathrm{ram}_R(\alpha) = 0$, $\hat{\alpha}$ maps to 0 in $H^i(G_{\bar{R}},\mathbb{Z})$ and hence comes from an element of $H^i(G_{\bar{R}},\mu) = H^i(\bar{R},\mu)$. We call this element α_R and think of it as the specialization of α at R. We claim:

Proposition 2.3. *Let R be a discrete valuation domain containing F and $K_1 = q(R)$. Suppose $\alpha \in H^i(K_1, \mu)_u$. Then $\alpha_R \in H^i(\bar{R}, \mu)_u$.*

Proof. Let $\hat{K}_1$ be the completion of K_1 at R. By 2.2, the image, $\hat{\alpha}$ is in $H^i(\hat{K}_1, \mu)_u$. By e.g. [Se] p. 33, $\hat{K}_1 \cong \bar{R}((t))$. Since $\mathrm{ram}_R(\alpha) = 0$, the above argument shows that $\hat{\alpha}$ is the image of $\alpha_R \in H^i(G_{\bar{R}}, \mu)$. Thus $\hat{\alpha}$ is the image of $\alpha_R \in H^i(\bar{R}, \mu)$, where $\bar{R}$ is viewed as a subfield of $\hat{K}_1$.

Suppose $S \subset \bar{R}$ is a discrete valuation domain containing F with $q(S) = \bar{R}$. Then S extends to a discrete valuation domain S_1 with residue field $\bar{S}((t))$ such that $e(S_1/S) = 1$ and $q(S_1) = \hat{K}_1$. Let $\beta = \mathrm{ram}_S(\alpha_R)$. By 2.1 a), the image of β in $H^{i-1}(\bar{S}((1)), \mu)$ equals $\mathrm{ram}_{S_1}(\hat{\alpha}) = 0$. Thus we are done once we show:

Lemma 2.4. *The map $H^j(k, \mu) \to H^j(k((t)), \mu)$ is injective.*

Proof. Since $H^i(k((t)), \mu) \cong H^i(G_k \oplus \hat{\mathbb{Z}}', \mu)$, it suffices to show the injectivity of $H^i(G_k, \mu) \to H^i(G_k \oplus \hat{\mathbb{Z}}', \mu)$, which is obvious. □

Suppose next that $R, \mathcal{M}$ is a regular local domain with $q(R) = K_1$ and residue field $\bar{R}$. Let $x_1, \ldots, x_n$ be a fixed R sequence generating $\mathcal{M}$. The the localization $R_{(x_1)}$ is a discrete valuation domain with residue field k_1 and the image of R in k_1, call it $R_1, \mathcal{M}_1$, is a regular local ring with $q(R_1) = k_1$. In addition, the images of the x_i, $i > 1$, form an R sequence generating $\mathcal{M}_1$. Thus by 2.3 and iteration, an $\alpha \in H^i(K_1, \mu)_u$ induces a $\alpha_R \in H^i(\bar{R}, \mu)_u$. More generally, we make the following recursive definition. Suppose $\alpha \in H^i(K_1, \mu)$. We say $\alpha_R \in H^i(\bar{R}, \mu)$ is *defined* if α is unramified at $R_{(x_1)}$ and the image of α in $H^i(k_1, \mu)$, call it α_1, satisfies that $(\alpha_1)_{\bar{R}}$ is defined where $\bar{R} = R/(x_1)$ with R sequence the images of $x_2, \ldots, x_n$. Of course we showed above that if $\alpha \in H^i(K_1, \mu)_u$, then α_R is defined.

From the definition, it looks like α_R depends on the choice of ordered R sequence. We do not know whether this is actually true. Luckily we can avoid this issue, at least in this paper. We do need other properties of this construction, as follows.

Lemma 2.5. *Let R be a regular domain with given R sequence containing the field L_1. Let $q(R) = K_1$.*

a) *Suppose $\alpha, \beta \in H^i(K_1, \mu)$ and α_R, β_R are both defined. Then $(\alpha + \beta)_R$ is defined and equals $\alpha_R + \beta_R$.*

b) *Suppose $\alpha \in H^i(K_1, \mu)$ is the image of $\alpha' \in H^i(L_1, \mu)$. Then $\alpha_R \in H^i(\bar{R}, \mu)$ is defined and is also the image of α'.*

c) *Suppose S is a regular domain lying over R and define specialization at S using the the same R sequence. Let $L_1 = q(S)$. Then the following diagram commutes, where the horizontal maps are specialization and the vertical maps restriction:*

$$\begin{array}{ccc} H^i(L_1,\mu)_u & \longrightarrow & H^i(\bar{S},\mu)_u \\ \uparrow & & \uparrow \\ H^i(K_1,\mu)_u & \longrightarrow & H^i(\bar{R},\mu)_u \end{array}$$

Proof. By induction we may assume R and S are discrete valuation domains. Part a) is now obvious and part b) follows from the fact that the composition $G_{\bar{R}} \subset G_{K_1} \to G_{L_1}$ is just the canonical map $G_{\bar{R}} \to G_{L_1}$. Part c) follows from the commutativity of:

$$\begin{array}{ccc} G_{L_1} & \longleftarrow & G_{\bar{S}} \\ \downarrow & & \downarrow \\ G_{K_1} & \longleftarrow & G_{\bar{R}} \end{array}$$

which is obvious. □

Unramified cohomology will not help with this paper's problem because of the next result. We say $S^{n-1}(K)$ has a smooth rational point if there is a smooth local domain with residue field F and $q(R) = S^{n-1}(K)$. Note this is always true if F is algebraically closed or K is unirational.

Theorem 2.6. *Suppose $S^{n-1}(K)$ has a smooth rational point. If*

$$H^i(S^n(K/F),\mu)_u = \mathrm{Im}(H^i(F,\mu)),$$

then $H^i(K,\mu)_u = \mathrm{Im}(H^i(F,\mu))$.

Proof. Since we actually prove the contrapositive, suppose $\alpha \in H^i(K,\mu)_u$ and $\alpha \notin \mathrm{Im}(H^i(F,\mu))$. The field extension $q(K^n)/S^n(K)$ is clearly S_n Galois, and we can set $L = q(K^n)^{S_{n-1}}$. Notice that $L = q(K \otimes_F S^{n-1}(K))$. In particular, $K \subset L$. Let $\alpha_L \in H^i(L,\mu)$ be the image of α and $\beta = \mathrm{Cor}^L_{S^n(K)}(\alpha_L)$. By 2.1, $\beta \in H^i(S^n(K/F),\mu)_u$. It will suffice to show that that

$$\beta \notin \mathrm{Im}(H^i(F,\mu)). \tag{1}$$

If $\beta_L \in H^i(L,\mu)_u$ is the image under restriction of β, it suffices to show

$$\beta_L \notin \mathrm{Im}(H^i(L,\mu)). \tag{2}$$

Recall that $L = KS^{n-1}(K)$. Now by the restriction–corestriction formula (e.g. [B] p. 82), $\beta_L = \alpha_L + \gamma$ where γ is a sum of elements which have the form

$$\mathrm{Cor}^{S_{n-1}}_{S_{n-1}\cap\sigma(S_{n-1})}(\mathrm{Res}^{\sigma(S_{n-1})}_{S_{n-1}\cap\sigma(S_{n-1})}(\sigma(\alpha_L)))$$

where $\sigma \notin S_{n-1}$. It follows that γ is in the image of $\gamma' \in H^i(S^{n-1}(K),\mu)_u$. Also note that since β_L and α_L lie in $H^i(L,\mu)_u$, then $\gamma \in H^i(L,\mu)_u$ also.

Let $R, \mathcal{M}$ be a smooth local domain with $q(R) = S^{n-1}(K)$ and $R/\mathcal{M} = F$. Let $S \subset L = KS^{n-1}(K)$ be the natural extension with the same R series, $q(S) = L$, and residue field K. By 2.5 c), γ_S is in the image of $H^i(F,\mu)$. If (2) fails, then $(\alpha_L)_S$ is in the image of $H^i(F,\mu)$. But by 2.5 b), $(\alpha_L)_S = \alpha$, contradicting the assumption. □

References

[BS] Borel, A., Springer, T.A.: Rationality properties for linear algebraic groups II. Tohuku Math. Journal **20**(2) 443–497 (1968)

[B] Brown, K.: *Cohomology of Groups* (Springer, Berlin, Heidelberg, New York 1982)

[C] Chevalley, C.: On algebraic group varieties. J. Math. Soc. Japan **6** 303–324 (1954)

[K] Krashen, D.: Birational isomorphisms between generalized Severi–Brauer varieties. Preprint arXiv:math.RA/0203117

[M] Merkurjev, A.: Closed points of Severi–Brauer varieties. Vestnik S.-Peterburg. Univ. Mat. Mekh. Astronom. **2** 51–53 (1993)

[S1] Saltman, D.J.: Brauer groups of invariant fields, geometrically negligible classes, and equivariant Chow group, and unramified H^3. In: Jacob, W.B., Rosenberg, A. (eds): *K-Theory and Algebraic Geometry: Connections with Quadratic Forms and Division Algebras.* Proc. Symp. Pure Math. **58** Part 1 (AMS, Providence, RI, 1995), pp. 189–246

[S2] Saltman, D.J.: Invariant fields of symplectic and orthogonal groups. J. Algebra **258**(2) 507–534 (2002)

[S3] Saltman, D.J.: Triality, cocycles, cross products, involutions, Clifford algebras and invariants. Preprint http://www.ma.utexas.edu/text/webpages/saltman.html

[Se] Serre, J.–P.: *Local Fields* (Springer, Berlin, Heidelberg, New York 1979)

Representation Theory and Projective Geometry

Joseph M. Landsberg[1] and Laurent Manivel[2]

[1] Georgia Institute of Technology, School of Mathematics, Atlanta, GA 30332-0160, USA. jml@math.gatech.edu

[2] Université Grenoble I, BP 74, Institut Fourier, UMR 5582 du CNRS, 38402 Saint Martin d'Hères cedex, France. Laurent.Manivel@ujf-grenoble.fr

Abstract. This article consists of three parts that are largely independent of one another. The first part deals with the projective geometry of homogeneous varieties, in particular their secant and tangential varieties. It culminates with an elementary construction of the compact Hermitian symmetric spaces and the closed orbits in the projectivization of the adjoint representation of a simple Lie algebra. The second part discusses division algebras, triality, Jordan agebras and the Freudenthal magic square. The third part describes work of Deligne and Vogel inspired by knot theory and several perspectives for understanding this work.

1 Overview

This article has two purposes. The first is to provide an elementary introduction to papers [59, 60, 61, 62, 63], related works and their historical context. Sections 2.1–2.3, 2.5–2.6, 3.1–3.5, 4.1–4.3 should be accessible to a general audience of mathematicians. The second is to provide generalizations, new perspectives, and complements to results in these papers, i.e., things we thought of after the papers were published. In particular, we mention 2.5, 2.7, 3.4 and 3.6–3.9. Each section begins with a description of its contents.

Simply put, our goals are to use geometry to solve questions in representation theory and to use representation theory to solve questions in geometry.

On the geometric side, the objects of interest are homogeneous varieties $X = G/P \subset \mathbb{P}V$. Here G is a complex semi-simple Lie group, V is an irreducible G-module, X is the unique closed orbit (projectivization of the orbit of a highest weight vector) and P is the stabilizer of a point. For example, let $W = \mathbb{C}^m$ and let $X = G(k, W) \subset \mathbb{P}(\Lambda^k W)$, the Grassmannian of k-planes through the origin in W. Here $G/P = \mathrm{SL}(W)/P$, $V = \Lambda^k W$. We are more generally interested in the geometry of orbit closures in $\mathbb{P}V$.

Basic questions one can ask about a variety are its dimension, its degree, and more generally its Hilbert function. We may ask the same questions for varieties associated to X. For example the degrees of the dual varieties of Grassmannians are still unknown in general (see [67, 26, 78]).

Other types of problems include recognition questions. For example, given a variety with certain geometric properties, are those properties enough to characterize the variety? For an example of this, see Zak's theorem on Severi varieties below in Section 2.6. For another example, Hwang and Mok characterize rational homogeneous varieties by the variety of tangent directions to minimal degree rational curves passing through a general point, see [44] for an overview. Some results along those lines are described in Section 2.5 below. We also mention the LeBrun–Salamon conjecture [68, 69] which states that any Fano variety equipped with a holomorphic contact structure must be the closed orbit in the adjoint representation $X_{\mathrm{ad}} \subset \mathbb{P}\mathfrak{g}$ for a complex simple Lie algebra. In this context also see [8, 52].

On the representation theory side, the basic objects are $\mathfrak{g}$, a complex semi-simple Lie algebra and V, an irreducible $\mathfrak{g}$-module (e.g., $\mathfrak{g} = \mathfrak{sl}(W)$, $V = \Lambda^k W$). Problems include the classification of orbit closures in $\mathbb{P}V$, to construct explicit models for the group action, to geometrically interpret the decomposition of $V^{\otimes k}$ into irreducible $\mathfrak{g}$-modules. We discuss these classical questions below, primarily for algebras occuring in "series".

Vassiliev theory points to the need for defining objects more general than Lie algebras. We have nothing to add about this subject at the moment, but the results of [62, 63] were partly inspired by work of Deligne [24] and Vogel [82, 83] in this direction.

For the mystically inclined, there are many strange formulas related to the exceptional groups. We present some such formulas in Sections 4.3, 4.6 below. Proctor and Gelfand–Zelevinski filled in "holes" in the classical formulas for the $\mathfrak{osp}_n$ series using the non-reductive *odd symplectic groups.* Our formulas led us to exceptional analogues of the odd symplectic groups. These analogues are currently under investigation (see [65]).

When not otherwise specified, we use the ordering of roots as in [9].

We now turn to details. We begin with some observations that lead to interesting rational maps of projective spaces.

2 Construction Of Complex Simple Lie Algebras Via Geometry

We begin in Section 2.1 with three ingredients that go into our study: local differential geometry (asymptotic directions), elementary algebraic geometry (rational maps of projective space) and homogeneous varieties (the correspondence between rational homogeneous varieties and marked Dynkin diagrams). We then, in Sections 2.2–2.4 describe two algorithms that construct new varieties from old that lead to new proofs of the classification of compact Hermitian symmetric spaces and the Cartan–Killing classification of complex simple Lie algebras. The proofs are constructive, via explicit rational maps and in

Section 2.5 we describe applications and generalizations of these maps. Our maps generalize maps used by Zak in his classification of Severi varieties and in Section 2.6 we describe his influence on our work. In Section 2.7 we return to a topic raised in Section 2.1, where we determined the parameter space of lines through a point of a homogeneous variety $X = G/P$. We explain *Tits' correspondences* which allow one to determine the parameter space of all lines on X and in fact parameter spaces for all G-homogeneous varieties on X. We explain how to use Tits correspondences to explicitly construct certain homogeneous vector bundles, an in turn to use the explicit construction to systematize Kempf's method for desingularizing orbit closures.

2.1 Differential Geometry, Algebraic Geometry and Representation Theory

2.1.1 Local Differential Geometry

Let $X^n \subset \mathbb{P}^{n+a}$ be an algebraic variety.

Question 2.1. How to study the geometry of X?

One answer: Take out a microscope. Fix a smooth point $x \in X$ and take derivatives at x. The first derivatives don't yield much. One gets $\tilde{T}_xX$, the embedded tangent projective space, the union of lines (linearly embedded $\mathbb{P}^1$'s) having contact to order one with X at x. (Contact to order zero means the line passes through x.)

Sometimes we work with vector spaces so for $Y \subset \mathbb{P}V$ we let $\hat{Y} \subset V$ denote the cone over Y and we let $\hat{T}_xX = \hat{\tilde{T}}_xX$. Let T_xX denote the Zariski (intrinsic) tangent space to X at x. We have $T_xX = \hat{x}^* \otimes \hat{T}_xX/\hat{x}$, see [59].

Taking second derivatives one obtains a variety $Y = Y_{X,x} \subset \mathbb{P}T_xX$ whose points are the *asymptotic directions*, the tangent directions to lines having contact to order two with X at x.

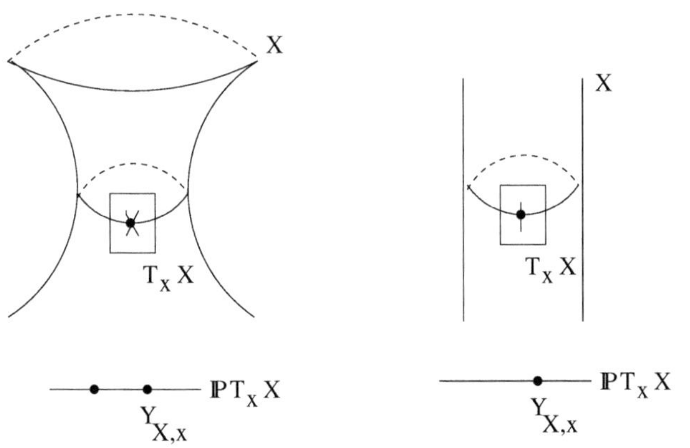

In an attempt to get global information from the asymptotic directions, restrict to the case where $x \in X$ is a *general* point

Question 2.2. How much of the geometry of X can be recovered from $Y_{X,x} \subset \mathbb{P}T_xX$?

Answer: Usually not much. For example, if X is a smooth hypersurface, then Y is always a smooth quadric hypersurface, i.e., all smooth hypersurfaces look the same to second order.

The set of asymptotic directions is the zero set of a system of quadratic equations generically of dimension equal to the codimension of X (unless the codimension of X is large, in which case it is generically the complete system of quadrics). If Y is sufficiently pathological one might hope to recover important information about X.

Consider the *Segre variety*, $\mathrm{Seg}(\mathbb{P}^k \times \mathbb{P}^l) \subset \mathbb{P}(\mathbb{C}^{k+1} \otimes \mathbb{C}^{l+1})$ of rank one matrices in the space of all $(k+1) \times (l+1)$ matrices. A short calculation shows $Y_{\mathrm{Seg},x} = \mathbb{P}^{k-1} \sqcup \mathbb{P}^{l-1}$, the disjoint union of a $\mathbb{P}^{k-1}$ with a $\mathbb{P}^{l-1}$. (Note that the codimension is sufficiently large here that one would expect Y to be empty based on dimension considerations.)

Griffiths and Harris [37] conjectured that if $Z \subset \mathbb{P}^8$ is a variety of dimension 4 and $z \in Z$ a general point, if $Y_{Z,z} \subset \mathbb{P}T_zZ$ is the disjoint union of two lines, then $Z = \mathrm{Seg}(\mathbb{P}^2 \times \mathbb{P}^2) \subset \mathbb{P}^8$.

Theorem 2.3. [58] *Let $k, l > 1$, let $Z^{k+l} \subset \mathbb{P}^N$ be a variety, and let $z \in Z$ be a general point. If $Y_{Z,z} = \mathbb{P}^{k-1} \sqcup \mathbb{P}^{l-1}$, then $Z = \mathrm{Seg}(\mathbb{P}^k \times \mathbb{P}^l)$.*

Moreover, the analogous rigidity is true for varieties having the same asymptotic directions as $X = G(2,m) \subset \mathbb{P}(\Lambda^2\mathbb{C}^m)$, the Grassmannian of two-planes in $\mathbb{C}^m$ in its Plucker embedding, with $m \geq 6$, and the complex Cayley plane $E_6/P_6 = \mathbb{OP}^2_{\mathbb{C}} \subset \mathbb{P}^{26}$ [58]. We recently prove similar rigidity results for homogeneous Legendrian varieties [64].

The rigidity results hold in the C^∞ category if one replaces the word "general" by the word "every".

If one adds the global assumption that Z is smooth, the results also hold for $X = G(2,5)$ and $X = \mathbb{S}_5 \subset \mathbb{P}^{15}$, the spinor variety using different methods.

We are unaware of any second order rigidity results for nonhomogeneous varieties. Thus, purely from the most naive differential geometry, one already encounters homogeneous varieties as examples of the most rigid projective varieties. In fact, so far we have just encountered the most homogeneous ones, the ones admitting Hermitian symmetric metrics. For more on this and the rigidity of other homogenous varieties see Section 2.5.

We define a homogeneous variety $X = G/P \subset \mathbb{P}V$ to be *minuscule* if G is simple, X admits a Hermitian symmetric metric and X is in its minimal homogeneous embedding. X is said to be *generalized minuscule* if it is homogeneously embedded and admits a Hermitian symmetric metric.

Remark 2.4. If $X \subset \mathbb{P}V$ is a variety cut out by quadratic polynomials, then the asymptotic directions $Y \subset \mathbb{P}T_xX$ are actually the tangent directions to lines (linearly embedded $\mathbb{P}^1$'s) on X. Rational homogeneous varieties are cut out by quadratic equations, in fact if $X \subset \mathbb{P}V_\lambda$, then the ideal is generated by $V_{2\lambda}{}^\perp \subset S^2V_\lambda^*$.

2.1.2 Lie Groups and Homogeneous Varieties

Let G be a complex (semi)-simple Lie group and let V be an irreducible G-module. Then there exists a unique closed orbit $X = G/P \subset \mathbb{P}V$.

Examples. 1. $G = \mathrm{SL}(n, \mathbb{C})$, the group preserving $\det \in \Lambda^n\mathbb{C}^{n*}$, $V = \Lambda^k\mathbb{C}^n$, $X = G(k, n)$, the Grassmannian of k-planes through the origin in $\mathbb{C}^n$.

2. $G = \mathrm{SO}(n, Q)$, the group preserving a nondegenerate $Q \in S^2\mathbb{C}^{n*}$, $V = \Lambda^k\mathbb{C}^n$, $X = G_Q(k, n)$, the Grassmannian of Q-isotropic k-planes through the origin in $\mathbb{C}^n$.

3. $G = \mathrm{Sp}(n, \omega)$, the group preserving a nondegenerate $\omega \in \Lambda^2\mathbb{C}^{n*}$, $V = \Lambda^k\mathbb{C}^n/(\Lambda^{k-2}\mathbb{C}^n \wedge \omega)$, $X = G_\omega(k, n)$, the Grassmannian of ω-isotropic k-planes through the origin in $\mathbb{C}^n$. Here n is usually required to be even (but see [73, 36], and Section 4.3).

Since linear algebra is easier than global geometry, we work with $\mathfrak{g} = T_{\mathrm{Id}}G$, the associated Lie algebra.

About a century ago, Killing and Cartan classified complex simple Lie algebras. Thanks to Coxeter and Dynkin, the classification can be expressed pictorially. (See [40] for a wonderful account of their work and the history surrounding it.)

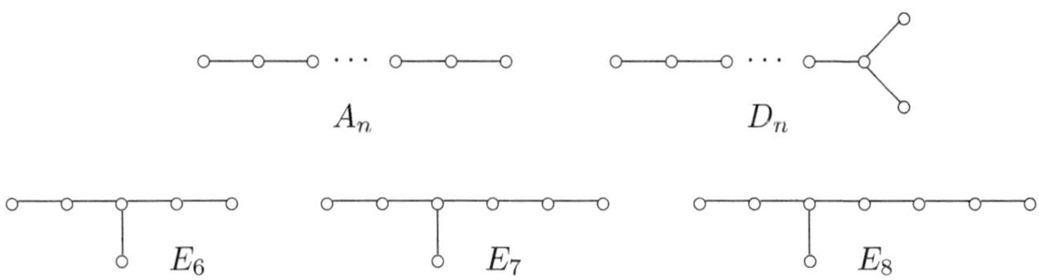

A_n D_n

E_6 E_7 E_8

If a diagram has symmetry, we are allowed to fold it along the symmetry and place an arrow pointing away from the hinge to get a new one:

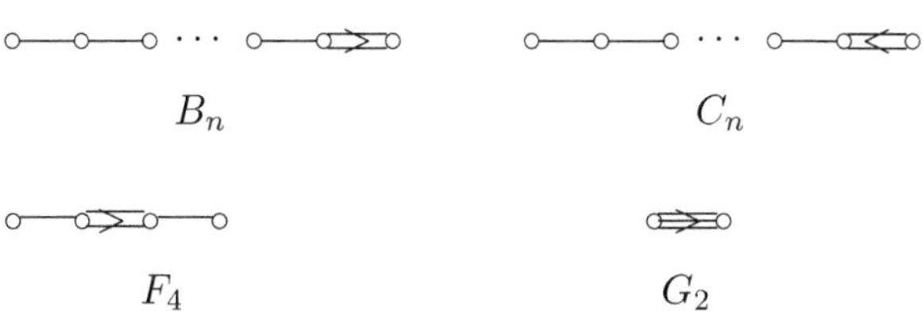

B_n C_n

F_4 G_2

For example, C_n is the fold of A_{2n-1}. Given a semi-simple Lie algebra $\mathfrak{g}$, we let $D(\mathfrak{g})$ denote its Dynkin diagram.

Returning to geometry, homogeneous varieties correspond to marked diagrams

$G(3,7)$

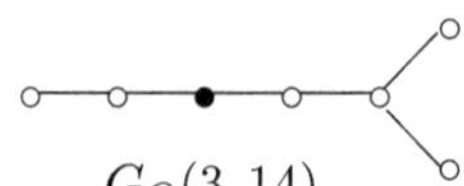

$G_Q(3,14)$

Note in particular that $\mathbb{P}^n$ has marked diagram

$\mathbb{P}^n$

If S is a subset of the simple roots of G, we use P_S to designate the parabolic subgroup of G corresponding to the simple roots in the complement of S. The Dynkin diagram for G/P_S is the diagram for G with roots in S marked. Also, if S consists of a single root $\{\alpha\}$ and $\alpha = \alpha_i$, we write $P_S = P_\alpha = P_i$.

Unless otherwise specified, we take $\mathbb{P}V$ to be ambient space for the the minimal homogeneous embedding of G/P. For example, if $P = P_\alpha$ is maximal, then $V = V_\omega$ where ω is the weight Killing-dual to the coroot of α.

Back to our question of how to study homogeneous varieties:

Idea: Study $X \subset \mathbb{P}V$ *via* $Y = Y_{X,x} \subset \mathbb{P}T_xX$.

For this idea to be a good one, Y should be a simpler space than X and it should be possible to determine Y systematically. Although Y is indeed cut out by quadratic equations, and has dimension strictly less than that of X, X is homogeneous, and Y need not be.

The following theorem originates with work of Tits [81], amplified by Cohen and Cooperstein [18]:

Theorem 2.5. [59] *Let $X = G/P_\alpha \subset \mathbb{P}V$ be a rational homogeneous variety such that α is not short (i.e., no arrow in $D(\mathfrak{g})$ points towards α).*

Then $Y \subset \mathbb{P}T_xX$ is homogeneous, in fact generalized minuscule (as defined above).

Moreover, Y can be determined pictorially: remove the node corresponding to α from $D(\mathfrak{g})$ and mark the nodes that were adjacent to α. One obtains a semi-simple Lie algebra $\mathfrak{h}$ with marked diagram. The resulting homogeneous space H/Q is Y.

Example 2.6.

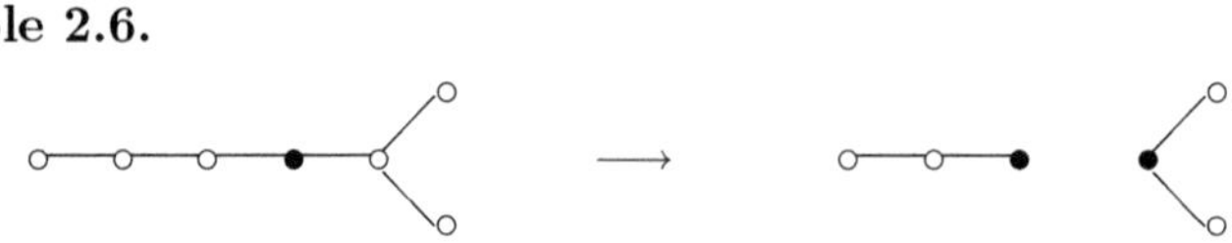

$X = G_Q(4,12)$ $\qquad$ $Y = \mathrm{Seg}(\mathbb{P}^3 \times G(2,4))$

The H-modules $\langle Y \rangle$ are studied in [49] where they are called *type-I θ-representations*.

The embedding of Y is minimal iff the diagram is simply laced. In the case of a double edge, one takes the quadratic Veronese embedding, for a triple edge, one takes the cubic Veronese (see the algorithms below).

Remark 2.7. In [59] we explicitly determine Y in the case of short roots and in fact arbitrary parabolics. We also give geometric models. In the case of $X = G/P_\alpha$ with α short, Y is the union of exactly two G-orbits, and the closed orbit has codimension one in Y.

If Y is still complicated, one can continue, studying the asymptotic directions of Y at a point. Eventually one gets (Segre products of re-embeddings of) $\mathbb{CP}^1$, the one homogeneous space we all pretend to understand.

We describe our discovery that one can reverse this infinitesimalization procedure below. First we need to review some elementary algebraic geometry.

2.1.3 Embeddings Of Projective Space

Recall the Veronese embeddings of projective space:

$$v_d : \mathbb{P}V \to \mathbb{P}(S^dV),$$
$$[w] \mapsto [w^d].$$

Dually, let $P_0, \ldots, P_N$ be a basis of S^dV^*. The map is

$$[w] \mapsto [P_0(w), \ldots, P_N(w)].$$

A remarkable fact is that all maps between projective spaces $\mathbb{P}V \to \mathbb{P}W$ are obtained by projecting a Veronese re-embedding.

If we project to $\mathbb{P}(S^dV/L)$ where $L^\perp \subset S^dV^*$ has basis $Q_0, \ldots, Q_m$, the map is

$$f : \mathbb{P}V \to \mathbb{P}(S^dV/L),$$
$$[w] \mapsto [Q_0(w), \ldots, Q_m(w)]$$

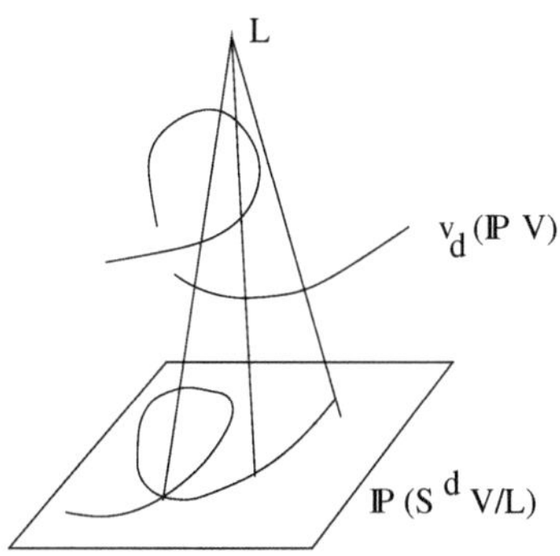

The image of f is smooth and isomorphic to $\mathbb{P}V$ iff no secant line of $v_d(\mathbb{P}V)$ intersects L.

In the following algorithm we will actually be interested in images that get squashed in the projection so that they are no longer isomorphic to $\mathbb{P}^n$ (but are still smooth).

2.2 First Algorithm

We are about to describe an algorithm, which you might like to think of as a game. Starting with $\mathbb{P}^1$ as initial input we build some new algebraic varieties subject to certain rules. The game has rounds, and in each new round, we are allowed to use the outputs from previous rounds as new inputs.

We fix some notation.

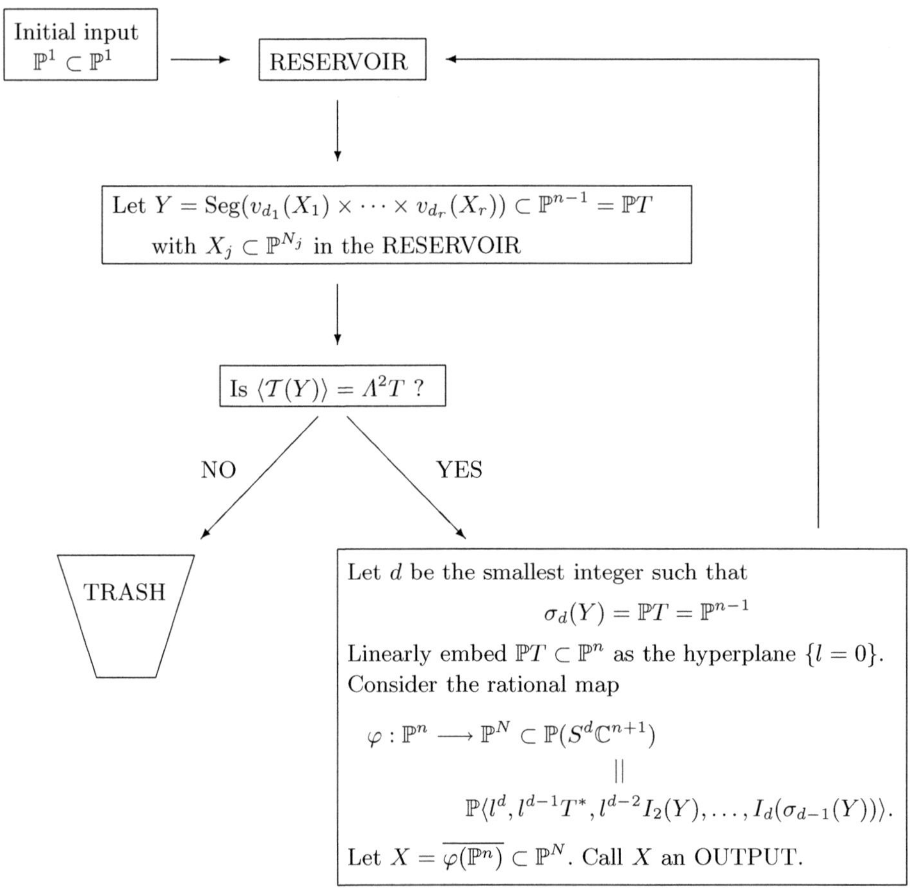

Fig. 1. First Algorithm

Let $X \subset \mathbb{P}V$, $Y \subset \mathbb{P}W$, be varieties.

$\mathrm{Seg}(X \times Y) \subset \mathbb{P}(V \otimes W)$ is the Segre product, $([x],[y]) \mapsto [x \otimes y]$. Sometimes we just write $X \times Y$ with the Segre product being understood.

$\sigma_k(X) \subset \mathbb{P}V$ is the union of all secant $\mathbb{P}^{k-1}$'s to X. We let $\sigma(X) = \sigma_2(X)$.

$\mathcal{T}(X) \subset G(2,V) \subset \mathbb{P}(\Lambda^2 \mathbb{P}V)$ denotes the union of all tangent lines to V. (Recall $G(k,V)$ is also the space of $\mathbb{P}^{k-1}$'s in $\mathbb{P}V$.)

$\hat{X} \subset V$ is the cone over X and $\langle X \rangle \subseteq V$ denotes the linear span of $\hat{X}$.

Given $Y \subset \mathbb{P}W$, we let $I_k(Y) \subset S^k W^*$ denote the component of the ideal in degree k. In the flowchart below we slightly abuse notation by letting $I_k(Y)$ denote a set of generators of $I_k(Y)$. Similarly, we let T^* be shorthand for a basis of T^*.

For the first run through the algorithm, the admissible varieties and their outputs are as follows

$$\begin{array}{rl} Y \subseteq \mathbb{P}^{n-1}, & X^n \subseteq \mathbb{P}^N, \\ \mathbb{P}^1 \subseteq \mathbb{P}^1, & \mathbb{P}^2 \subseteq \mathbb{P}^2, \\ \mathbb{P}^1 \times \mathbb{P}^1 \subset \mathbb{P}^3, & \mathbb{Q}^4 \subset \mathbb{P}^5, \\ v_2(\mathbb{P}^1) \subset \mathbb{P}^2, & \mathbb{Q}^3 \subset \mathbb{P}^4. \end{array}$$

Here and below, $\mathbb{Q}^m \subset \mathbb{P}^{m+1}$ denotes the smooth quadric hypersurface.

For the second round,

$$\begin{array}{rr} Y \subseteq \mathbb{P}^{n-1}, & X^n \subseteq \mathbb{P}^N, \\ \mathbb{P}^2 \subseteq \mathbb{P}^2, & \mathbb{P}^3 \subseteq \mathbb{P}^3, \\ v_2(\mathbb{P}^2) \subset \mathbb{P}^5, & G_\omega(3,6) \subset \mathbb{P}^{11}, \\ \mathbb{P}^1 \times \mathbb{P}^2 \subset \mathbb{P}^5, & G(2,5) \subset \mathbb{P}^9, \\ \mathbb{P}^2 \times \mathbb{P}^2 \subset \mathbb{P}^8, & G(3,6) \subset \mathbb{P}^{19}, \\ \mathbb{Q}^4 \subset \mathbb{P}^5, & \mathbb{Q}^6 \subset \mathbb{P}^8, \\ \mathbb{Q}^3 \subset \mathbb{P}^4, & \mathbb{Q}^5 \subset \mathbb{P}^7. \end{array}$$

Here $G(k,l)$ denotes the Grassmannian of k-planes in $\mathbb{C}^l$ and $G_\omega(k,2k)$ denotes the Grassmannian of Lagrangian k-planes for a given symplectic form.

Question 2.8. What comes out? Since the algorithm goes on forever, is it even possible to answer this question?

Proposition 2.9. [60] *The algorithm is effective. A priori, $r, d_j \leq 2$ and the algorithm stabilizes after six rounds.*

So at least our question is reasonable. Now for the answer:

Theorem 2.10. [60] *OUTPUTS = MINUSCULE VARIETIES.*

Corollary 2.11. [60] *A new proof of the classification of the compact Hermitian symmetric spaces without any reference to Lie groups.*

We have the following stable round:

$$\begin{array}{rr} Y \subseteq \mathbb{P}^{n-1}, & X^n \subseteq \mathbb{P}^N, \\ \mathbb{P}^{n-1} \subseteq \mathbb{P}^{n-1}, & \mathbb{P}^n \subseteq \mathbb{P}^n, \\ v_2(\mathbb{P}^{m-1}) \subset \mathbb{P}^{\binom{m+1}{2}-1}, & G_\omega(m,2m) \subset \mathbb{P}^{C_{m+1}-1}, \\ \mathbb{P}^{k-1} \times \mathbb{P}^{l-1} \subset \mathbb{P}^{kl-1}, & G(k,k+l) \subset \mathbb{P}^{\binom{k+l}{k}-1}, \\ \mathbb{Q}^{2m-2} \subset \mathbb{P}^{2m-1}, & \mathbb{Q}^{2m} \subset \mathbb{P}^{2m+1}, \\ \mathbb{Q}^{2m-1} \subset \mathbb{P}^{2m}, & \mathbb{Q}^{2m+1} \subset \mathbb{P}^{2m+2}, \\ G(2,m) \subset \mathbb{P}^{\binom{m}{2}-1}, & \mathbb{S}_m \subset \mathbb{P}^{2^{m-1}-1}. \end{array}$$

Here $C_{m+1} = \frac{1}{m+2}\binom{2m+2}{m+1}$ is the $(m+1)$-st Catalan number. The spinor variety $\mathbb{S}_m$ of D_m consists of one family of maximal isotropic subspaces of $\mathbb{C}^{2m}$ endowed with a nondegenerate quadratic form and embedded in the projectivization of one of the two half-spin representations.

The most interesting (but terminal) path is:

$$\begin{array}{rcl} \mathrm{Seg}(\mathbb{P}^1 \times \mathbb{P}^2) & = & A_1/P_1 \times A_2/P_2 \\ & \downarrow & \\ G(2,5) & = & A_4/P_2 \\ & \downarrow & \\ \mathbb{S}_5 & = & D_5/P_5 \\ & \downarrow & \\ \mathbb{OP}^2_{\mathbb{C}} & = & E_6/P_6 \\ & \downarrow & \\ G_w(\mathbb{O}^3, \mathbb{O}^6) & = & E_7/P_7. \end{array}$$

Here $\mathbb{OP}^2_{\mathbb{C}} = E_{6(-14)}$ is the complexification of $\mathbb{OP}^2 = F_4/\mathrm{Spin}_9$, the celebrated Cayley plane discovered by Chevalley. As a topological space, $\mathbb{OP}^2$ is built out of three cells of dimension $0, 8$ and 16. It is $F_{4(-20)}$ in the notation of Tits (see, e.g., [70]). The notation $G_w(\mathbb{O}^3, \mathbb{O}^6)$ is discussed in Section 4.6.

2.3 Second Algorithm

Now that was fun, but it was a shame to throw away some of our favorite varieties like $v_3(\mathbb{P}^1)$ and $\mathrm{Seg}(\mathbb{P}^1 \times \mathbb{P}^1 \times \mathbb{P}^1)$. Let's revise the algorithm slightly and have just one consolation round. Given a smooth variety $Y \subset \mathbb{P}V$, we let $\tau(Y) \subset \mathbb{P}V$ denote the union of all points on all embedded tangent lines to Y.

This algorithm is also effective. We show *a priori* that $r, d_j \leq 3$.

So, what do we get???

Theorem 2.12. [60] *OUTPUTS = Fundamental adjoint varieties $X \subset \mathbb{P}\mathfrak{g}$.*

The fundamental adjoint varieties are the closed orbits in the adjoint representation when it is fundamental. The Lie algebras whose adjoint varieties are not fundamental are the pathological A_n and its fold (when foldable) C_m.

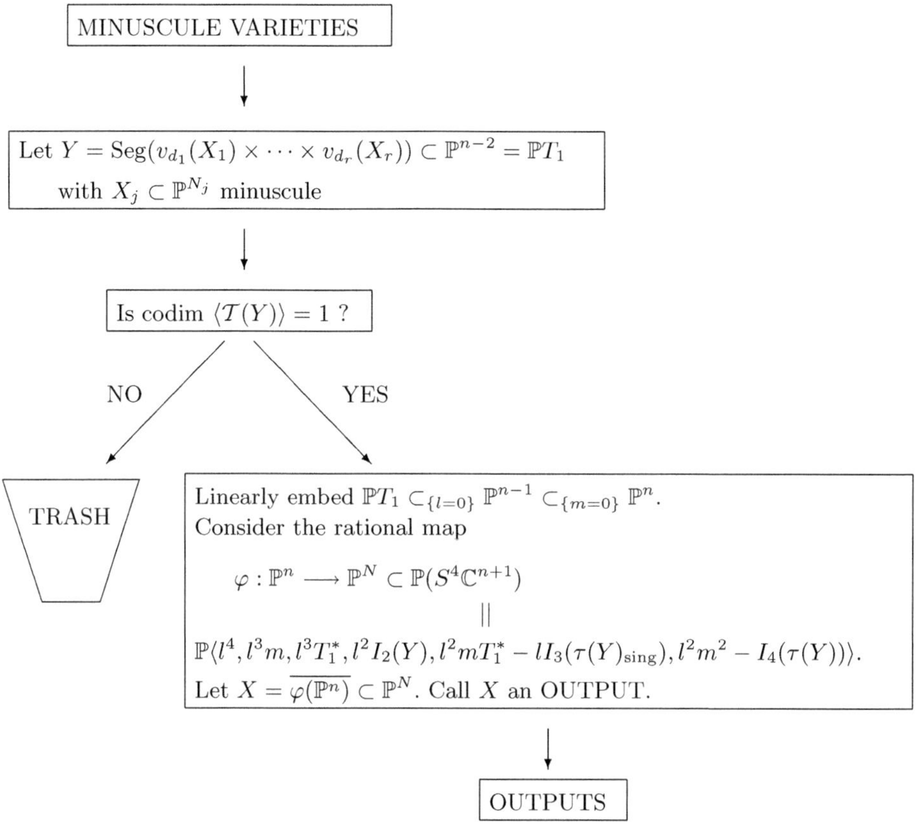

Fig. 2. Consolation prize (one round)

Our algorithm again provides a beautiful and easy proof of the classification of fundamental adjoint varieties. We also account for these two pathological cases, but it involves some less elegant work. We do obtain:

Corollary 2.13. [60] *A new proof of the Cartan–Killing classification of complex simple Lie algebras.*

The examples for the adjoint algorithm are as follows:

$$\begin{array}{rl}
Y \subset \mathbb{P}^{n-2}, & \mathfrak{g}, \\
v_3(\mathbb{P}^1) \subset \mathbb{P}^3, & \mathfrak{g}_2, \\
\mathbb{P}^1 \times \mathbb{Q}^{m-4} \subset \mathbb{P}^{2m-5}, & \mathfrak{so}_m, \\
G_\omega(3,6) \subset \mathbb{P}^{13}, & \mathfrak{f}_4, \\
G(3,6) \subset \mathbb{P}^{19}, & \mathfrak{e}_6, \\
\mathbb{S}_6 \subset \mathbb{P}^{31}, & \mathfrak{e}_7, \\
G_w(\mathbb{O}^3, \mathbb{O}^6) \subset \mathbb{P}^{55}, & \mathfrak{e}_8.
\end{array}$$

The two exceptional (i.e., non-fundamental) cases are

$$\mathbb{P}^{k-3} \sqcup \mathbb{P}^{k-3} \subset \mathbb{P}^{2k-3}, \quad \mathfrak{sl}_k,$$
$$\emptyset \subset \mathbb{P}^{2m-1}, \quad \mathfrak{sp}_{2m}.$$

Note that the algorithm works in these two cases, we just didn't have the varieties in our reservoir. (In the case of $\mathfrak{sp}_{2m}$, it may have been there and just difficult to see.)

2.4 Outline Of the Proofs

The proofs have three ingredients: differential invariants, local Lie algebras, and relating Casimirs to geometry.

2.4.1 Differential Invariants

Given a variety $X \subset \mathbb{P}V$ and $x \in X$, we can recover X from its Taylor series at x. The projective differential invariants comprise a series of tensors encoding the geometric (i.e., invariant under $GL(V)$) information in the Taylor series. We prove *a priori* facts about the differential invariants of any putative minuscule variety or fundamental adjoint variety.

2.4.2 Local Lie Algebras

A local Lie algebra is a graded vector space

$$\mathfrak{g}_{-1} \oplus \mathfrak{g}_0 \oplus \mathfrak{g}_1$$

equipped with a bracket for which $\mathfrak{g}_0$ is a Lie algebra. The bracket must respect the grading but the Jacobi identity need not hold. If the Jacobi identity fails, one can construct a (unique) $\mathbb{Z}$-graded Lie algebra from the local Lie algebra. The traditional way to do this (see, e.g., [48]) is to take the free algebra generated by the brackets and then mod out by the relations. Note that if one does that, one has no idea how many factors (if any at all) one will be adding on to obtain the final result.

The data $Y = H/Q \subseteq \mathbb{P}T_1$ furnishes (up to scale) a local Lie algebra with $\mathfrak{g}_0 = \mathfrak{h} \oplus \mathbb{C}$, $\mathfrak{g}_1 = T_1$, $\mathfrak{g}_{-1} = T_1^*$. Here the action of $\mathbb{C}$ is as a scalar times the identity and we do not initially specify the scalar.

Since this does give rise to a unique $\mathbb{Z}$-graded Lie algebra, (in particular, a Lie algebra equipped with a representation V supported on one fundamental weight), we can study the resulting homogeneous variety $X' = G/P \subset \mathbb{P}V$, and calculate its differential invariants.

Note that if X' is minuscule, $\mathfrak{g} = \mathfrak{g}_{-1} \oplus \mathfrak{g}_0 \oplus \mathfrak{g}_1$ and if X' is adjoint, $\mathfrak{g} = \mathfrak{g}_{-2} \oplus \mathfrak{g}_{-1} \oplus \mathfrak{g}_0 \oplus \mathfrak{g}_1 \oplus \mathfrak{g}_2$ with $\mathfrak{g}_{\pm 2} = \mathbb{C}$.

2.4.3 Compare X' With the Constructed Variety X

We compare the differential invariants of X and X' and show that they are the same. The key point in the minuscule case is that a local Lie algebra is already a Lie algebra iff the Jacobi identity holds. We show this is the case iff $\Lambda^2\mathfrak{g}_1$ is an eigenspace for the Casimir operator of $\mathfrak{h}$. We then connect this to the geometry by showing that $\langle \mathcal{T}(Y)\rangle$ is a Casimir eigenspace!

For the adjoint case the idea is similar: one shows that one can get away with a one-dimensional correction.

2.5 Applications, Generalizations and Related Work

2.5.1 More General Algorithms

The adjoint algorithm (with a different test for admission) also constructs all homogeneous varieties and representations corresponding to a five step grading. M. Dillon is currently formulating more general constructions for gradings including extensions to affine Lie algebras. (The algorithms have no hope of being effective for graded Lie algebras with exponential growth.) An amusing exercise is to construct the grading of $\mathfrak{e}_8{}^{(1)}$ associated to α_5. Here $\mathfrak{h} = \mathfrak{a}_4 + \mathfrak{a}_4$, and

$$\begin{aligned} Y = Y_1 = G(2,5)\times\mathbb{P}^4 = \mathrm{Seg}(G(2,V)\times\mathbb{P}W) &\subset \mathbb{P}(\Lambda^2V\otimes W) = \mathbb{P}\mathfrak{g}_1,\\ Y_2 = \mathrm{Seg}(\mathbb{P}V^*\times G(2,W)) &\subset \mathbb{P}(V^*\otimes\Lambda^2W) = \mathbb{P}\mathfrak{g}_2,\\ Y_3 = \mathrm{Seg}(\mathbb{P}V\times G(2,W^*)) &\subset \mathbb{P}(V\otimes\Lambda^2W^*) = \mathbb{P}\mathfrak{g}_3,\\ Y_4 = \mathrm{Seg}(G(2,V)\times\mathbb{P}W^*) &\subset \mathbb{P}(\Lambda^2V\otimes W^*) = \mathbb{P}\mathfrak{g}_4, \end{aligned}$$

and then the cycle repeats. Note that all the varieties are isomorphic as varieties and they appear in all possible combinations in terms of the $\mathfrak{h}$-action.

In this context, a result of Kostant [56] (Theorem 1.50) is interesting. He determines, given a reductive Lie algebra $\mathfrak{r}$ and an $\mathfrak{r}$-module V, when $\mathfrak{r}+V$ can be given the structure of a Lie algebra compatible with the $\mathfrak{r}$-actions (and in how many different ways).

2.5.2 Constructions In an Algebraic Context

Let $\mathcal{A}$ be an algebra with unit and involution $a\mapsto\overline{a}$. B. Allison [2] defines $\mathcal{A}$ to be *structurable* if the following graded vector space is actually a graded Lie algebra. Let

$$\mathfrak{g}(\mathcal{A}) = \mathfrak{g}_{-2}\oplus\mathfrak{g}_{-1}\oplus\mathfrak{g}_0\oplus\mathfrak{g}_1\oplus\mathfrak{g}_2$$

with $\mathfrak{g}_{\pm1}\simeq\mathcal{A}$, $\mathfrak{g}_{\pm2}\simeq\mathcal{S}(\mathcal{A}) := \{a\in\mathcal{A}\mid \overline{a}=-a\}$ and $\mathfrak{g}_0$ the set of derivations of $\mathcal{A}$ generated by linear maps $V_{a,b}:\mathcal{A}\to\mathcal{A}$, where $V_{a,b}(c) = (a\overline{b})c + (c\overline{b})a - (c\overline{a})b$.

$\mathfrak{g}(\mathcal{A})$ always has a natural bracket, and is a Lie algebra iff the Jacobi identity holds. In analogy with our situation, there is essentially one identity to check, which, in Allison's notation becomes

$$[V_{a,1}, V_{b,c}] = V_{V_{a,1}(b),c} - V_{b,V_{\overline{a},1}(c)}.$$

Allison shows there is a one to one correspondence between simple structurable algebras and certain symmetric five step gradings of simple Lie algebras. His constructions work over all fields and one motivation for his constructions was to determine simple Lie algebras over arbitrary fields. Our more general 5-step constructions correspond to algebraic structures termed *Kantor triple systems.*

Jordan triple systems form a special class of Kantor triple systems, those with trivial involution $\overline{a} = a$ and thus $\mathcal{S}(\mathcal{A}) = 0$, and they give rise to the minuscule gradings our minuscule algorithm produces.

2.5.3 Rigidity

In contrast to the rigidity theorems in Section 2.1.1 we have:

Theorem 2.14. *There exist "fake" adjoint varieties. That is, for each adjoint variety $X \subset \mathbb{P}\mathfrak{g}$, there exists $Z \subset \mathbb{P}^N$, not isomorphic to X, $U \subset Z$ a Zariski open subset, and a holomorphic map $\phi : U \to X$ such that asymptotic directions are preserved (in fact the entire projective second fundamental form is preserved, see [60]).*

Moreover, the same is true for all non-minuscule homogeneous varieties.

This local flexibility fails globally for adjoint varieties. Consider the following result of Hong [42]:

Theorem 2.15. *A Fano manifold with a geometric structure modeled after a fundamental adjoint variety Z is biholomorphic to Z and the geometric structure is locally isomorphic to the standard geometric structure on Z.*

A geometric structure on a variety X modeled after a fundamental adjoint variety Z may be understood as follows. Tor all $x \in X$, one has a subvariety $Y_x \subset \mathbb{P}T_xX$ isomorphic to the asymptotic directions $Y_z \subset \mathbb{P}T_zZ$. Note that here the subvariety Y_x is not required to play any particular role, but it must be present at every point so in particular it determines a family of distributions on X.

Hong's theorem is a variant of an earlier rigidity result of Hwang and Mok [45], where a geometric structure may be understood in the analogous way.

Theorem 2.16. *A Fano manifold with a geometric structure modeled after a compact irreducible Hermitian symmetric space S of rank ≥ 2 is biholomorphic to S and the geometric structure is locally isomorphic to the standard geometric structure on S.*

The work of Hwang and Mok (also see [44]) relies on studying an intrinsic analog of the set of asymptotic directions. Namely a Fano manifold X is uniruled by rational curves, and, fixing a reference line bundle, one has a subvariety $Y_x \subset \mathbb{P}T_xX$ of tangent directions to minimal degree rational curves. They also study deformation rigidity of homogeneous varieties with Picard number one. It is worth remarking that in the case of Hermitian symmetric spaces [46], a key fact used in their proof is that $\mathcal{T}(Y_x) \subset \mathbb{P}T_xX$ is linearly nondegenerate, i.e., the same condition we use in the minuscule algorithm.

2.5.4 Normal Forms For Singularities

In [4], Arnold classified the simple singularities. Like many interesting things in life they are in correspondence with Dynkin diagrams, in fact just the simply laced ones. He also gave normal forms in a minimal set of variables (two). It was also known (but evidently unpublished) that the simple singularities can be realized as degree three hypersurface singularities if one allows the number of variables to grow with the Milnor number. Holweck [41] has found a nice realization of these hypersurface singularities using a theorem of Knop [54] and the construction in Section 2.3.

2.6 Why Secant and Tangent Lines?

The idea that secant and tangent lines should so strongly control the geometry of homogeneous varieties was inspired by Zak's theorem on Severi varieties.

Theorem 2.17. (Zak's theorems on linear normality and Severi varieties) *Let $X^n \subset \mathbb{P}^{n+a}$ be a smooth variety, not contained in a hyperplane and such that $\sigma(X) \neq \mathbb{P}^{n+a}$ Then*

(i) $a \geq \frac{n}{2} + 2$.

(ii) *If $a = \frac{n}{2} + 2$ then X is one of* $v_2(\mathbb{P}^2) \subset \mathbb{P}^5$, $\mathrm{Seg}(\mathbb{P}^2 \times \mathbb{P}^2) \subset \mathbb{P}^8$, $G(2,6) \subset \mathbb{P}^{14}$, $\mathbb{OP}^2_{\mathbb{C}} \subset \mathbb{P}^{26}$.

The four critical varieties are called the *Severi* varieties after Severi who proved the $n = 2$ case of the theorem.

It is not known if there is a bound on the secant defect $\delta_\sigma(X) := 2n + 1 - \dim\sigma(\mathrm{X})$ for smooth subvarieties of projective space with degenerate secant varieties. On the other hand, Zak established an upper bound on the codimension of a smooth variety of a given secant defect. He then went on to classify the varieties achieving this bound, which he calls the *Scorza varieties*. They are all closed orbits $G/P \subset \mathbb{P}V$, namely $v_2(\mathbb{P}^n)$, $\mathrm{Seg}(\mathbb{P}^n \times \mathbb{P}^n)$, $G(2,n)$ and $\mathbb{OP}^2_{\mathbb{C}}$.

Using Zak's result, Ein and Shepherd–Baron proved these four varieties also classify quadro-quadro Cremona transformations [29]. There are numerous other characterization problems where the answer is the Severi varieties.

Zak's proofs of his theorems rely on looking at *entry loci*. Namely let $X \subset \mathbb{P}V$ be a variety and let $y \in \mathbb{P}V \backslash X$. Define the entry locus of y to be

$$\Sigma_y := \overline{\{x \in X \mid \exists z \in X, y \in \mathbb{P}^1_{xz}\}}.$$

Here $\mathbb{P}^1_{xz}$ denotes the projective line spanned by x and z. Zak shows that for a Severi variety, the entry locus of a general point of $\sigma(X)$ is a quadric hypersurface in X. He then goes on to show that a variety so uniruled by quadric hypersurfaces and satisfying the dimension requirements must satisfy further dimension restrictions and eventually must be one of the four Severi varieties. More precisely, he shows that each Severi variety is the image of a rational map of a projective space. Our algorithms generalize his construction of the Severi varieties.

Recently Chaput [15] has shown that such a uniruling by quadrics immediately implies homogeneity, which gives a quicker proof of Zak's theorem.

The rigidity results for the Severi varieties other than $v_2(\mathbb{P}^2)$, are an outgrowth of a different proof of Zak's theorem, where one first shows that any putative Severi variety infinitesimally looks like an actual Severi variety to second order at a general point, and then uses the rigidity to finish the proof. ($v_2(\mathbb{P}^2)$ requires special treatment as its tangential variety is nondegenerate.) See [57] and [58].

Many other important problems are related to the geometry of secant varieties. Consider the classical Waring problem: given a generic homogeneous polynomial of degree d in $n+1$ variables, what is the minimal k such that the polynomial may be written as a sum of k d-th powers? Phrased geometrically, the problem is to find the minimal k such that $\sigma_k(v_d(\mathbb{P}^n)) = \mathbb{P}(S^d\mathbb{C}^{n+1})$. This problem was solved by Alexander and Hirschowitz [1]. Generalizations and variants are still open, see [13, 47, 75] for recent progress.

2.7 Tits Correspondences and Applications

In Section 2.7.1 we describe a construction, *Tits correspondences* [81], to determine the homogeneous (in the sense described below) unirulings of a rational homogeneous variety $X \subset \mathbb{P}V$. We explain how to use Tits correspondences to explicitly construct the homogeneous vector bundles over X in Section 2.7.2. In Section 2.7.3 we relate Tits correspondences to other orbit closures in $\mathbb{P}V$ and systematize Kempf's method for desingularizing orbit closures. Tits correspondences led us to the decomposition formulas described in Section 4.2 and we believe they will have more applications in the future.

2.7.1 Tits Correspondences

Tits associated to any simple group a full set of geometries, encoded by the parabolic subgroups and their relative positions. This culminated in the definition of *buildings*, which has since known formidable developments. The Tits correspondences we now describe come from this perspective.

Let G be a simple Lie group, let S, S' be two subsets of simple roots of G. Consider the diagram

$$G/P_{S\cup S'}$$
$$\pi \swarrow \qquad \searrow \pi'$$
$$X = G/P_S \qquad\qquad X' = G/P_{S'}\ .$$

Let $x' \in X'$ and consider $Y := \pi(\pi'^{-1}(x')) \subset X$. We call $Y = Y_{x'}$ the *Tits transform* of x'. The variety X is covered by such varieties Y. Tits shows that $Y = H/Q$ where $\mathcal{D}(H) = \mathcal{D}(G)\backslash(S\backslash S')$, and $Q \subset H$ is the parabolic subgroup corresponding to $S'\backslash S$. We call such subvarieties Y of X, *G-homogeneous subvarieties.*

2.7.2 Unirulings

Unirulings by lines. Let $X = G/P_\alpha \subset \mathbb{P}V$ be a homogeneous variety. In Theorem 2.5 above we determined the parameter space of lines through a point of X. Here we determine the space of all lines on X. Let $\mathbb{F}_1(X)^G \subset G(2,V) \subset \mathbb{P}\Lambda^2 V$ denote the variety parametrizing the G-homogeneous $\mathbb{P}^1$'s on X. Then $\mathbb{F}_1(X)^G = G/P_S$ where S is the set of simple roots adjacent to α in $D(\mathfrak{g})$. If P is not maximal, there is one family of G-homogeneous $\mathbb{P}^1$'s for each marked node. If α is not a short root, all lines are G-homogeneous.

Unirulings by $\mathbb{P}^k$'s. The diagram for $\mathbb{P}^k$ is $(\mathfrak{a}_k, \omega_1)$. Here there may be several G-homogeneous families of $\mathbb{P}^k$'s on $X = G/P_\alpha$, each arising from a subdiagram of $D(\mathfrak{g}, \omega)$ isomorphic to a $D(\mathfrak{a}_k, \omega_1)$. If α is not short, then all $\mathbb{P}^k$'s on X are G-homogeneous.

Example 2.18. The largest linear space on E_n/P_1 is a $\mathbb{P}^{n-1}$, via the chain terminating with α_n, so E_n/P_1 is maximally uniruled by $\mathbb{P}^{n-1}$'s. There is a second chain terminating with α_2, so E_n/P_1 is also maximally uniruled by $\mathbb{P}^4$'s. (The unirulings by the $\mathbb{P}^4$'s are maximal in the sense that none of the $\mathbb{P}^4$'s of the uniruling are contained in any $\mathbb{P}^5$ on E_n/P_1.) The varieties parametrizing these two unirulings of E_n/P_1 are respectively E_n/P_2 and E_n/P_5.

Unirulings by quadrics. Here we look for subdiagrams isomorphic to the diagram of a quadric hypersurface, i.e., $D(\mathfrak{so}_n, \omega_1)$. Note that there are at most two possible such subdiagrams except for the case of $(\mathfrak{g}, V) = (\mathfrak{so}_8, V_{\omega_2})$ where there are three isomorphic ones.

Example 2.19. $(\mathfrak{g}, V) = (\mathfrak{so}_{2m}, V_{\omega_k})$ has two such unirulings, by $\mathbb{Q}^4$'s parametrized by $D_m/P_{\alpha_{k-2},\alpha_{k+2}}$ and by $\mathbb{Q}^{2(m-k-1)}$'s parametrized by $D_m/P_{\alpha_{k-1}}$.

Example 2.20. (The Severi varieties) The G-homogeneous quadrics uniruling the Severi varieties are parametrized by the same Severi variety in the dual projective space.

Unirulings by quadrics give rise to subvarieties of $\sigma(X)$ as follows. The union of the spans of the quadrics produce a subvariety of $\sigma(X)$ whose entry loci contain the quadrics (and in all examples we know, are equal to the quadrics). It also appears that, when X is homogeneous, these give the subvarieties of $\sigma(X)$ with maximal entry loci. As explained below, these orbit closures admit uniform desingularizations by Kempf's method.

2.7.3 Homogeneous Vector Bundles

The G-homogeneous vector bundles over rational homogeneous varieties $Z = G/Q$ are defined by Q-modules. (We change notation to reserve G/P for a different role below.) If W is a Q-module, one obtains the vector bundle $E_W := G\times_Q W \to G/Q$ where $(gp, w) \simeq (g, pw)$ for all $p \in Q$. Note that $D(\mathfrak{f})$ is $D(\mathfrak{g})$ with the nodes corresponding to Q deleted.

Some homogeneous vector bundles over rational homogeneous varieties $Z = G/Q$ can be understood in terms of Tits fibrations. Let $X = G/P \subset \mathbb{P}V$. For each $z \in Z$ we obtain a variety $Y_z \subset X$ and thus a linear space $\langle Y_z\rangle \subset V$. As we vary z, we obtain a vector bundle $E \to Z$ whose fibers are the $\langle Y_z\rangle$'s. In particular, E is a subbundle of the trivial bundle $V \otimes \mathcal{O}_Z \to Z$.

2.7.4 Collapses à la Kempf

Let $X = G/P \subset \mathbb{P}V$ be a rational homogeneous variety. In [53], Kempf defined a method for desingularizing orbit closures $\overline{\mathcal{O}} \subset \mathbb{P}V$ by finding a vector bundle over a different homogeneous space $Z = G/Q$. He calls this technique the *collapsing of a vector bundle*. He gave some examples that appear to be found ad hoc. In [61] we gave new examples of collapsing occurring in series. Using the above discussion, we now give another description of Kempf's method.

Recipe For Kempf's Method

Let V be a G-module with closed orbit $X \subset \mathbb{P}V$. Let $Z = G/Q$ be a rational homogeneous variety of G, defining a Tits correspondence with X.

Consider the G-variety, which often appears to be an orbit closure

$$\overline{\mathcal{O}} = \cup_{z\in Z}\langle Y_z\rangle \subset \mathbb{P}V.$$

If $\overline{\mathcal{O}} \neq \mathbb{P}V$, then $\overline{\mathcal{O}}$ is a singular variety. It admits a desingularization by $\mathbb{P}E$, where $E \to Z$ is the vector bundle constructed above as the natural map $\mathbb{P}E \to Z$ is generically one to one.

Note that the G-variety $\tau(X)$ is of course desingularized by $\mathbb{P}TX$, which is a special case of the above with $Q = P$.

3 Triality and Exceptional Lie Algebras

We begin with a review of division algebras via the Cayley-Dickson process in Section 3.1 and explicitly describe the derivations of $\mathbb{O}$ in Section 3.2 for later use. In Section 3.3 we review the triality principle for the octonions and its extension to all structurable algebras. We pause for a detour in Section 3.4, exploring the triality model for $\mathfrak{so}_{4,4}$ in detail, which ends up being a 4-ality. We review various constructions of the magic square in Section 3.5. We present a new result regarding the compatiblity of the Cayley–Dickson process and inclusions of triality algebras in Section 3.6, which at first glance appears to be trivial, but whose "obvious" generalization to structurable algebras is false. In Section 3.7 we study automorphisms related to the algebras in the magic square and the resulting magic squares of symmetric and trisymmetric spaces. In Section 3.8 we use our result in Section 3.6 to show how some towers of dual pairs naturally occur in series. In Section 3.9 we discuss quaternionic symmetric spaces, their relation to adjoint varieties and the conjecture of LeBrun and Salamon.

3.1 Division Algebras

A classical theorem of Hurwitz, published in 1898, asserts that there are only four normed division algebras over the field of real numbers: $\mathbb{R}$ itself, $\mathbb{C}$, the quaternions $\mathbb{H}$, and the octonions $\mathbb{O}$. A *division algebra* is an algebra without zero divisor and a *normed division algebra* is a division algebra endowed with a norm such that the norm of a product is the product of the norms.

Quaternions were discovered by Hamilton in 1843, and octonions shortly afterwards by Graves and Cayley. A nice way to define these algebras is to imitate the definition of complex numbers by a pair of real numbers. Let $\mathbb{A}$ be a real algebra endowed with a conjugation $x \mapsto \overline{x}$, such that $x + \overline{x}$ and $x\overline{x} = \overline{x}x$ are always real numbers (more precisely scalar multiples of the unit element), the later being positive when x is nonzero. If $\mathbb{A}$ is alternative, i.e., any subalgebra generated by two elements is associative [5], then $\mathbb{A}$ is a normed algebra for the norm $\|x\|^2 = x\overline{x}$.

One can then define a new algebra with conjugation $\mathbb{B} = \mathbb{A}\oplus\mathbb{A}$ by letting $(x, y)(z, t) = (xz - t\overline{y}, \overline{x}t + zy)$ with conjugation $\overline{(x, y)} = (\overline{x}, -y)$. This is the *Cayley–Dickson* process. The new algebra $\mathbb{B}$ will be alternative, hence a normed algebra, iff $\mathbb{A}$ is associative.

In particular, $\mathbb{A}=\mathbb{C}$ gives the quaternion algebra $\mathbb{B}=\mathbb{H}$, which is associative, so the Cayley–Dickson process can be applied once more and $\mathbb{A}=\mathbb{H}$ gives the octonion algebra $\mathbb{B}=\mathbb{O}$. This algebra is no longer associative (although it is alternative, hence normed), and the process then fails to produce new normed algebras.

A useful variant of the Cayley–Dickson process is obtained by changing a sign in the formula for the product in $\mathbb{B}$, letting $(x,y)(z,t)=(xz+t\overline{y},\overline{x}t+zy)$. The resulting algebra $\tilde{\mathbb{B}}$ is called *split*. It is no longer normed, but endowed with a nondegenerate quadratic form of signature (a,a) compatible with the product. (Here a denotes the dimension of $\mathbb{A}$.) In particular, it is a *composition algebra*. The algebra $\tilde{\mathbb{C}}$ of split complex numbers is the algebra $\mathbb{R}\oplus\mathbb{R}$ with termwise multiplication, while the split quaternion algebra $\tilde{\mathbb{H}}$ is equivalent to $M_2(\mathbb{R})$.

3.2 Derivations

Any automorphism of $\mathbb{H}$ is inner, so that any derivation of $\mathbb{H}$ is of the form L_a-R_a for some imaginary quaternion $a\in\mathrm{Im}\mathbb{H}$, where L_a and R_a denote the operators of left and right mulplication by a. In particular $\mathrm{Der}\mathbb{H}=\mathrm{Im}\mathbb{H}=\mathfrak{so}_3$.

The derivation algebra of $\mathbb{O}$ is the compact Lie algebra $\mathfrak{g}_2$. The algebra $\mathrm{Der}\mathbb{H}=\mathfrak{so}_3$ does not imbed in $\mathrm{Der}\mathbb{O}$ in a canonical way, but there is one prefered embedding for each decomposition $\mathbb{O}=\mathbb{H}\oplus e\mathbb{H}$, where the product in $\mathbb{O}$ is deduced from that in $\mathbb{H}$ through the Cayley–Dickson process. Indeed, for $\phi\in Der\mathbb{H}$, the endomorphism $\tilde{\phi}$ of $\mathbb{O}$ defined by $\tilde{\phi}(x+ey)=\phi(x)+e\phi(y)$ is a derivation. Note that the subalgebra $\mathrm{Der}(\mathbb{O},\mathbb{H})$ of the derivations of $\mathbb{O}$ stabilizing $\mathbb{H}$ is strictly greater than $\mathrm{Der}\mathbb{H}$. It contains, for each imaginary quaternion h, the map $\psi_h(x+ey)=e(yh)$. This gives another copy of $\mathfrak{so}_3$ in $\mathrm{Der}(\mathbb{O},\mathbb{H})=\mathfrak{so}_3\times\mathfrak{so}_3$, and the restriction of these derivations to $e\mathbb{H}$ gives the full $\mathfrak{so}_4=\mathfrak{so}_3\times\mathfrak{so}_3$. Explicitly, choose a standard basis $e_0=1$, e_1, e_2, $e_3=e_1e_2$ of $\mathbb{H}$, and let $e_4=e$, $e_5=ee_1$, $e_6=ee_2$, $e_7=ee_3$. Using this basis of $\mathbb{O}$, we obtain a matrix representation of $\mathfrak{g}_2=\mathrm{Der}(\mathbb{O})$:

$$\mathfrak{g}_2=\begin{pmatrix}
0 & 0 & 0 & 0 & 0 & 0 & 0 & 0\\
0 & 0 & -\alpha_2 & -\alpha_3 & -\alpha_4 & -\alpha_5 & -\alpha_6 & -\alpha_7\\
0 & \alpha_2 & 0 & -\beta_3 & -\beta_4 & -\beta_5 & -\beta_6 & -\beta_7\\
0 & \alpha_3 & \beta_3 & 0 & \alpha_6+\beta_5 & -\alpha_5-\beta_4 & -\alpha_4+\beta_7 & \alpha_7-\beta_6\\
0 & \alpha_4 & \beta_4 & -\alpha_6-\beta_5 & 0 & -\gamma_5 & -\gamma_6 & -\gamma_7\\
0 & \alpha_5 & \beta_5 & \alpha_5+\beta_4 & \gamma_5 & 0 & -\alpha_2+\gamma_7 & -\alpha_3-\gamma_6\\
0 & \alpha_6 & \beta_6 & \alpha_4-\beta_7 & \gamma_6 & \alpha_2-\gamma_7 & 0 & \beta_3+\gamma_5\\
0 & \alpha_7 & \beta_7 & -\alpha_7+\beta_6 & \gamma_7 & \alpha_3+\gamma_6 & -\beta_3-\gamma_5 & 0
\end{pmatrix}.$$

It follows that the Lie sub-algebra $\mathrm{Der}(\mathbb{O},\mathbb{H})$ may be written:

$$\mathrm{Der}(\mathbb{O},\mathbb{H}) = \begin{pmatrix} 0 & 0 & 0 & 0 & 0 & 0 & 0 & 0 \\ 0 & 0 & -\alpha_2 & -\alpha_3 & 0 & 0 & 0 & 0 \\ 0 & \alpha_2 & 0 & -\beta_3 & 0 & 0 & 0 & 0 \\ 0 & \alpha_3 & \beta_3 & 0 & 0 & 0 & 0 & 0 \\ 0 & 0 & 0 & 0 & 0 & -\gamma_5 & -\gamma_6 & -\gamma_7 \\ 0 & 0 & 0 & 0 & \gamma_5 & 0 & -\alpha_2+\gamma_7 & -\alpha_3-\gamma_6 \\ 0 & 0 & 0 & 0 & \gamma_6 & \alpha_2-\gamma_7 & 0 & \beta_3+\gamma_5 \\ 0 & 0 & 0 & 0 & \gamma_7 & \alpha_3+\gamma_6 & -\beta_3-\gamma_5 & 0 \end{pmatrix}.$$

Our first type of derivations in $\mathrm{Der}(\mathbb{O},\mathbb{H})$ corresponds to those matrices for which $\gamma_5 = \gamma_6 = \gamma_7 = 0$, and the second type to those for which $\alpha_2 = \alpha_3 = \beta_3 = 0$. These two copies of $\mathfrak{so}_3$ commute. Finally, the south-east corner of the matrix above gives a copy of $\mathfrak{so}_4$.

3.3 Triality

3.3.1 Cartan's Triality

The Dynkin diagram of type D_4 is the only one with a threefold symmetry:

D_4

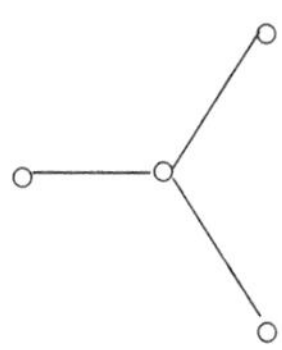

Symmetries of Dynkin diagrams detect outer automorphisms of the corresponding complex (or split) Lie algebras, and in the case of D_4 this is closely related with the algebra structure of the octonions. Indeed, let

$$T(\mathbb{O}) = \{(\mathrm{U}_1,\mathrm{U}_2,\mathrm{U}_3) \in \mathrm{SO}(\mathbb{O})^3 \mid \mathrm{U}_1(xy) = \mathrm{U}_2(x)\mathrm{U}_3(y) \ \ \forall x,y \in \mathbb{O}\}.$$

This group contains $\mathrm{Aut}(\mathbb{O}) = G_2$ and has three projections π_1, π_2, π_3 on $\mathrm{SO}(\mathbb{O})$, hence three representations of dimension eight. The following theorem is due to Elie Cartan and can be found page 370 of his paper [10] from 1925. Define Spin_8 to be the compact, simply connected Lie group with Lie algebra $\mathfrak{so}_8$.

Theorem 3.1. *The group $T(\mathbb{O})$ is Spin_8, and each projection π_i is a twofold covering of SO_8. The three corresponding eight dimensional representations of Spin_8 are not equivalent.*

With this description of Spin_8 in hand, it is easy to define an action of $\mathfrak{S}_3$ by outer automorphisms. Indeed, since $\mathbb{O}$ is alternative and $\overline{x} = -x$ modulo the unit element of $\mathbb{O}$, the relation $\overline{x}(xy) = (\overline{x}x)y$ holds. (This and similar identities are due to Moufang, see [39].) A little computation then shows that the identities

$$\mathrm{U}_1(xy) = \mathrm{U}_2(x)\mathrm{U}_3(y), \quad \mathrm{U}_2(xy) = \mathrm{U}_1(x)\overline{\mathrm{U}_3(\overline{y})}, \quad \mathrm{U}_3(xy) = \overline{\mathrm{U}_2(\overline{x})}\mathrm{U}_1(y)$$

are equivalent. If we denote by $\tau \in \mathrm{SO}(\mathbb{O})$ the conjugation anti-automorphism, this implies that we can define automorphisms s and t of Spin_8 by letting

$$\begin{aligned} s(\mathrm{U}_1, \mathrm{U}_2, \mathrm{U}_3) &= (\mathrm{U}_2, \mathrm{U}_1, \tau \circ \mathrm{U}_3 \circ \tau), \\ t(\mathrm{U}_1, \mathrm{U}_2, \mathrm{U}_3) &= (\tau \circ \mathrm{U}_1 \circ \tau, \tau \circ \mathrm{U}_3 \circ \tau, \tau \circ \mathrm{U}_2 \circ \tau). \end{aligned}$$

These automorphisms are not inner, since they exchange the three inequivalent eight dimensional representations of Spin_8. Moreover, we have $s^2 = t^2 = 1$ and $sts = tst$, so that s and t generate a group isomorphic to $\mathfrak{S}_3$, which is the full group of outer automorphisms of Spin_8.

Note that at the level of Lie algebras, Cartan's theorem above implies the *infinitesimal triality principle*, which we state as

$$\mathfrak{t}(\mathbb{O}) := \{(u_1, u_2, u_3) \in \mathfrak{so}(\mathbb{O})^3 \mid u_1(xy) = u_2(x)y + xu_3(y) \ \forall x, y \in \mathbb{O}\} = \mathfrak{so}_8.$$

This means that if we fix u_1 for example, there is a unique pair of skew-symmetric endomorphism of $\mathbb{O}$ such that the above relation holds.

3.3.2 Triality Algebras

Allison and Faulkner [3], following work of Allison [2] and Kantor [50] define the following vast generalization of the triality principle. Let $\mathcal{A}$ be any algebra with unit and involution $a \mapsto \overline{a}$ over a ring of scalars. Let $\mathrm{Im}\mathcal{A} = \{a \in \mathcal{A}, \ \overline{a} = -a\}$. Define the *triality algebra of* $\mathcal{A}$

$$\mathfrak{t}(\mathcal{A}) := \{T = (T_1, T_2, T_3) \in \mathfrak{gl}(\mathcal{A})^{\oplus 3} \mid \overline{T_1}(ab) = T_2(a)b + aT_3(b) \ \ \forall a, b \in \mathcal{A}\},$$

with $\overline{T_1}(a) = \overline{T_1(\overline{a})}$. This is a Lie subalgebra of $\mathfrak{gl}(\mathcal{A})^{\oplus 3}$. In [3] it is call the set of *partially related Lie triples of* $\mathcal{A}$. If 3 is invertible in the ring of scalars, they provide (Corollary 3.5) a general description of $\mathfrak{t}(\mathcal{A})$, which in the case where $\mathcal{A}$ is alternative gives

$$\begin{aligned} \mathfrak{t}(\mathcal{A}) &= \{(D + L_s - R_t, D + L_t - R_r, D + L_r - R_s) \mid D \in \mathrm{Der}(\mathcal{A}), \\ &\qquad\qquad\qquad\qquad r, s, t \in \mathrm{Im}(\mathcal{A}), \ r + s + t = 0\} \\ &= \mathrm{Der}(\mathcal{A}) \oplus \mathrm{Im}(\mathcal{A})^{\oplus 2}. \end{aligned}$$

If D is a derivation, $(D, D, D) \in \mathfrak{t}(\mathcal{A})$ because $\overline{D} = D$. Also, recall that $\mathcal{A}$ being alternative means that any subalgebra generated by two elements is associative, or equivalently, that the *associator*

$$\{x, y, z\} := (xy)z - x(yz), \quad x, y, z \in \mathcal{A},$$

is a skew-symmetric trilinear function. In particular, we have $\{x, y, z\} = \{z, x, y\}$, which can be rewritten as $(xy)z + z(xy) = x(yz) + (zx)y$. If z is imaginary, this shows that $(-L_z - R_z, L_z, R_z)$ is in $\mathfrak{t}(\mathcal{A})$. Using the triality automorphisms defined above, one gets that $(L_s - R_t, L_t - R_r, L_r - R_s)$ belongs to $\mathfrak{t}(\mathcal{A})$ as soon as r, s, t are imaginary with $r + s + t = 0$.

For composition algebras, we get

$$\begin{gathered}\mathfrak{t}(\mathbb{R}) = 0,\\ \mathfrak{t}(\mathbb{C}) = \mathbb{R}^2,\ \mathfrak{t}(\tilde{\mathbb{C}}) = \mathbb{R}^2,\\ \mathfrak{t}(\mathbb{H}) = (\mathfrak{so}_3)^{\oplus 3},\ \mathfrak{t}(\tilde{\mathbb{H}}) = (\mathfrak{sl}_2)^{\oplus 3},\\ \mathfrak{t}(\mathbb{O}) = \mathfrak{so}_8,\ \mathfrak{t}(\tilde{\mathbb{O}}) = \mathfrak{so}_{4,4}.\end{gathered}$$

3.4 Triality and $\mathfrak{so}_{4,4}$: 4-ality

An interesting version of triality holds for $\mathfrak{so}_{4,4}$, the split version of $\mathfrak{so}_8$. We use a model is related to the fourfold symmetry of the extended Dynkin diagram $\tilde{D}_4$:

$\tilde{D}_4$

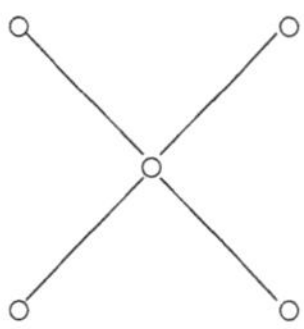

Choose four real two-dimensional vector spaces A, B, C, D and non-degenerate two-forms on each of them. Then

$$\mathfrak{so}_{4,4} = \mathfrak{sl}(A) \times \mathfrak{sl}(B) \times \mathfrak{sl}(C) \times \mathfrak{sl}(D) \oplus (A \otimes B \otimes C \otimes D).$$

In this model, $\mathfrak{sl}(A) \times \mathfrak{sl}(B) \times \mathfrak{sl}(C) \times \mathfrak{sl}(D)$ is a sub-Lie algebra and acts on $A \otimes B \otimes C \otimes D$ in the obvious way. The bracket of two vectors in $A \otimes B \otimes C \otimes D$ can be defined as follows. Choose non-degenerate skew-symmetric two forms on A, B, C and D. We denote these forms by the same letter ω. We have then natural isomorphisms $S^2 A \to \mathfrak{sl}(A)$, and so on, by sending a square a^2, where $a \in A$, to the endomorphism $a' \mapsto \omega(a, a')a$. With this understood, we let

$$\begin{aligned}&[a \otimes b \otimes c \otimes d, a' \otimes b' \otimes c' \otimes d']\\ &\quad = \omega(b, b')\omega(c, c')\omega(d, d')aa' + \omega(a, a')\omega(c, c')\omega(d, d')bb' +\\ &\qquad + \omega(a, a')\omega(b, b')\omega(d, d')cc' + \omega(a, a')\omega(b, b')\omega(c, c')dd'.\end{aligned}$$

One can then check that the Jacobi identity holds, so that we obtain a Lie algebra which can easily be identified with $\mathfrak{so}_{4,4}$. One could also take in $\mathfrak{so}_{4,4}$, once a Cartan subalgebra and a set of positive roots have been chosen, the product of the four copies of $\mathfrak{sl}_2$ generated by the three simple roots and by the highest root. These four copies commute, and the sum of the other root spaces forms a module over their product. It is then routine to identify this module with the tensor product of the natural representations.

What is particularily nice with this model is that we can explicitly define three inequivalent eight dimensional representations, which we denote by

$$\begin{aligned}\tilde{\mathbb{O}}_1 &= (A\otimes B)\oplus(C\otimes D),\\ \tilde{\mathbb{O}}_2 &= (A\otimes C)\oplus(B\otimes D),\\ \tilde{\mathbb{O}}_3 &= (A\otimes D)\oplus(B\otimes C).\end{aligned}$$

The action of $\mathfrak{so}_{4,4}$ is easy to describe: the action of $\mathfrak{sl}(A)\times\mathfrak{sl}(B)\times\mathfrak{sl}(C)\times\mathfrak{sl}(D)$ is the obvious one , while that of $A\otimes B\otimes C\otimes D$ on $\tilde{\mathbb{O}}_1$, for example, is defined by

$$a\otimes b\otimes c\otimes d.(a'\otimes b'+c'\otimes d')=\omega(c,c')\omega(d,d')a\otimes b+\omega(a,a')\omega(b,b')c\otimes d.$$

To see that we get three inequivalent representations, we can describe the root system of $\mathfrak{so}_{4,4}$ thanks to the previous model, and check that the highest weights are distinct. Moreover, once isomorphisms of A, B, C and D with a fixed two-dimensional vector space have been chosen, there is an obvious action of the symmetric group $\mathfrak{S}_4$ on $\mathfrak{so}_{4,4}$ and its triplet of representations. The resulting group of outer automorphisms is only $\mathfrak{S}_3$, of course, which in this model appears as the quotient of $\mathfrak{S}_4$ by the normal subgroup of order four whose three nontrivial elements are products of disjoint two-cycles (recall that the alternating group $\mathcal{A}_4$ is not simple!). (See [55] for a geometric interpretation of the $\mathfrak{so}_{4,4}$ triality, and an application to the construction of a minimal unitary representation.) Note also that the subalgebra of fixed points of the $\mathfrak{S}_3$-action (defined by the permutations of A, B, C) on $\mathfrak{so}_{4,4}$ is

$$\tilde{\mathfrak{g}}_2=\mathfrak{sl}(A)\times\mathfrak{sl}(D)\oplus(S^3A\otimes D),$$

the split form of $\mathfrak{g}_2$.

As pointed out by J. Wolf in [34], one can twist this $\mathfrak{S}_3$-action so as to get a different fixed point subalgebra. We define automorphisms τ and τ' of $\mathfrak{so}_{4,4}$, acting on $(M,N,P,Q)\in\mathfrak{sl}(A)\times\mathfrak{sl}(B)\times\mathfrak{sl}(C)\times\mathfrak{sl}(D)$ and on $a\otimes b\otimes c\otimes d\in A\otimes B\otimes C\otimes D$ in the following way:

$$\begin{aligned}\tau(M,N,P,Q)&=(P,M,N,Q), & \tau(a\otimes b\otimes c\otimes d)&=a\otimes b\otimes c\otimes d,\\ \tau'(M,N,P,Q)&=(P,M,N,ad(T)Q), & \tau'(a\otimes b\otimes c\otimes d)&=c\otimes a\otimes b\otimes T(d),\end{aligned}$$

where T is a rotation matrix of order three (so that $T^2+T+1=0$). The fixed point set of τ' is $\mathfrak{sl}(A)\times\mathbb{R}H\oplus M$, where $H=1+2T$ and M is the space of fixed points of τ' inside $A\otimes B\otimes C\otimes D$. Write such a tensor as $X\otimes e+Y\otimes Te$ for some nonzero vector e in D, where X and Y belong to $A\otimes B\otimes C$. It is fixed by τ' if and only if $X=-\tau(Y)$ and $(1+\tau+\tau^2)(Y)=0$, which means that $Y=Z-\tau(Z)$ for some Z, and then $X=\tau^2(Z)-\tau(Z)$. In particular M has dimension 4 and it is easy to see that the space of fixed points of τ' is a copy of $\mathfrak{sl}_3$.

3.5 The Magic Square

3.5.1 Jordan Algebras

The relation between exceptional Lie groups and normed division algebras, especially the octonion algebra, was first noticed by E. Cartan, at least as early as 1908 (see the encyclopedia article for the French version of the Mathematische Wissenschaften). Cartan observed that the automorphism group of $\mathbb{O}$ is the compact group G_2. In 1950, Chevalley and Schafer obtained the compact group F_4 as the automorphism group of the Jordan algebra

$$\mathcal{J}_3(\mathbb{O}) := \left\{ \begin{pmatrix} r_1 & x_3 & x_2 \\ \overline{x_3} & r_2 & x_1 \\ \overline{x_2} & \overline{x_1} & r_3 \end{pmatrix} \mid r_i \in \mathbb{R},\ x_j \in \mathbb{O} \right\}.$$

This is an algebra with product $A.B = \frac{1}{2}(AB+BA)$. It is a commutative but nonassociative algebra, in which the characteristic identity $A^2(AB) = A(A^2B)$ of Jordan algebras holds. By remarks in [12], it appears Cartan was aware of this construction at least as early as 1939.

Jordan algebras were introduced in the 1930's by Jordan, von Neumann and Wigner as a natural mathematical framework for quantum theory. Any associative algebra (over a field of characteristic not equal to 2) can be considered as a Jordan algebra for the symmetrized matrix product $a.b = \frac{1}{2}(ab + ba)$. A Jordan algebra is said to be *exceptional* if it cannot be embedded as a Jordan subalgebra of an associative algebra. The algebra $\mathcal{J}_3(\mathbb{O})$ is an exceptional Jordan algebra. Moreover, if A is an algebra such that $\mathcal{J}_n(A)$, with its symmetrized product, is a Jordan algebra, then A must be alternative if $n \geq 3$, and associative if $n \geq 4$.

The closed orbits $X \subset \mathcal{J}_n(\mathbb{A})$, with $\mathbb{A}$ the complexification of a division algebra are exactly the Scorza varieties while the Severi varieties are exactly the varieties $X \subset \mathcal{J}_3(\mathbb{A})$, see [85]. The close relationship between Scorza varieties and Jordan algebras, which is not explained in [85], is investigated in [16].

3.5.2 The Freudenthal–Tits Construction

The discovery that $F_4 = \mathrm{Aut}\mathcal{J}_3(\mathbb{O})$ led several authors (including Freudenthal, Tits, Vinberg, Jacobson, Springer) to study the relations between exceptional groups and Jordan algebras. Freudenthal and Tits obtained a unified construction of all compact exceptional Lie algebras in the following way: take two normed division algebras $\mathbb{A}$ and $\mathbb{B}$, and let

$$\mathfrak{g}(\mathbb{A},\mathbb{B}) := \mathrm{Der}\mathbb{A} \oplus (\mathrm{Im}\mathbb{A} \otimes \mathcal{J}_3(\mathbb{B})_0) \oplus \mathrm{Der}\mathcal{J}_3(\mathbb{B}),$$

where $\mathcal{J}_3(\mathbb{B})_0 \subset \mathcal{J}_3(\mathbb{B})$ is the hyperplane of traceless matrices. One can then define a Lie algebra structure on $\mathfrak{g}(\mathbb{A},\mathbb{B})$, with $\mathrm{Der}\mathbb{A} \oplus \mathrm{Der}\mathcal{J}_3(\mathbb{B})$ as a Lie subalgebra acting on $\mathrm{Im}\mathbb{A} \otimes \mathcal{J}_3(\mathbb{B})_0$ in a natural way [79]. The result of this construction is the *Freudenthal-Tits magic square of Lie algebras*:

	$\mathbb{R}$	$\mathbb{C}$	$\mathbb{H}$	$\mathbb{O}$
$\mathbb{R}$	$\mathfrak{su}_2$	$\mathfrak{su}_3$	$\mathfrak{sp}_6$	$\mathfrak{f}_4$
$\mathbb{C}$	$\mathfrak{su}_3$	$\mathfrak{su}_3 \times \mathfrak{su}_3$	$\mathfrak{su}_6$	$\mathfrak{e}_6$
$\mathbb{H}$	$\mathfrak{sp}_6$	$\mathfrak{su}_6$	$\mathfrak{so}_{12}$	$\mathfrak{e}_7$
$\mathbb{O}$	$\mathfrak{f}_4$	$\mathfrak{e}_6$	$\mathfrak{e}_7$	$\mathfrak{e}_8$

Let $\mathbb{B} = \underline{0}$ denote the zero algebra, so $\mathcal{J}_n(\underline{0}) \simeq \mathbb{C}^n$ is the diagonal matrices. Let Δ be something slightly smaller, so that $\mathcal{J}_n(\Delta) \subset \mathcal{J}_n(\underline{0})$ consists of the scalar matrices. With these conventions, we obtain a magic rectangle:

	Δ	$\underline{0}$	$\mathbb{R}$	$\mathbb{C}$	$\mathbb{H}$	$\mathbb{O}$
$\mathbb{R}$	0	0	$\mathfrak{su}_2$	$\mathfrak{su}_3$	$\mathfrak{sp}_6$	$\mathfrak{f}_4$
$\mathbb{C}$	0	0	$\mathfrak{su}_3$	$\mathfrak{su}_3 \times \mathfrak{su}_3$	$\mathfrak{su}_6$	$\mathfrak{e}_6$
$\mathbb{H}$	$\mathfrak{su}_2$	$(\mathfrak{su}_2)^{\oplus 3}$	$\mathfrak{sp}_6$	$\mathfrak{su}_6$	$\mathfrak{so}_{12}$	$\mathfrak{e}_7$
$\mathbb{O}$	$\mathfrak{g}_2$	$\mathfrak{so}_8$	$\mathfrak{f}_4$	$\mathfrak{e}_6$	$\mathfrak{e}_7$	$\mathfrak{e}_8$

3.5.3 The Allison Construction

While it is impressive that the vector space $\mathfrak{g}(\mathbb{A}, \mathbb{B})$ is a Lie algebra, the symmetry of the square appears to be miraculous. Vinberg, in 1966, obtained a symmetric construction (see [72] for an exposition). We present a variant of the Vinberg construction first discovered by Allison [2], then rediscovered independently in more geometric form by Dadok and Harvey [23] (who, after discovering Allison's work decided not to publish their manuscript), Barton and Sudbery [7], and ourselves [62].

Allison shows that given an arbitrary structurable algebra $\mathcal{A}$, one can put a Lie algebra structure on

$$\mathfrak{t}(\mathcal{A}) \oplus \mathcal{A}^{\oplus 3}.$$

Allison observes moreover that the algebra $\mathcal{A} = \mathbb{A} \otimes \mathbb{B}$ with $\mathbb{A}, \mathbb{B}$ division algebras is structurable and his construction applied to them yields the triality model, see [2, 3].

Using the Allison construction, the magic square may be described as follows:

$$\mathfrak{g}(\mathbb{A}, \mathbb{B}) = \mathfrak{t}(\mathbb{A}) \times \mathfrak{t}(\mathbb{B}) \oplus (\mathbb{A}_1 \otimes \mathbb{B}_1) \oplus (\mathbb{A}_2 \otimes \mathbb{B}_2) \oplus (\mathbb{A}_3 \otimes \mathbb{B}_3).$$

This bracket is defined so that $\mathfrak{t}(\mathbb{A}) \times \mathfrak{t}(\mathbb{B})$ is a Lie subalgebra, acting on each $\mathbb{A}_i \otimes \mathbb{B}_i$ in the natural way. The bracket of an element $a_1 \otimes b_1 \in \mathbb{A}_1 \otimes \mathbb{B}_1$ with $a_2 \otimes b_2 \in \mathbb{A}_2 \otimes \mathbb{B}_2$ is simply $a_1 a_2 \otimes b_1 b_2$, considered as an element of $\mathbb{A}_3 \otimes \mathbb{B}_3$. This is the general rule for taking the bracket of an element of $\mathbb{A}_i \otimes \mathbb{B}_i$ with one of $\mathbb{A}_j \otimes \mathbb{B}_j$, although there are some slight twists whose details can be found in [3] or [62]. Finally, the bracket of two elements in $\mathbb{A}_i \otimes \mathbb{B}_i$ is defined by the quadratic forms on $\mathbb{A}_i$, $\mathbb{B}_i$ and the natural maps $\Psi_i : \wedge^2 \mathbb{A}_i \to \mathfrak{t}(\mathbb{A})$

obtained by dualizing the action of $\mathfrak{t}(\mathbb{A})$ on $\mathbb{A}_i$, which can be seen as a map $\mathfrak{t}(\mathbb{A}) \to \wedge^2 \mathbb{A}_i$. (Note that $\mathfrak{t}(\mathbb{A})$ is always reductive, hence isomorphic to its dual as a $\mathfrak{t}(\mathbb{A})$-module.) The key formulas that ensure that the Jacobi identities hold in $\mathfrak{g}(\mathbb{A}, \mathbb{B})$ are then the following, which we state only for Ψ_1:

$$\begin{aligned}
\Psi_1(u \wedge v)_1 x &= Q(u, x)v - Q(v, x)u, \\
\Psi_1(u \wedge v)_2 x &= \tfrac{1}{2}(\overline{v}(ux) - \overline{u}(vx)), \\
\Psi_1(u \wedge v)_3 x &= \tfrac{1}{2}((xu)\overline{v} - (xv)\overline{u}).
\end{aligned}$$

These formulas are classical in the study of triality, that is for $\mathbb{A} = \mathbb{O}$. It is easy to check that in the other cases, which are much easier, we can arrange so that they also hold.

Proposition 3.2. *The real Lie algebra* $\mathfrak{g}(\mathbb{A}, \mathbb{B})$ *is compact, while* $\mathfrak{g}(\tilde{\mathbb{A}}, \tilde{\mathbb{B}})$ *is split.*

Proof. Let $K_{\mathbb{A}}$ be the non-degenerate quadratic form on $\mathfrak{t}(\mathbb{A})$ which we used in the construction of $\mathfrak{g}(\mathbb{A}, \mathbb{B})$. For $\mathbb{A} = \mathbb{O}$, this form is the Killing form on $\mathfrak{so}_8$ and is therefore negative definite; this is also true for the other division algebras. Now it is easy to check that the quadratic form

$$\mathcal{Q} := K_{\mathbb{A}} + K_{\mathbb{B}} - \sum_{i=1}^{3} Q_{\mathbb{A}_i} \otimes Q_{\mathbb{B}_i}$$

is invariant and negative definite. Since $\mathfrak{g}(\mathbb{A}, \mathbb{B})$ is simple (except for $\mathbb{A} = \mathbb{B} = \mathbb{C}$), this invariant form must be proportional to the Killing form, which is therefore negative definite as well (it cannot be positive definite!). Thus $\mathfrak{g}(\mathbb{A}, \mathbb{B})$ is compact.

Finally $\mathfrak{g}(\tilde{\mathbb{A}}, \tilde{\mathbb{B}})$ is clearly split since both $\mathfrak{t}(\tilde{\mathbb{A}})$ and $\mathfrak{t}(\tilde{\mathbb{B}})$ are split, and their product is a maximal rank subalgebra of $\mathfrak{g}(\mathbb{A}, \mathbb{B})$. □

3.6 Inclusions

It is not at all obvious that an inclusion $\mathbb{B} \subset \mathbb{B}'$ of composition algebras induces an inclusion $\mathfrak{g}(\mathbb{A}, \mathbb{B}) \subset \mathfrak{g}(\mathbb{A}, \mathbb{B}')$ of Lie algebras. In fact the Allison-Faulkner process applied to structurable algebras $\mathcal{A} \subset \mathcal{B}$ does *not* imply inclusions $\mathfrak{g}(\mathcal{A}) \subset \mathfrak{g}(\mathcal{B})$. To discuss this problem in our situation we may assume that $\mathbb{B}' = \mathbb{B} \oplus e\mathbb{B}$ is deduced from $\mathbb{B}$ via the Cayley–Dickson process.

Theorem 3.3. *There is a unique embedding of Lie algebras* $\mathfrak{t}(\mathbb{B}) \hookrightarrow \mathfrak{t}(\mathbb{B}')$ *that makes* $\mathfrak{g}(\mathbb{A}, \mathbb{B})$ *a Lie subalgebra of* $\mathfrak{g}(\mathbb{A}, \mathbb{B}')$.

If $\mathbb{B}'$ *is deduced from* $\mathbb{B}$ *by the Cayley–Dickson process, the relevant embedding of* $\mathfrak{t}(\mathbb{B})$ *inside* $\mathfrak{t}(\mathbb{B}')$ *is*

$$\mathfrak{t}(\mathbb{B}) \simeq \{(U_1, U_2, U_3) \in \mathfrak{t}(\mathbb{B}') \mid U_i(\mathbb{B}) \subset \mathbb{B}\} \subset \mathfrak{t}(\mathbb{B}').$$

Proof. We want the inclusion $\mathfrak{g}(\mathbb{A},\mathbb{B}) \hookrightarrow \mathfrak{g}(\mathbb{A},\mathbb{B}')$ to be a morphism of Lie algebras. It is easy to see that the only non-obvious verification to be made is that the bracket of two elements in $\mathbb{A}_i \otimes \mathbb{B}_i$ is the same in $\mathfrak{g}(\mathbb{A},\mathbb{B})$ as in $\mathfrak{g}(\mathbb{A},\mathbb{B}')$, which amounts to proving that we have a commutative diagram

$$\begin{array}{ccc} \wedge^2(\mathbb{A}_i \otimes \mathbb{B}_i) & \longrightarrow & \mathfrak{t}(\mathbb{A}) \times \mathfrak{t}(\mathbb{B}) \\ \downarrow & & \downarrow \\ \wedge^2(\mathbb{A}_i \otimes \mathbb{B}_i') & \longrightarrow & \mathfrak{t}(\mathbb{A}) \times \mathfrak{t}(\mathbb{B}'). \end{array}$$

Since the norm on $\mathbb{B}$ is the restriction of the norm on $\mathbb{B}'$, it is actually enough to prove the commutativity of the following diagram:

$$\begin{array}{ccc} \wedge^2\mathbb{B}_i & \xrightarrow{\Psi_i} & \mathfrak{t}(\mathbb{B}) \\ \downarrow & & \downarrow \\ \wedge^2\mathbb{B}_i' & \xrightarrow{\Psi_i'} & \mathfrak{t}(\mathbb{B}'). \end{array}$$

We must check this for all i, so we consider the sum $\Psi_{\mathbb{B}} : \wedge^2\mathbb{B}_1 \oplus \wedge^2 \mathbb{B}_2 \oplus \wedge^2 \mathbb{B}_3 \longrightarrow \mathfrak{t}(\mathbb{B})$. Note that $\Psi_{\mathbb{B}}$ is dual to the action, hence surjective since the action map is injective by definition. So the commutativity of the diagram above for $i = 1, 2, 3$ completely determines the inclusion of $\mathfrak{t}(\mathbb{B})$ inside $\mathfrak{t}(\mathbb{B}')$. For this inclusion map to be well-defined, we need a compatibility condition, namely that $\mathrm{Ker}\Psi_{\mathbb{B}} \subset \mathrm{Ker}\Psi_{\mathbb{B}'}$. Everything being trivial for $\mathbb{B} = \mathbb{R}$, we will check that $\mathrm{Ker}\Psi_{\mathbb{C}} \subset \mathrm{Ker}\Psi_{\mathbb{H}} \subset \mathrm{Ker}\Psi_{\mathbb{O}}$ and deduce the relevant embeddings $\mathfrak{t}(\mathbb{C}) \subset \mathfrak{t}(\mathbb{H}) \subset \mathfrak{t}(\mathbb{O})$.

We begin with $\Psi_{\mathbb{C}}$, which is dual to the inclusion map

$$\mathfrak{t}(\mathbb{C}) = \{(a,b,c) \in \mathbb{R}^3 \mid a = b + c\} \subset \wedge^2\mathbb{C}_1 \oplus \wedge^2 \mathbb{C}_2 \oplus \wedge^2 \mathbb{C}_3,$$

where for example $a \in \mathbb{R}$ maps to the skew-symmetric endomorphism $\left(\begin{smallmatrix} 0 & a \\ -a & 0 \end{smallmatrix}\right)$ of $\mathbb{C}_1$. On $\mathfrak{t}(\mathbb{C})$ we use the restriction of the canonical norm of $\mathbb{R}^3$, and we find that the dual map is

$$\Psi_{\mathbb{C}} : \left(\begin{pmatrix} 0 & a \\ -a & 0 \end{pmatrix}, \begin{pmatrix} 0 & b \\ -b & 0 \end{pmatrix}, \begin{pmatrix} 0 & c \\ -c & 0 \end{pmatrix} \right) \mapsto \left(\frac{2a+b+c}{3}, \frac{a+2b-c}{3}, \frac{a-b+2c}{3} \right),$$

whose kernel is generated by $\left(\left(\begin{smallmatrix} 0 & 1 \\ -1 & 0 \end{smallmatrix}\right), \left(\begin{smallmatrix} 0 & -1 \\ 1 & 0 \end{smallmatrix}\right), \left(\begin{smallmatrix} 0 & -1 \\ 1 & 0 \end{smallmatrix}\right) \right)$.

Now to compute $\Psi_{\mathbb{H}}$, using that $\mathrm{Der}(\mathbb{H}) \simeq \mathrm{Im}\mathbb{H}$ we may describe $\mathfrak{t}(\mathbb{H})$ as

$$\mathfrak{t}(\mathbb{H}) = \{(L_a + R_b, L_a - R_c, L_c + R_b) \mid a, b, c \in \mathrm{Im}\mathbb{H}\}.$$

Every skew-symmetric endomorphism of $\mathbb{H}$ is of the form $L_x + R_y$ for some imaginary quaternions x and y, and the norm defined by the Killing form is $||L_x + R_y||^2 = ||x||^2 + ||y||^2$. With the induced norm on $\mathfrak{t}(\mathbb{H})$ we compute that

$$\begin{aligned} \Psi_{\mathbb{H}} : &(L_{x_1} + R_{y_1}, L_{x_2} + R_{y_2}, L_{x_3} + R_{y_3}) \\ &\mapsto (L_{x_1+x_3} + R_{y_1+y_2}, L_{x_1+x_3} - R_{x_2-y_3}, L_{x_2-y_3} + R_{y_1+y_2}). \end{aligned}$$

Its kernel is $\{(L_x + R_y, L_z - R_y, -L_x + R_z) \mid x, y, z \in \mathrm{Im}\mathbb{H}\}$.

Finally, the computation of $\Psi_{\mathbb{O}}$ is an immediate consequence of the triality principle. Indeed, we know that each projection $\mathfrak{t}(\mathbb{O}) \to \wedge^2 \mathbb{O}_i$ is an isomorphism. If we denote by (U, U', U'') the elements of $\mathfrak{t}(\mathbb{O})$, this means that such a triplet is uniquely determined by U, U' or U''. We deduce that

$$\Psi_{\mathbb{O}} : (U_1, U_2', U_3'') \mapsto (U_1 + U_2 + U_3, U_1' + U_2' + U_3', U_1'' + U_2'' + U_3'').$$

We are ready to check our compatibility conditions. First note that the endomorphism $\left(\begin{smallmatrix} 0 & 1 \\ -1 & 0 \end{smallmatrix}\right)$ of $\mathbb{C}$, when extended to $\mathbb{H}$ in the obvious way, is $1 \wedge i = \frac{1}{2}(L_i + R_i)$. Taking $x = y = -z = i$ in our description of Ker$\Psi_{\mathbb{H}}$, we see that it contains Ker$\Psi_{\mathbb{C}}$, as required, and that the induced inclusion $\mathfrak{t}(\mathbb{C}) \subset \mathfrak{t}(\mathbb{H})$ is given by

$$(b, c) \in \mathbb{R}^2 \mapsto (L_{(b+2c)i} + R_{(2b+c)i}, L_{(b+2c)i} + R_{(c-b)i}, L_{(b-c)i} + R_{(2b+c)i}).$$

Now take $(L_x + R_y, L_z - R_y, -L_x + R_z)$ in Ker$\Psi_{\mathbb{H}}$, where $x, y, z \in \mathrm{Im}\mathbb{H}$, and consider it as an element of $\wedge^2\mathbb{O}_1 \oplus \wedge^2 \mathbb{O}_2 \oplus \wedge^2 \mathbb{O}_3$, which we denote by $((L_x + R_y, 0), (L_z - R_y, 0), (-L_x + R_z, 0))$. Here $(L_z, 0)$, for example, denotes the endomorphism of $\mathbb{O}$ defined on elements of $\mathbb{H}$ by left multiplication by $z \in \mathbb{H}$, and by zero on $\mathbb{H}^\perp$. Note that with these conventions, it follows from the multiplication rules in the Cayley-Dickson process that for $z \in \mathrm{Im}\mathbb{H}$, we have $L_z = (L_z, -L_z)$ and $R_z = (R_z, L_z)$.

It is easy to check that every skew-symmetric endomorphism of $\mathbb{H}$ is of the form $L_a + R_b$. In matrix terms, if $a = \alpha_1 i + \alpha_2 j + \alpha_3 k$ and $b = \beta_1 i + \beta_2 j + \beta_3 k$, we have

$$L_a = \begin{pmatrix} 0 & -\alpha_1 & -\alpha_2 & -\alpha_3 \\ \alpha_1 & 0 & -\alpha_3 & \alpha_2 \\ \alpha_2 & \alpha_3 & 0 & -\alpha_1 \\ \alpha_3 & -\alpha_2 & \alpha_1 & 0 \end{pmatrix}, \quad R_b = \begin{pmatrix} 0 & -\beta_1 & -\beta_2 & -\beta_3 \\ \beta_1 & 0 & \beta_3 & \beta_2 \\ \beta_2 & -\beta_3 & 0 & \beta_1 \\ \beta_3 & \beta_2 & -\beta_1 & 0 \end{pmatrix}.$$

Under the natural isomorphism $\mathfrak{so}(\mathbb{H}) \simeq \wedge^2\mathbb{H}$ induced by the norm on $\mathbb{H}$, this means that

$$\begin{aligned} L_a &= \alpha_1(1{\wedge}i + j{\wedge}k) + \alpha_2(1{\wedge}j + k{\wedge}i) + \alpha_3(1{\wedge}k + i{\wedge}j), \\ R_b &= \beta_1(1{\wedge}i - j{\wedge}k) + \beta_2(1{\wedge}j - k{\wedge}i) + \beta_3(1{\wedge}k - i{\wedge}j). \end{aligned}$$

Now we can ask what is the element of $\mathfrak{t}(\mathbb{O})$ defined by $u_1 = 1{\wedge}i \in \mathfrak{so}(\mathbb{H}) \subset \mathfrak{so}(\mathbb{O})$? To answer this question, we just need to note that $1{\wedge}i = \frac{1}{2}(L_i + R_i)$, so $u_2 = \frac{1}{2}L_i$ and $u_3 = \frac{1}{2}R_i$. Also if we take $u_1 = j{\wedge}k$, a straightforward computation gives $u_2 = j{\wedge}k + \frac{1}{2}R_i = \frac{1}{2}(L_i, L_i)$ and $u_3 = j{\wedge}k - \frac{1}{2}L_i = \frac{1}{2}(-R_i, L_i)$. Letting automorphisms of $\mathbb{H}$ act, we deduce that the following triplets (u_1, u_2, u_3) belong to $\mathfrak{t}(\mathbb{O})$, where z, u, v denote imaginary quaternions:

u_1	$1{\wedge}z$	$u{\wedge}v$	$(L_z, 0)$	$(R_z, 0)$	$(0, L_z)$
u_2	$\frac{1}{2}L_z$	$u{\wedge}v + \frac{1}{4}R_{[u,v]}$	$(L_z, 0)$	$(0, -L_z)$	$(R_z, 0)$
u_3	$\frac{1}{2}R_z$	$u{\wedge}v - \frac{1}{4}L_{[u,v]}$	$(0, L_z)$	$(R_z, 0)$	$(-L_z, 0)$

Thus we have

$$\begin{aligned} \Psi_{\mathbb{O}}((L_x,0),0,0) &= ((L_x,0),(0,-L_x),(L_x,0)) &= \Psi_{\mathbb{O}}(0,0,(L_x,0)),\\ \Psi_{\mathbb{O}}((R_y,0),0,0) &= ((R_y,0),(R_y,0),(0,L_y)) &= \Psi_{\mathbb{O}}(0,(R_y,0),0),\\ \Psi_{\mathbb{O}}(0,(L_z,0),0) &= ((0,-L_z),(L_z,0),(-R_z,0)) &= \Psi_{\mathbb{O}}(0,0,(-R_z,0)). \end{aligned}$$

This immediately implies that $((L_x+R_y,0),(L_z-R_y,0),(-L_x+R_z,0))$ belongs to $\mathrm{Ker}\Psi_{\mathbb{O}}$, and we deduce from the preceding formulas that the induced imbedding of $\mathfrak{t}(\mathbb{H}) \subset \mathfrak{t}(\mathbb{O})$ is given by

$$\begin{aligned} &(L_a+R_b, L_a-R_c, L_c+R_b) \in \mathfrak{t}(\mathbb{H})\\ &\qquad\mapsto ((L_a+R_b,L_c),(L_a-R_c,-L_b),(L_c+R_b,L_a)) \in \mathfrak{t}(\mathbb{O}). \end{aligned}$$

In particular, one can check that as a subalgebra of $\mathfrak{t}(\mathbb{O})$, $\mathfrak{t}(\mathbb{C})$ is generated by (L_i+R_i,L_i,R_i) and $(R_i,-R_i,L_i+R_i)$.

A more careful inspection of these formulas yields the explicit conclusion stated in the theorem. □

3.7 Automorphisms, Symmetric and Trisymmetric Spaces

The symmetry in the triality model in the roles of $\mathbb{A}$ and $\mathbb{B}$ allows one to exhibit automorphisms of exceptional Lie algebras with interesting properties. For example:

Proposition 3.4. [62] *The endomorphism of $\mathfrak{g}(\mathbb{A},\mathbb{B})$ defined as the identity on $\mathfrak{h}(\mathbb{A},\mathbb{B}) = \mathfrak{t}(\mathbb{A}) \times \mathfrak{t}(\mathbb{B}) \oplus \mathbb{A}_1 \otimes \mathbb{B}_1$, and minus the identity on $\mathbb{A}_2 \otimes \mathbb{B}_2 \oplus \mathbb{A}_3 \otimes \mathbb{B}_3$, is a Lie algebra involution.*

Exhibiting a Lie algebra involution or a symmetric space is more or less equivalent, and we conclude that there exists a "magic square" of symmetric spaces of dimensions $2ab$ where a and b denote the dimensions of $\mathbb{A}$ and $\mathbb{B}$. In particular we always get powers of 2! These are not all uniquely defined by our Lie algebra involution, but a version of this magic square is the following:

	$\mathbb{R}$	$\mathbb{C}$	$\mathbb{H}$	$\mathbb{O}$
$\mathbb{R}$	$\mathbb{RP}^2$	$\mathbb{CP}^2$	$\mathbb{HP}^2$	$\mathbb{OP}^2$
$\mathbb{C}$	$\mathbb{CP}^2$	$\mathbb{CP}^2 \times \mathbb{CP}^2$	$G_{\mathbb{C}}(2,6)$	$\mathbb{OP}^2_{\mathbb{C}}$
$\mathbb{H}$	$\mathbb{HP}^2$	$G_{\mathbb{C}}(2,6)$	$G_{\mathbb{R}}(4,12)$	$E_{7(-5)}$
$\mathbb{O}$	$\mathbb{OP}^2$	$\mathbb{OP}^2_{\mathbb{C}}$	$E_{7(-5)}$	$E_{8(8)}$

The symmetric space $E_{8(8)} = E_8/\mathrm{SO}_{16}$, of dimension 128, is particularly intriguing and it would be very nice to have a direct construction of it. It is claimed in [76] that it can be interpreted as a projective plane over $\mathbb{O}\otimes\mathbb{O}$, and that in fact the whole square above of symmetric spaces can be obtained by taking projective planes, in a suitable sense, over the tensor products $\mathbb{A}\otimes\mathbb{B}$.

It is also easy to describe Lie algebra automorphisms of order three, reflecting the trialitarian origin of the magic square. Recall that $\mathfrak{t}(\mathbb{A})$ has a natural automorphism $\tau_{\mathbb{A}}$ or order three.

Proposition 3.5. *The endomorphism of* $\mathfrak{g}(\mathbb{A},\mathbb{B})$ *defined by*

$$\tau_{\mathbb{A},\mathbb{B}}(X+Y+U_1+U_2+U_3)=\tau_{\mathbb{A}}(X)+\tau_{\mathbb{B}}(Y)+U_2+U_3+U_1$$

for $X\in\mathfrak{t}(\mathbb{A})$, $Y\in\mathfrak{t}(\mathbb{B})$, $\mathrm{U}_i\in\mathbb{A}_i\otimes\mathbb{B}_i$, *is a Lie algebra automorphism of order three, whose fixed points set is the subalgebra*

$$\mathfrak{k}(\mathbb{A},\mathbb{B})=Der\mathbb{A}\times Der\mathbb{B}\oplus\mathbb{A}\otimes\mathbb{B}.$$

When $\mathbb{B}=\mathbb{O}$, *the conclusion also holds when we replace the usual triality automorphism* $\tau_{\mathbb{O}}$ *by its twisted version* $\tau'_{\mathbb{O}}$. *In this case, the fixed point set is the subalgebra* $\mathfrak{k}'(\mathbb{A},\mathbb{B})=Der\mathbb{A}\times\mathfrak{su}_3\oplus\mathbb{A}\otimes\mathbb{B}$.

As in the case of involutions, we deduce from this statement a "magic square" of homogeneous spaces which are no longer symmetric, but are sometimes called trisymmetric: they are quotients of a semisimple Lie group G by a subgroup K which is, up to a finite group, the fixed point set of an automorphism of G of order three. Their dimensions are equal to $2ab+2a+2b-4$.

	$\mathbb{R}$	$\mathbb{C}$	$\mathbb{H}$	$\mathbb{O}$
$\mathbb{R}$	$\mathrm{U}_2/\mathrm{U}_1\times\mathrm{U}_1$	$\mathrm{U}_3/(\mathrm{U}_1)^3$	$\mathrm{Sp}_6/\mathrm{U}_2\times\mathrm{Sp}_2$	$F_4/(\mathrm{Spin}_7\times\mathrm{S}^1/\mathbb{Z}_2)$
$\mathbb{C}$	$\mathrm{U}_3/(\mathrm{U}_1)^3$	$\mathrm{U}_3\times\mathrm{U}_3/(\mathrm{U}_1)^6$	$\mathrm{U}_6/(\mathrm{U}_2)^3$	$E_6/\mathrm{SO}_8\times(\mathrm{SO}_2)^2$
$\mathbb{H}$	$\mathrm{Sp}_6/\mathrm{U}_2\times\mathrm{Sp}_2$	$\mathrm{U}_6/(\mathrm{U}_2)^3$	$\mathrm{SO}_{12}/\mathrm{U}_4\times\mathrm{SO}_4$	$E_7/S(\mathrm{U}_7\times\mathrm{U}_1)/\mathbb{Z}_2$
$\mathbb{O}$	$F_4/(\mathrm{Spin}_7\times\mathrm{S}^1/\mathbb{Z}_2)$	$E_6/\mathrm{SO}_8\times(\mathrm{SO}_2)^2$	$E_7/\mathrm{S}(\mathrm{U}_7\times\mathrm{U}_1)/\mathbb{Z}_2$	$E_8/\mathrm{SO}_{14}\times\mathrm{SO}_2$

Here we denoted by F_4,E_6,E_7,E_8 the compact centerless groups of these types. Note that for example, $\mathrm{U}_3/\mathrm{U}_1\times\mathrm{U}_1\times\mathrm{U}_1$ is the variety of complete flags in $\mathbb{C}^3$. Similarly, $\mathrm{U}_6/\mathrm{U}_2\times\mathrm{U}_2\times\mathrm{U}_2$ is the flag variety $\mathbb{F}(2,4,6)$. Also $\mathrm{SO}_{12}/\mathrm{U}_4\times\mathrm{SO}_4$ is the space of 8-dimensional subspaces in $\mathbb{R}^{12}$ endowed with an orthogonal complex structure. This is because SO_{2n} acts transitively on the space of orthogonal complex structures. Moreover, the stabilizer of a point is the group of orthogonal transformations commuting with the corresponding complex structure, and this subgroup of SO_{2n} is a copy of U_n.

In the twisted case, for $\mathbb{B}=\mathbb{O}$, we replace the last line by

$$F_4/(\mathrm{SU}_3\times\mathrm{SU}_3/\mathbb{Z}_3)\quad E_6/(\mathrm{SU}_3\times\mathrm{SU}_3\times\mathrm{SU}_3/\mathbb{Z}_3)\quad E_7/(\mathrm{SU}_3\times\mathrm{SU}_6/\mathbb{Z}_3)\quad E_8/\mathrm{SU}_3\times E_6,$$

a series of trisymmetric spaces of dimension $18a+18$. Note that at the infinitesimal level, we obtain the quotients $\mathfrak{g}(\mathbb{A},\mathbb{O})/\mathfrak{su}_3\times\mathfrak{g}(\mathbb{A},\mathbb{C})$.

Automorphisms of order three were studied and classified by Wolf and Gray [34]. In the exceptional cases the construction above is not far from giving the whole classification.

3.8 Dual Pairs

The triality model allows one to identify series of *dual pairs* in the exceptional Lie algebras. Recall that a pair $(\mathfrak{h},\mathfrak{h}')$ of Lie subalgebras of a Lie algebra $\mathfrak{g}$ is a dual pair if $\mathfrak{h}$ is the centralizer of $\mathfrak{h}'$ and vice versa. Dual pairs of Lie algebras, or of Lie groups (with the same definition), have been extensively studied after the discovery by R. Howe that certain dual pairs of groups inside the real symplectic group $\mathrm{Sp}(2n,\mathbb{R})$ have very special properties with respect to the (infinite dimensional) *metaplectic representation* [43].

First recall that, as we saw in 4.7, the triality model is compatible with inclusions, in the sense that $\mathfrak{g}(\mathbb{A},\mathbb{B})$ is naturally embedded in $\mathfrak{g}(\mathbb{A}',\mathbb{B}')$ when $\mathbb{A}\subset\mathbb{A}'$ and $\mathbb{B}\subset\mathbb{B}'$.

Proposition 3.6. *The centralizers in $\mathfrak{g}(\mathbb{A},\mathbb{O})$ of the subalgebras $\mathfrak{g}(\mathbb{A},\mathbb{H})$, $\mathfrak{g}(\mathbb{A},\mathbb{C})$, $\mathfrak{g}(\mathbb{A},\mathbb{R})$ and $\mathfrak{t}(\mathbb{A})$, are isomorphic to $\mathfrak{su}_2$, $\mathfrak{su}_3$, $\mathfrak{g}_2$ and $\mathfrak{z}(\mathfrak{t}(\mathbb{A}))\times\mathfrak{so}_8$, respectively. Moreover, the centralizers of these centralizers are the subalgebras themselves.*

Here we denoted by $\mathfrak{z}(\mathfrak{t}(\mathbb{A}))$ the center of $\mathfrak{t}(\mathbb{A})$, which is $\mathfrak{t}(\mathbb{A})$ itself when this algebra is commutative, that is for $\mathbb{A}=\mathbb{R}$ or $\mathbb{C}$, and zero otherwise.

Proof. It is easy to see that the centralizer of $\mathfrak{g}(\mathbb{A},\mathbb{B})$ inside $\mathfrak{g}(\mathbb{A},\mathbb{O})$ is a subalgebra of $\mathfrak{t}(\mathbb{O})$, equal to the subalgebra of the centralizer of $\mathfrak{t}(\mathbb{B})$ inside $\mathfrak{t}(\mathbb{O})$ acting trivially on each $\mathbb{B}_i\subset\mathbb{O}_i$.

If $\mathbb{B}=\mathbb{R}$ we get the subalgebra of elements $(u_1,u_2,u_3)\in\mathfrak{t}(\mathbb{O})$ killing the unit element in each $\mathbb{O}_i$. Making $x=1$ and $y=1$ in the triality relation $u_1(xy)=u_2(x)y+xu_3(y)$, we see that $u_1=u_2=u_3$ is a derivation of $\mathbb{O}$, so that the center of $\mathfrak{g}(\mathbb{A},\mathbb{R})$ inside $\mathfrak{g}(\mathbb{A},\mathbb{O})$ is $\mathrm{Der}\mathbb{O}=\mathfrak{g}_2$.

When $\mathbb{B}=\mathbb{C}$, the condition that (u_1,u_2,u_2) kills $\mathbb{C}_i\subset\mathbb{O}_i$ implies that $u=u_1=u_2=u_3$ is a derivation of $\mathbb{O}$ such that $u(i)=0$ ($i=e_1$ in the notations of 5.1). But then $u(ix)=iu(x)$, which means that u is a complex endomorphism of $\mathbb{C}^\perp\simeq\mathbb{C}^3$, endowed with the complex structure defined by L_i (and also that defined by R_i, which is the conjugate complex structure, since $u(xi)=u(x)i$). Note that the corresponding element of $\mathfrak{t}(\mathbb{O})$ automatically centralizes $\mathfrak{t}(\mathbb{C})$, since u commutes with L_i,R_i and we saw that $\mathfrak{t}(\mathbb{C})$ is generated as a subalgebra of $\mathfrak{t}(\mathbb{O})$ by (L_i+R_i,L_i,R_i) and $(R_i,-R_i,L_i+R_i)$. Finally, take the general matrix of $\mathfrak{g}_2$ as written in Section 3.2. The condition that $u(i)=0$ means that $\alpha_2=\alpha_3=\alpha_4=\alpha_5=\alpha_6=\alpha_7=0$. A complex basis of the space $\mathbb{C}^\perp$ is (e_2,e_4,e_6), in terms of which u is the complex endomorphism given by the matrix

$$\begin{pmatrix} i\beta_3 & -\beta_4+i\beta_5 & -\beta_6+i\beta_7\\ \beta_4+i\beta_5 & i\gamma_5 & -\gamma_6+i\gamma_7\\ \beta_6+i\beta_7 & \gamma_6+i\gamma_7 & -i(\beta_3+\gamma_5)\end{pmatrix},$$

which is a general element of $\mathfrak{su}_3$.

Finally let $\mathbb{B}=\mathbb{H}$. Again, our description of $\mathfrak{g}_2$ in Section 3.2 shows that the derivations of $\mathbb{O}$ vanishing on $\mathbb{H}$ form a copy of $\mathfrak{so}_3$, which we can describe as the space of endomorphisms of the form $(0,R_z)$ with $z\in\mathrm{Im}\mathbb{H}$. But we saw that $\mathfrak{t}(\mathbb{H})$, as a subalgebra of $\mathfrak{t}(\mathbb{C})$, was generated by the triplets $((L_a+R_b,L_c),(L_a-R_c,-L_b),(L_c+R_b,L_a))$, where $a,b,c\in\mathrm{Im}\mathbb{H}$. But these elements of $\mathfrak{t}(\mathbb{O})$ commute with those of the form $((0,R_z),(0,R_z),(0,R_z))$. This proves that the centralizer of $\mathfrak{g}(\mathbb{A},\mathbb{H})$ in $\mathfrak{g}(\mathbb{A},\mathbb{O})$ is $\mathfrak{so}_3$.

Alternatively we can identify this centralizer to the subalgebra of the centralizer of $\mathfrak{t}(\mathbb{C})$ annihilating $e_2 = j$, which is clearly $\mathfrak{su}_2$.

The rest of the claim is straightforward. □

In particular, we have pairs of reductive subalgebras in $\mathfrak{g}(\mathbb{A},\mathbb{O})$, each of which is the centralizer of the other: this is a *reductive dual pair*. The classification of reductive dual pairs inside reductive complex Lie algebras was obtained in [77]. We deduce from the proposition above the existence of uniform series of what Rubenthaler calls *towers* of reductive dual pairs:

$$\begin{array}{ccc}
\mathfrak{g}(\mathbb{A},\mathbb{O}) & = & \mathfrak{g}(\mathbb{A},\mathbb{O}) \\
\cup & & \cup \\
\mathfrak{g}(\mathbb{A},\mathbb{H}) & & \mathfrak{so}_8 \times \mathfrak{z}(\mathfrak{t}(\mathbb{A})) \\
\cup & & \cup \\
\mathfrak{g}(\mathbb{A},\mathbb{C}) & & \mathfrak{g}_2 \\
\cup & & \cup \\
\mathfrak{g}(\mathbb{A},\mathbb{R}) & & \mathfrak{su}_3 \\
\cup & & \cup \\
\mathfrak{t}(\mathbb{A}) & & \mathfrak{su}_2
\end{array}$$

We conclude that $\mathfrak{so}_8 \times \mathfrak{t}(\mathbb{A})$, $\mathfrak{g}_2 \times \mathfrak{g}(\mathbb{A},\mathbb{R})$, $\mathfrak{su}_3 \times \mathfrak{g}(\mathbb{A},\mathbb{C})$ and $\mathfrak{so}_3 \times \mathfrak{g}(\mathbb{A},\mathbb{H})$ are maximal rank reductive subalgebras of $\mathfrak{g}(\mathbb{A},\mathbb{O})$. A natural question to ask is to describe the module structure of $\mathfrak{g}(\mathbb{A},\mathbb{O})$ over each of these subalgebras. For the first one, the answer is given by the triality construction. For the second one, we note that $\mathfrak{g}(\mathbb{A},\mathbb{R}) = \mathrm{Der}(\mathcal{J}_3(\mathbb{A}))$ and $\mathfrak{g}_2 = \mathrm{Der}\mathbb{O}$, so that we are back to the original Tits construction of the exceptional series of Lie algebras. We obtain:

$$\begin{aligned}
\mathfrak{g}(\mathbb{A},\mathbb{O}) &= \ \mathfrak{g}_2 \times \mathfrak{g}(\mathbb{A},\mathbb{R}) \ \oplus \ \mathbb{R}^7 \otimes \mathcal{J}_3(\mathbb{A})_0, \\
&= \mathfrak{su}_3 \times \mathfrak{g}(\mathbb{A},\mathbb{C}) \ \oplus \ \mathbb{R}^3 \otimes \mathcal{J}_3(\mathbb{A}) \oplus (\mathbb{R}^3 \otimes \mathcal{J}_3(\mathbb{A}))^*, \\
&= \mathfrak{su}_2 \times \mathfrak{g}(\mathbb{A},\mathbb{H}) \ \oplus \ \mathbb{R}^2 \otimes \mathcal{Z}_3(\mathbb{A}).
\end{aligned}$$

Here $\mathcal{Z}_3(\mathbb{A})$ denotes the space of *Zorn matrices*, which is a $\mathfrak{g}(\mathbb{A},\mathbb{H})$-module, see Section 4.6. Note that $\mathfrak{su}_2$ acts not on $\mathbb{R}^2$ but on $\mathbb{C}^2$. Therefore there must exist a complex structure on $\mathcal{Z}_3(\mathbb{A}) = \mathcal{Z}_3^0(\mathbb{A}) \oplus i\mathcal{Z}_3^0(\mathbb{A})$, which is not $\mathfrak{g}(\mathbb{A},\mathbb{H})$-invariant, so that $\mathbb{R}^2 \otimes \mathcal{Z}_3(\mathbb{A}) = \mathbb{C}^2 \otimes \mathcal{Z}_3^0(\mathbb{A})$, the $\mathfrak{su}_2$-action coming from $\mathbb{C}^2$.

These towers of dual pairs in series were discovered in joint work with B. Westbury.

3.9 The Quaternionic Form

The algebras $\mathfrak{g}(\mathbb{A},\tilde{\mathbb{B}})$ are also interesting, especially when $\mathbb{B} = \mathbb{O}$. Following Wolf [84], each simple complex Lie algebra has a unique real form such that the associated compact or non-compact symmetric spaces have an invariant quaternionic structure. We will say that this real form is *quaternionic*.

Recall that the *rank* of a symmetric space G/K, corresponding to some Cartan decomposition $\mathfrak{g} = \mathfrak{k} \oplus \mathfrak{p}$, is the maximal dimension of a subalgebra of $\mathfrak{g}$ contained in $\mathfrak{p}$. Such a subalgebra is automatically abelian and is called a Cartan subspace.

Theorem 3.7. *The real Lie algebra $\mathfrak{g}(\mathbb{A}, \tilde{\mathbb{O}})$ is quaternionic. A Cartan subspace of $\mathfrak{t}(\tilde{\mathbb{O}}) = \mathfrak{so}_{4,4}$ embeds as a Cartan subspace of $\mathfrak{g}(\mathbb{A}, \tilde{\mathbb{O}})$, which in particular has real rank four (independent of $\mathbb{A}$).*

Proof. Recall that each simple complex Lie algebra $\mathfrak{g}$, once a Cartan subalgebra and a set of positive roots has been chosen, has a canonical 5-graduation defined by the highest root $\tilde{\alpha}$, more precisely by the eigenspaces of $\mathrm{ad}(H_{\tilde{\alpha}})$, where $H_{\tilde{\alpha}}$ denotes the coroot of α.

If $\mathfrak{g}$ is the complexification of the compact real Lie algebra $\mathfrak{g}(\mathbb{A}, \mathbb{O})$, we can arrange so that the highest root comes from $\mathfrak{t}(\mathbb{O})_{\mathbb{C}} = \mathfrak{so}_8(\mathbb{C})$. Note that the space of highest root vectors in $\mathfrak{so}_8(\mathbb{C})$ is the cone over the Grassmannian of isotropic planes in $\mathbb{C}^8$. Using this, we then check that we can arrange so that

$$iH_{\tilde{\alpha}} = \begin{pmatrix} 0 & 0 & 0 & 0 & 0 & 0 & 0 & 0 \\ 0 & 0 & 0 & 0 & 0 & 0 & 0 & 0 \\ 0 & 0 & 0 & 0 & 0 & 0 & 0 & 0 \\ 0 & 0 & 0 & 0 & 0 & 0 & 0 & 0 \\ 0 & 0 & 0 & 0 & 0 & -1 & 0 & 0 \\ 0 & 0 & 0 & 0 & 1 & 0 & 0 & 0 \\ 0 & 0 & 0 & 0 & 0 & 0 & 0 & -1 \\ 0 & 0 & 0 & 0 & 0 & 0 & 1 & 0 \end{pmatrix}.$$

The eigenvalues of $\mathrm{ad}(H_{\tilde{\alpha}})$ acting on $\mathfrak{g}$ are $0, \pm 1, \pm 2$. The corresponding eigenspaces are

$$\begin{aligned} \mathfrak{g}_0 &= \mathfrak{t}(\mathbb{A}_{\mathbb{C}}) \times \mathfrak{t}(\mathbb{O}_{\mathbb{C}})_0 \oplus \mathbb{A}_{\mathbb{C},1} \otimes \mathbb{O}^0_{\mathbb{C},1} \oplus \mathbb{A}_{\mathbb{C},2} \otimes \mathbb{O}^0_{\mathbb{C},2} \oplus \mathbb{A}_{\mathbb{C},3} \otimes \mathbb{O}^0_{\mathbb{C},3}, \\ \mathfrak{g}_{\pm 1} &= \mathfrak{t}(\mathbb{O}_{\mathbb{C}})_{\pm 1} \oplus \mathbb{A}_{\mathbb{C},1} \otimes \mathbb{O}^{\pm 1}_{\mathbb{C},1} \oplus \mathbb{A}_{\mathbb{C},2} \otimes \mathbb{O}^{\pm 1}_{\mathbb{C},2} \oplus \mathbb{A}_{\mathbb{C},3} \otimes \mathbb{O}^{\pm 1}_{\mathbb{C},3}, \\ \mathfrak{g}_{\pm 2} &= \mathfrak{t}(\mathbb{O}_{\mathbb{C}})_{\pm 2}, \end{aligned}$$

where we denoted by $\mathbb{O}^t_i$ the t-eigenspace for the action of $H_{\tilde{\alpha}}$ on $\mathbb{O}_i$. Note that $iH_{\tilde{\alpha}}$ belongs to $\mathrm{Der}\mathbb{O} = \mathfrak{g}_2$ (see our description of $\mathfrak{g}_2$ above), so we can let $i = 1$ and consider the standard action of $H_{\tilde{\alpha}}$. Then it is clear that the kernel of $H_{\tilde{\alpha}}$ is the complexification of the standard quaternion subalgebra $\mathbb{H}$, and that the sum of the two other eigenspaces is the complexification of $e\mathbb{H}$.

Now, let θ denote the involution of $\mathfrak{g}$ associated to its 5-graduation. By definition, θ acts by $(-1)^k$ on $\mathfrak{g}_k$. Then θ stabilizes $\mathfrak{g}(\mathbb{A}, \mathbb{O})$, and the corresponding Cartan decomposition $\mathfrak{g}(\mathbb{A}, \mathbb{O}) = \mathfrak{k} \oplus \mathfrak{p}$ into eigenspaces of θ is given, in terms of the Cartan decomposition $\mathfrak{t}(\mathbb{O}) = \mathfrak{k}_{\mathbb{O}} \oplus \mathfrak{p}_{\mathbb{O}}$, by

$$\begin{aligned} \mathfrak{k} &= \mathfrak{t}(\mathbb{A}) \times \mathfrak{k}_{\mathbb{O}} \oplus \mathbb{A}_1 \otimes \mathbb{H}_1 \oplus \mathbb{A}_2 \otimes \mathbb{H}_2 \oplus \mathbb{A}_3 \otimes \mathbb{H}_3, \\ \mathfrak{p} &= \mathfrak{p}_{\mathbb{O}} \oplus \mathbb{A}_1 \otimes e\mathbb{H}_1 \oplus \mathbb{A}_2 \otimes e\mathbb{H}_2 \oplus \mathbb{A}_3 \otimes e\mathbb{H}_3. \end{aligned}$$

Finally (see [84, 17]), we obtain the quaternionic form $\mathfrak{g}(\mathbb{A},\mathbb{O})_{\mathbb{H}}$ of $\mathfrak{g}$ by twisting this decomposition, that is, letting $\mathfrak{g}(\mathbb{A},\mathbb{O})_{\mathbb{H}} = \mathfrak{k} \oplus i\mathfrak{p}$, which amounts to multiplying the brackets in $\mathfrak{g}(\mathbb{A},\mathbb{O})$ of two elements of $\mathfrak{p}$ by -1. From the description of $\mathfrak{k}$ and $\mathfrak{p}$ we have just given, we see that this twist amounts to doing two things. First, twist the Cartan decomposition of $\mathfrak{t}(\mathbb{O})$, which means that we replace $\mathfrak{t}(\mathbb{O}) = \mathfrak{so}_8$ by its quaternionic form $\mathfrak{so}_{4,4} = \mathfrak{t}(\tilde{\mathbb{O}})$. Second, twist the Cayley–Dickson process of construction of $\mathbb{O}$ from $\mathbb{H}$ by multiplying by -1 the product of two elements in $e\mathbb{H}$, which amounts to replace $\mathbb{O}$ by the split Cayley algebra $\tilde{\mathbb{O}}$. This proves the first part of the theorem. The rest of the proof is straightforward. □

Adjoint Varieties and Quaternionic Symmetric Spaces

Symmetric spaces with quaternionic structures are closely related to adjoint varieties. Recall that on a Riemannian manifold M, a quaternionic structure is defined as a parallel field of quaternion algebras $\mathbb{H}_x \subset \mathrm{End}(T_xM)$, $x \in M$, such that the unit sphere of $\mathbb{H}_x$ is contained in the orthogonal group $\mathrm{SO}(T_xM)$. The dimension of M is then equal to $4m$ for some $m > 0$, and its reduced holonomy group is contained in $\mathrm{Sp}(m)\mathrm{Sp}(1) \subset \mathrm{SO}(4m)$, where $\mathrm{Sp}(m)$ denotes the group of quaternionic unitary matrices of order m (in particular, $\mathrm{Sp}(1) \simeq S^3$ is the unit sphere in $\mathbb{H}$). Such matrices act on $\mathbb{H}^m = \mathbb{R}^{4m}$ by multiplication on the left, which commutes with the scalar multiplication on the right by unitary quaternions: the resulting group of orthogonal transformations of $\mathbb{R}^{4m}$ is $\mathrm{Sp}(m)\mathrm{Sp}(1)$.

This group appears in Berger's classification as one of the few possible reduced holonomy groups of nonsymmetric Riemannian manifolds. The case of symmetric manifolds was discussed in detail by J. Wolf in [84], who proved that there exists exactly one G-homogeneous symmetric space M with a quaternionic structure for each simple compact Lie group G. When G is simply connected, we can write $M = G/K.\mathrm{Sp}(1)$. Choosing a complex plane in $\mathbb{H}$ amounts to choosing a circle $\mathrm{S}^1 \subset \mathrm{Sp}(1)$, and this induces a fibration

$$X = G/K.\mathrm{S}^1 \longrightarrow M = G/K.\mathrm{Sp}(1)$$

whose fibers are two-spheres. Wolf proved that X is a complex variety, homogeneous under the complexification $G^{\mathbb{C}}$ of the real compact group G, and endowed with an invariant contact structure. Such varieties were previously classified by Boothby, who showed that they are in correspondance with simple complex Lie groups of adjoint type. In fact, X is the *adjoint variety* $(G^{\mathbb{C}})^{\mathrm{ad}}$, the closed orbit in $\mathbb{P}\mathfrak{g}^{\mathbb{C}}$.

In modern terminology, X is the *twistor space* of M, and Wolf's work is at the origin of this twistor theory, which assigns to each Riemannian manifold M with a quaternionic structure a complex variety X which is a S^2-bundle over M, defined at each point $x \in M$ by the unit sphere in $\mathrm{Im}\mathbb{H}_x$.

The adjoint varieties are endowed with natural contact structures (induced by the Kostant–Kirillov symplectic structures on the minimal nilpotent coadjoint orbits, and of which the adjoint varieties are the projectivizations). In fact, any compact Riemannian manifold with exceptional holonomy $\mathrm{Sp}(n)\mathrm{Sp}(1)$ and positive scalar curvature has a twistor space that is a contact Fano manifold. Our understanding of contact Fano manifolds has recently improved very much. First of all, there are not very many:

Theorem 3.8. (LeBrun–Salamon [69]) *Up to biholomorphism, there are only finitely many contact Fano manifolds of any given dimension.*

This lead to the following conjecture:

Conjecture. (LeBrun–Salamon) *Let X be a smooth complex projective contact variety with $b_2(X) = 1$. Then X must be the adjoint variety of a simple complex Lie group.*

Both the theorem and conjecture can be rephrased purely on the Riemannian side using the twistor transform.

In [52], S. Kebekus proved that the space of minimal rational curves through a fixed point of X is a Legendre variety (except for the case of $\mathbb{P}^{2n-1}$), just as is the case for the adjoint varieties.

4 Series Of Lie Algebras Via Knot Theory and Geometry

4.1 From Knot Theory To the Universal Lie Algebra

One of the main achievements of the last ten years in topology has been the definition of the *Kontsevich integral*, which is a universal invariant of finite type for knots. The Kontsevich integral associates to each knot a formal series in a space of chord diagrams. The space is a quotient of the space of formal combinations of certain types of graphs by the AS and IHX relations. From the Kontsevitch integral, it is possible to deduce more standard invariants, say numerical or polynomial invariants, if one is given a *quadratic Lie algebra*, i.e., a finite dimensional Lie algebra endowed with an invariant nondegenerate quadratic form. An essential point is that the AS and IHX relations can be interpreted respectively as the antisymmetry of the Lie bracket and the Jacobi identity.

The main reference for the Kontsevich integral is the beautiful article by Bar–Natan [6]. In it he defines and proves (Theorem 4) the mapping from quadratic Lie algebras equipped with a representation to a functional on chord diagrams (called a *weight system*) and makes the conjecture (Conjecture 1) that all weight systems come from this mapping. This conjecture was disproved by P. Vogel in [82] which raised the problem of constructing a category more general than that of quadratic Lie algebras that would give rise to all weight systems.

In [83] Vogel defines a candidate for such an object, which he calls the *universal Lie algebra.* The universal Lie algebra is a category $\mathcal{D}$ with the following property: for any quadratic Lie (super)algebra $\mathfrak{g}$ over a field k, there is a natural functor from $\mathcal{D}$ to the category $\mathrm{Mod}_k\mathfrak{g}$ of $\mathfrak{g}$-modules. The objects of the category $\mathcal{D}$ are finite sets $[n] = \{1, \dots, n\}$, $n \geq 0$. Morphisms in $\mathcal{D}$ are defined as follows: $\mathrm{Hom}_{\mathcal{D}}([p], [q])$ is the $\mathbb{Z}$-module of formal combinations of uni-trivalent abstract graphs with univalent vertices $[p] \cup [q]$, modulo isotopy, and modulo the AS and IHX relations. If $\mathfrak{g}$ is a quadratic Lie algebra, a functor from $\mathcal{D}$ to $\mathrm{Mod}_k\mathfrak{g}$ is defined at the level of objects by sending $[n]$ to $\mathfrak{g}^{\otimes n}$. To determine the morphisms, consider some uni-trivalent graph defining an element of $\mathrm{Hom}_{\mathcal{D}}([p], [q])$. Possibly after isotopy, such a graph can be seen as made of elementary pieces of the form

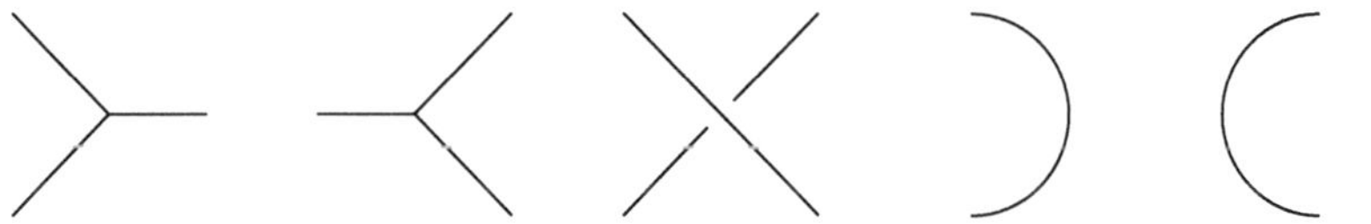

These pieces should be interpreted as indicated, respectively as the Lie bracket $\mathfrak{g}^{\otimes 2} \to \mathfrak{g}$, the cobracket $\mathfrak{g} \to \mathfrak{g}^{\otimes 2}$ (the dual to the bracket), the symmetry $\mathfrak{g}^{\otimes 2} \to \mathfrak{g}^{\otimes 2}$, the quadratic form $\mathfrak{g}^{\otimes 2} \to \mathfrak{g}^{\otimes 0} = k$, and the Casimir element $\mathfrak{g}^{\otimes 0} \to \mathfrak{g}^{\otimes 2}$ (the dual to the quadratic form). Putting together the contributions of all its elementary pieces, one associates a morphism from $\mathfrak{g}^{\otimes p}$ to $\mathfrak{g}^{\otimes q}$ to the graph. This morphism depends only on the isotopy class of the graph, and factors through the AS and IHX relations, because of the skew-symmetry and Jacobi relation for the Lie bracket in $\mathfrak{g}$.

Vogel proves a stronger result for simple quadratic Lie algebras. One can then replace the category $\mathcal{D}$ by another one $\mathcal{D}'$, which is deduced from $\mathcal{D}$ by forcing an action of the algebra Λ of skew-symmetric connected uni-trivalent graphs with exactly three univalent vertices. The functor from $\mathcal{D}'$ to $\mathrm{Mod}_k\mathfrak{g}$ is then compatible with the action of Λ through a character $\chi_{\mathfrak{g}} : \Lambda \to k$.

From the point of view of representation theory, the existence and the properties of the categories $\mathcal{D}$ and $\mathcal{D}'$ have very interesting consequences. Indeed, each "decomposition" of $[p]$ in $\mathcal{D}'$ will imply the existence of a decomposition of $\mathfrak{g}^{\otimes p}$ of the same type, *for every simple quadratic Lie algebra.* Here, by a "decomposition" of $[p]$, we must understand a decomposition of the identity endomorphism of $[p]$ as a sum of idempotents, which will correspond to projectors onto submodules of $\mathfrak{g}^{\otimes p}$.

Describing such idempotents is far from easy and requires quite subtle computations with uni-trivalent graphs. For $p = 2$, Vogel obtained the following result. (Vogel's proof relied on conjecture 3.6 as stated in [83], but Vogel himself realized that it is not correct as Λ, which is conjectured to be integral, has zero divisors. Nevertheless, a case by case verification shows that Vogel's conclusions are indeed correct.)

Let $\mathfrak{g}$ be a simple Lie algebra, say over the complex numbers. Then there are decompositions

$$\begin{aligned} \wedge^2\mathfrak{g} &= X_1 \oplus X_2, \\ S^2\mathfrak{g} &= X_0 \oplus Y_2 \oplus Y_2' \oplus Y_2'', \end{aligned}$$

into simple $\mathfrak{g}$-modules (some of which may be zero), with $X_0 = \mathbb{C}$ and $X_1 = \mathfrak{g}$. Moreover, there exists α, β, γ such that one half the Casimir operator acts on X_1, X_2, Y_2, Y_2' and Y_2'' by multiplication by $t, 2t, 2t-\alpha, 2t-\beta$ and $2t-\gamma$ respectively, where $t = \alpha+\beta+\gamma$. (In fact the Casimir organizes the representations in series, e.g., the representations X_k have Casimir eigenvalue kt.) Finally, the dimensions of these $\mathfrak{g}$-modules are given by rational functions in α, β, γ (the dimensions of Y_2' and Y_2'' are deduced from that of Y_2 by cyclic permutations of α, β, γ):

$$\begin{aligned} \dim\mathfrak{g} &= \frac{(\alpha-2t)(\beta-2t)(\gamma-2t)}{\alpha\beta\gamma}, \\ \dim X_2 &= -\frac{(\alpha-2t)(\beta-2t)(\gamma-2t)(\alpha+t)(\beta+t)(\gamma+t)}{\alpha^2\beta^2\gamma^2}, \\ \dim Y_2 &= -\frac{t(\beta-2t)(\gamma-2t)(\beta+t)(\gamma+t)(3\alpha-2t)}{\alpha^2\beta\gamma(\alpha-\beta)(\alpha-\gamma)}. \end{aligned}$$

The scalars α, β, γ are readily computed for each simple complex Lie algebra. Note that they are defined only up to permutation and multiplication by a same scalar (this is because the Casimir operator has not been normalized). We can therefore consider (α, β, γ) as a point in $\mathbb{P}^2/\mathfrak{S}_3$.

Series	Lie algebra	α	β	γ
SL	$\mathfrak{sl}_n$	-2	2	n
OSP	$\mathfrak{so}_n, \mathfrak{sp}_{-n}$	-2	4	$n-4$
EXC	$\mathfrak{sl}_3$	-2	3	2
	$\mathfrak{g}_2$	-3	5	4
	$\mathfrak{so}_8$	-2	6	4
	$\mathfrak{f}_4$	-2	5	6
	$\mathfrak{e}_6$	-2	6	8
	$\mathfrak{e}_7$	-2	8	12
	$\mathfrak{e}_8$	-2	12	20

Note that up in $\mathbb{P}^2$, the points of the three series SL, OSP and EXC are located on three lines, respectively $\alpha+\beta=0$, $2\alpha+\beta=0$, $2\alpha+2\beta-\gamma=0$. Hence the slogan that *there exists only three complex simple Lie algebras,* SL, OSP, *and* EXC. At the level of representations, it is a classical and well-known fact that the categories of modules over the $\mathfrak{sl}_n$, or the $\mathfrak{so}_n$ and $\mathfrak{sp}_{2n}$, have close and precise relationships. This is more surprising for the exceptional series.

In Section 4.6 we discuss another collection of Lie algebras that form a "line" in Vogel's plane.

Remark 4.1. In a spirit similar to Vogel's, Cvitanovic [21] has proposed a proof of the Killing–Cartan classification. Using notation inspired by the Feynman diagrams of quantum field theory, he represents the invariant tensors of a Lie algebra by diagrams, and by requring that irreducible representations be of integer dimension, he deduces severe limits on what simple groups could possibly exist. His approach dates back to 1976 [19, 20]. The method provides the correct Killing–Cartan list of all possible simple Lie algebras, but fails to prove existence.

4.2 Vogel's Decompositions and Tits Correspondences

As remarked above, Tits correspondences aid one in decomposing the tensor powers of nice (e.g., fundamental) representations into irreducible factors. We illustrate this by showing how to recover Vogel's decompositions of $S^2\mathfrak{g}$ and $\Lambda^2\mathfrak{g}$. In fact we can recover Vogel's higher decompositions as well, but we need to use a slightly more general technique which we call *diagram induction*, see [63]. Using diagram induction, we are also able to recover the Casimir functions for the representations occuring in series such as the X_k.

If V is a fundamental representation corresponding to a root that is not short (e.g., the fundamental adjoint representations) and $X = G/P_\alpha \subset \mathbb{P}V$ the closed orbit, then the Fano variety $\mathbb{F}_1(X)$ of $\mathbb{P}^1$'s on X is G-homogeneous by Theorem 2.5, and according to Tits fibrations, $\mathbb{F}_1(X) = G/P_S$ where S is the set of roots joined to α in $D(\mathfrak{g})$. Let $V_2 = \langle \mathbb{F}_1(X) \rangle$ denote the linear span of the cone over $\mathbb{F}_1(X)$ in its minimal homogeneous embedding. Note that $\mathbb{F}_1(X) \subset G(2,V) \subset \Lambda^2 V$, and we conclude that $V_2 \subset \Lambda^2 V$. (To do this, one needs to make sure the embedding of $\mathbb{F}_1(X)$ is indeed the minimal one, which is one reason we need the stronger technique of diagram induction.)

In the case of the adjoint representation, we also know, because of the cobracket, that $\mathfrak{g} \subset \Lambda^2\mathfrak{g}$. In fact, we have $\Lambda^2\mathfrak{g} = \mathfrak{g} \oplus \mathfrak{g}_2$. (In the case of $\mathfrak{a}_n$, $(\mathfrak{a}_n)_2$ is the direct sum of two dual irreducible representations, which is also predicted by Tits fibrations.)

Similiarly, inside $S^2\mathfrak{g}$, we know there is a trivial representation X_0 corresponding to the Killing form, and the Cartan power $\mathfrak{g}^{(2)}$, which, breaking Vogel's symmetry, we set $Y_2 = \mathfrak{g}^{(2)}$ so the problem becomes to recover the remaining modules, which we think of as the "primitive" ones. These are provided in most cases by Tits fibrations of homogeneous quadrics contained in X. There are up to two such families, we call the larger one $\mathfrak{g}_Q = Y_2'$ and if there is a second we call it $\mathfrak{g}_{Q'} = Y_2''$, see [63]. Thus, using Tit's correspondences to find the primitive factors, one recovers the Vogel decomposition of $\mathfrak{g}^{\otimes 2}$ for $\mathfrak{g}$ simple (except for $\mathfrak{c}_n$, which is tractible using diagram induction). Even when our methods predict all factors, we would like to emphasize that they do *not* explain why these are the only factors.

While our observations enable one to recover uniform decompositions of plethysms in series, the questions of why such series exist and how to find them systematically without looking at the list of simple Lie algebras is mysterious using our present techniques. It would be wonderful and challenging to generalize our perspective sufficiently to approach Vogel's. One indication that such a generalization might be possible is that, with the above choices, we obtain a geometric interpretation of Vogel's parameter β, namely β is the dimension of the largest quadric hypersurface contained in X_{ad} (see [63]). If there is a second unextendable quadric in X_{ad}, then γ is its dimension.

4.3 The Exceptional Series

Inspired by Vogel's work, Deligne [24], investigated the decompositions of the tensor powers of the exceptional Lie algebras into irreducible components, going up to degree four (with the help of Cohen, deMan and the computer program LiE, [14]) and giving more explicit dimension formulas.

4.3.1 Decomposition Formulas

As with the Vogel decompositions, these decomposition formulas are all *uniform*, in the sense, for example, that the numbers of irreducible components are always the same when the algebra varies in the exceptional series. This assertion has to be understood with some care: it happens that some components vanish, or should be taken with a minus sign. In addition, Deligne and Vogel deal with the algebra twisted by the symmetry of the marked (for the adjoint representation) diagram.

In degree two, one has the decomposition of Vogel, but the decomposition of $S^2\mathfrak{g}$ simplifies because the equation $2\alpha+2\beta-\gamma=0$ implies Y_2'' is zero. In degree three we have

$$\begin{aligned} S^3\mathfrak{g} &= \mathfrak{g}\oplus X_2\oplus A\oplus Y_3\oplus Y_3', \\ \wedge^3\mathfrak{g} &= \mathbb{C}\oplus X_2\oplus Y_2\oplus Y_2'\oplus X_3, \\ S_{21}\mathfrak{g} &= 2\mathfrak{g}\oplus X_2\oplus Y_2\oplus Y_2'\oplus A\oplus C\oplus C' \end{aligned}$$

where the (normalized) Casimir operator acts on $X_2, X_3, A, Y_2, Y_2', Y_3, Y_3', C, C'$ by multiplication by $2, 3, 8/3, (2\beta+5\gamma)/3\gamma, (2\alpha+5\gamma)/3\gamma, 2(\beta+\gamma)/\gamma, 2(\alpha+\gamma)/\gamma, (2\beta+5\gamma)/2\gamma, (2\alpha+5\gamma)/2\gamma$ respectively. Again, we are able to account for nearly all the factors and Casimir functions appearing using diagram induction.

4.3.2 Dimension Formulas

For each exceptional Lie algebra $\mathfrak{sl}_2$, $\mathfrak{sl}_3$, $\mathfrak{so}_8$ $\mathfrak{g}_2$, $\mathfrak{f}_4$, $\mathfrak{e}_6$, $\mathfrak{e}_7$, $\mathfrak{e}_8$, let $\lambda = -3, -2, -1, -3/2, -2/3, -1/2, -1/3, -1/5$ respectively. λ is a linear function of the length of the longest root (with the Casimir normalized to act as the identity on $\mathfrak{g}$). In Vogel's parametrization this gives $(\alpha, \beta, \gamma) = (\lambda, 1-\lambda, 2)$.

From the decomposition formulas stated above and the knowledge of the Casimir eigenvalue on each irreducible component, it is easy (at least with a computer) to calculate the dimensions of these components as rational functions of λ, if one knows the dimension of $\mathfrak{g}$ itself in terms of λ. For example,

$$\dim\mathfrak{g} = -2\frac{(\lambda+5)(\lambda-6)}{\lambda(\lambda-1)},$$

$$\dim X_2 = 5\frac{(\lambda+3)(\lambda+5)(\lambda-4)(\lambda-6)}{\lambda^2(\lambda-1)^2},$$

$$\dim Y_2 = -90\frac{(\lambda+5)(\lambda-4)}{\lambda^2(\lambda-1)(2\lambda-1)},$$

$$\dim Y_3 = -10\frac{(\lambda+5)(5\lambda-6)(\lambda-4)(\lambda-5)(\lambda-6)}{\lambda^3(\lambda-1)^2(2\lambda-1)(3\lambda-1)}.$$

A *miracle*, Deligne says, is that the numerators and denominators of these rational functions all factor out into products of linear forms with simple integers coefficients! Note that this miracle does not occur for the Vogel dimension formulas starting with degree three.

In addition, there is a striking duality property: the involution $\lambda \mapsto 1-\lambda$ takes the dimension formula for Y_2, Y_3 into those for Y_2', Y_3', while it leaves unchanged the dimension formulas for $\mathfrak{g}$, X_2, X_3, A.

One can attempt to continue the decompositions in degree five, but there for the first time there appear different representations with the same Casimir eigenvalue and Deligne's method cannot be used to compute dimensions. This suggests that if one wants to continue the formulas, it might be better to restrict to decompositions into Casimir eigenspaces rather than irreducible modules.

In the last part of [24], Deligne conjectured the existence of a category, supposed (just as the universal Lie algebra defined by Vogel, but with much stronger properties) to map to the categories of modules over each exceptional Lie algebra. The properties of this hypothetic category would then explain at least some of the phenomena discovered by Vogel and Deligne for the exceptional series. At the moment, it seems no progress has been made on this conjecture.

4.3.3 Deligne Dimension Formulas Via Triality

Deligne's conjecture is elegant while his proof is brute force computer based. In [62] we reprove and extend his formulas using methods whose level of elegance is somewhere between the conjecture and proof of Deligne. The starting point was our observation that the parameter λ could be written as $\lambda = -2/(a+2)$, with $a = -2/3, 1, 2, 4, 8$. (This implies that $(\alpha, \beta, \gamma) = (-2, a+4, 2a+4)$.)

This indicated that the relation between exceptional Lie algebras and normed division algebras should be exploited. The idea was to use this relation to find a good description of the exceptional root systems, and then apply the Weyl dimension formula.

The Tits construction is not convenient to describe the exceptional root systems and we were led to rediscover the triality model. Recall that for the exceptional series,

$$\mathfrak{g}(\mathbb{A},\mathbb{O}) = \mathfrak{t}(\mathbb{A}) \times \mathfrak{t}(\mathbb{O}) \oplus (\mathbb{A}_1 \otimes \mathbb{O}_1) \oplus (\mathbb{A}_2 \otimes \mathbb{O}_2) \oplus (\mathbb{A}_3 \otimes \mathbb{O}_3),$$

can be identified with the compact form of $\mathfrak{f}_4, \mathfrak{e}_6, \mathfrak{e}_7, \mathfrak{e}_8$ respectively for $\mathbb{A} = \mathbb{R}, \mathbb{C}, \mathbb{H}, \mathbb{O}$. In what follows we *complexify* this construction, without changing notation. In particular $\mathfrak{t}(\mathbb{O})$ is now the complex orthogonal Lie algebra $\mathfrak{so}_8(\mathbb{C})$. The point is that $\mathfrak{t}(\mathbb{A}) \times \mathfrak{t}(\mathbb{O})$ is now a maximal rank reductive subalgebra of $\mathfrak{g}(\mathbb{A},\mathbb{O})$. In particular, if we choose Cartan subalgebras $\mathfrak{h}(\mathbb{A}) \subset \mathfrak{t}(\mathbb{A})$ and $\mathfrak{h}(\mathbb{O}) \subset \mathfrak{t}(\mathbb{O})$, we obtain a Cartan subalgebra $\mathfrak{h}(\mathbb{A}) \times \mathfrak{h}(\mathbb{O}) \subset \mathfrak{g}(\mathbb{A},\mathbb{O})$. This provides a nice description of the root system of $\mathfrak{g}(\mathbb{A},\mathbb{O})$. There are three kinds of roots:

- the *heart* of the root system is the set of roots of $\mathfrak{t}(\mathbb{O}) = \mathfrak{so}_8(\mathbb{C})$,
- the *linear part* is made of the roots of the form $\mu + \nu$, where μ is a weight of $\mathbb{O}_i$ and ν a weight of $\mathbb{A}_i$ for some i,
- the *residue* is the set of roots of $\mathfrak{t}(\mathbb{A})$.

When $\mathbb{A} = \underline{0}$, the root system is of type D_4. When $\mathbb{A} = \mathbb{R}$, the root system is of type F_4. There is no residue in this case. The heart is the set of long roots, and the linear part the set of short roots. Both are root systems of type D_4. When we change $\mathbb{A}$ into $\mathbb{C}, \mathbb{H}, \mathbb{O}$, the linear part *expands*, is size being each time multiplied by two, and its roots become long. The residue also increases, but in limited proportions.

Next we specify the positive roots inside the exceptional root systems. They are the roots on which some linear form takes positive values, and we choose a linear form that takes very large values on the set Δ_+ of positive roots of $\mathfrak{so}_8(\mathbb{C})$. Then the positive roots in the linear part of the root system are the $\mu+\nu$ for μ belonging to an explicit set Σ of weights. This implies that the half-sum of the positive roots can be expressed as $\rho = \rho_{\mathfrak{t}(\mathbb{A})} + \rho_{\mathfrak{t}(\mathbb{O})} + a\gamma_{\mathfrak{t}(\mathbb{O})}$, where $\rho_{\mathfrak{t}(\mathbb{A})} \in \mathfrak{h}(\mathbb{A})^*$ denotes the half sum of the positive roots in $\mathfrak{t}(\mathbb{A})$ and $\gamma_{\mathfrak{t}(\mathbb{O})} \in \mathfrak{h}(\mathbb{O})^*$ is the sum of the weights belonging to Σ.

We are now close to being able to apply the Weyl dimension formula to $\mathfrak{g}(\mathbb{A},\mathbb{O})$-modules. What remains to do is to describe the dominant integral weights in $\mathfrak{h}(\mathbb{A})^* \times \mathfrak{h}(\mathbb{O})^*$. At least we can describe the set $C(\mathbb{O}) \subset \mathfrak{h}(\mathbb{O})^*$ of weights that are dominant and integral for each $\mathfrak{g}(\mathbb{A},\mathbb{O})$. Of course, such weights are be dominant and integral in $\mathfrak{so}_8(\mathbb{C})$, but this is not sufficient.

Proposition 4.2. *The set $C(\mathbb{O}) \subset \mathfrak{h}(\mathbb{O})^*$ is the simplicial cone of nonnegative integer linear combinations of the four following weights:*

$$\omega(\mathfrak{g}) = 1 - 2 \begin{array}{l} \diagup 1 \\ \diagdown 1 \end{array}, \quad \omega(X_2) = 2 - 3 \begin{array}{l} \diagup 2 \\ \diagdown 2 \end{array},$$

$$\omega(X_3) = 3 - 4 \begin{array}{l} \diagup 2 \\ \diagdown 3 \end{array}, \quad \omega(Y_2') = 2 - 2 \begin{array}{l} \diagup 1 \\ \diagdown 1 \end{array}.$$

The lattice in $\mathfrak{t}(\mathbb{O})^*$ generated by $C(\mathbb{O})$ is not the weight lattice, but the root lattice of $\mathfrak{so}_8(\mathbb{C})$.

These weights, that we expressed above in terms of simple roots, occur respectively, $\omega(\mathfrak{g}) = \omega_2$ as the highest weight of $\mathfrak{g}$, $\omega(X_2) = \omega_1 + \omega_3 + \omega_4$ as the highest weight of $\Lambda^2\mathfrak{g}$, $\omega(X_3) = 2\omega_1 + 2\omega_3$ as the highest weight of $\Lambda^3\mathfrak{g}$, and $\omega(Y_2') = 2\omega_1$ as the highest weight of $S^2\mathfrak{g} - \mathfrak{g}^{(2)}$.

We can now apply the Weyl dimension formula to weights in $C(\mathbb{O})$, considered for each choice of $\mathbb{A}$ as integral dominant weights of $\mathfrak{g}(\mathbb{A},\mathbb{O})$. It is essential that the residue of the root system will not contribute. We obtain (see [62] for details):

Theorem 4.3. *The dimension of the irreducible $\mathfrak{g}(\mathbb{A},\mathbb{O})$-module of highest weight $\omega \in C(\mathbb{O})$ is given by the following formula:*

$$\dim V_\omega = \prod_{\alpha \in \Delta_+ \cup \Sigma} \frac{(a\gamma_{\mathfrak{t}(\mathbb{O})} + \rho_{\mathfrak{t}(\mathbb{O})} + \omega, \alpha^\vee)}{(a\gamma_{\mathfrak{t}(\mathbb{O})} + \rho_{\mathfrak{t}(\mathbb{O})}, \alpha^\vee)}$$

$$\times \prod_{\beta \in \Sigma} \frac{\begin{pmatrix} (a\gamma_{\mathfrak{t}(\mathbb{O})} + \rho_{\mathfrak{t}(\mathbb{O})} + \omega, \beta^\vee) + \frac{a}{2} - 1 \\ (\omega, \beta^\vee) \end{pmatrix}}{\begin{pmatrix} (a\gamma_{\mathfrak{t}(\mathbb{O})} + \rho_{\mathfrak{t}(\mathbb{O})} + \omega, \beta^\vee) - \frac{a}{2} \\ (\omega, \beta^\vee) \end{pmatrix}}.$$

If $\omega = p\omega(\mathfrak{g}) + q\omega(X_2) + r\omega(X_3) + s\omega(Y_2^*)$, this formula gives a rational function of a, whose numerator and denominator are products of $6p + 12q + 16r + 10s + 24$ linear forms in a with integer coefficients. Since $\lambda = -2/(a+2)$, we obtain formulas of the type of those of Deligne, and an infinite family of such. For example, the k-th Cartan power $\mathfrak{g}(\mathbb{A},\mathbb{O})^{(k)}$ is Y_k in Deligne's notations, and we get

$$\dim Y_k = \frac{(2k-1)\lambda - 6}{k!\lambda^k(\lambda+6)} \prod_{j=1}^{k} \frac{((j-1)\lambda - 4)((j-2)\lambda - 5)((j-2)\lambda - 6)}{(j\lambda - 1)((j-1)\lambda - 2)}.$$

Also we can understand why we can expect good dimension formulas for all irreducible components of $\mathfrak{g}(\mathbb{A},\mathbb{O})^{\otimes k}$ only for small k. This is because when k increases, we will unavoidably get components whose highest weight does not come only from $\mathfrak{so}_8(\mathbb{C})$, but has a contribution from $\mathfrak{h}(\mathbb{A})^*$. Then the Weyl dimension formula does not give a well behaved expression.

While this takes some of the mystery out of the dimension formulas, the remarkable symmetry $\lambda \mapsto 1 - \lambda$ noticed by Deligne, which in our parameter a is $a \mapsto -4(a+3)/(a+4)$ remains beyond our understanding.

4.3.4 Numerology

In the formula above for dim Y_k, there are very few values of λ for which dimY_k is an integer for all k. We are currently, with B. Westbury [65] investigating these extra numbers that produce integers and Lie algebras that go with them. At least in the case $a = 6$, one is led to an algebra of dimension six, the *sextonions*, that leads to a nonreductive row of the magic square between the third and fourth. Like the *odd symplectic groups* of Proctor [73] and Gelfand and Zelevinsky [36], this series has certain behavior as if it were a series of reductive Lie algebras.

4.4 Freudenthal Geometries

The discovery of the Cayley plane $\mathbb{OP}^2$ and its automorphism group by Chevalley, Schafer and others led to a period of intense activity around the exceptional groups and their geometric interpretations in the 1950's involving Freudenthal, Tits and Rozenfeld.

Freudenthal and Rozenfeld defined explicit varieties whose automorphism groups (or groups related to the automorphism groups) were the exceptional groups. The starting point is the Jordan algebras $\mathcal{J}_3(\mathbb{A})$. The projective plane $\mathbb{AP}^2$ is the space of rank one idempotents of $\mathcal{J}_3(\mathbb{A})$. The subgroup of $GL(\mathcal{J}_3(\mathbb{A}))$ preserving the determinant of order three matrices (which is well defined even for $\mathbb{A} = \mathbb{O}$) then acts transitively on $\mathbb{AP}^2$. In particular, one recovers the transitive action of E_6 on $\mathbb{OP}^2$. The subgroup of $\mathrm{GL}(\mathcal{J}_3(\mathbb{A}))$ preserving the determinant and the quadratic form $(A, B) = \mathrm{trace}(AB)$ is the automorphism group of the Jordan algebra $\mathcal{J}_3(\mathbb{A})$. It acts irreducibly on $\mathcal{J}_3(\mathbb{A})_0$, the subspace of trace zero matrices. Tits and Freudenthal defined elliptic and projective geometries on $\mathbb{AP}^2$ with these groups of isometries. In the elliptic geometry, a *point* is the same as a *line*, and is defined by an element of $\mathbb{AP}^2_0$. In the projective geometry, a point is defined to be an element of $\mathbb{AP}^2$ and a line is determined by an element $[\alpha]$ of the dual projective plane (the rank one idempotents in the dual Jordan algebra corresponding to the dual vector space), namely $\{X \in \mathbb{AP}^2 \mid \alpha(X) = 0\}$.

A synthetic geometry Freudenthal terms *symplectic* may be associated to each of the groups in the third row. A *point* is an element $[X] \in \mathbb{P}\mathfrak{g}(\mathbb{A}, \mathbb{H})$, such that $\mathrm{ad}(X)^2 = 0$. This condition is equivalent to requiring that $[X]$ is in the adjoint variety $X_{\mathrm{ad}} \subset \mathbb{P}\mathfrak{g}(\mathbb{A}, \mathbb{H})$. A *plane* is is determined by an element $[P] \in \mathbb{P}\mathcal{Z}_3(\mathbb{A})$, $P \times P = 0$. (See Section 4.6 for the definition of the cross product.) This condition is equivalent to saying that $[P] \in G_w(\mathbb{A}^3, \mathbb{A}^6) \subset \mathbb{P}\mathcal{Z}_3(\mathbb{A})$, the closed orbit. The corresponding plane is $\{[X] \in X_{\mathrm{ad}} \mid XP = 0\}$. A *line* is an intersection of two planes with at least two points. It is also determined by a point $[\alpha] \in \mathbb{F}_1(X) \subset \mathbb{P}V_2$, where V_2 is the representation defined in Section 4.6 below and $\mathbb{F}_1(G_w(\mathbb{A}^3, \mathbb{A}^6))$ is the closed orbit (which, as the notation suggests, parametrizes the lines on the closed orbit $G_w(\mathbb{A}^3, \mathbb{A}^6) \subset$

$\mathbb{P}\mathcal{Z}_3(\mathbb{A})$). One can then define incidence rules for these geometric elements which generalize the symplectic geometry in $\mathbb{P}^5$.

For the fourth line of Freudenthal's magic square Freudenthal defines a *metasymplectic geometry*. There are now four types of elements, points, lines, planes and *symplecta*, with incidence rules explained in detail by Freudenthal. As above, each geometric element is a point in the closed orbit of the projectivization of a $\mathfrak{g}(\mathbb{A}, \mathbb{O})$-module. Here is the table of types of elements:

Geometric element	Variety	Dimension
Point	$X_{\text{point}} \subset \mathbb{P}\mathfrak{g}_Q$	$9a + 6$
Line	$X_{\text{line}} \subset \mathbb{P}\mathfrak{g}_3$	$11a + 9$
Plane	$X_{\text{plane}} \subset \mathbb{P}\mathfrak{g}_2$	$9a + 11$
Symplecta	$X_{\text{symplecta}} \subset \mathbb{P}\mathfrak{g}(\mathbb{A}, \mathbb{O})$	$6a + 9$

We see the four generators of the cone $C(\mathbb{O})$ in Proposition 4.2 define the four types of elements of Freudenthal's metasymplectic geometry! The analogous results hold for the second and third rows of the magic chart, that is, Freudenthals elements are the closed orbits in the projectivizations of the modules whose highest weights define the fixed cones for our dimension formulas in [62].

4.5 A Geometric Magic Square

Consider the following square of complex homogeneous varieties, which we investigate in [61]:

	$\mathbb{R}$	$\mathbb{C}$	$\mathbb{H}$	$\mathbb{O}$	
$\mathbb{R}$	$v_2(Q^1)$	$\mathbb{P}(T\mathbb{P}^2)$	$G_\omega(2,6)$	$\mathbb{OP}^2_{\mathbb{C},0}$	section of Severi
$\mathbb{C}$	$v_2(\mathbb{P}^2)$	$\mathbb{P}^2 \times \mathbb{P}^2$	$G(2,6)$	$\mathbb{OP}^2_{\mathbb{C}}$	Severi
$\mathbb{H}$	$G_\omega(3,6)$	$G(3,6)$	$\mathbb{S}_{12}$	E_7/P_7	Legendre
$\mathbb{O}$	F_4^{ad}	E_6^{ad}	E_7^{ad}	E_8^{ad}	adjoint

This *geometric magic square* is obtained as follows. One begins with the adjoint varieties for the exceptional groups: this is the fourth line of the square. Taking the varieties of lines through a point and applying Theorem 2.5, one obtains the third line. The second line is deduced from the third by the same process. Then take hyperplane sections to get the first line. We obtain projective varieties that are homogeneous under groups given by Freudenthal's magic square.

4.6 The Subexceptional Series

In this final section we consider the series corresponding to the third row of the extended Freudenthal chart:

$$\mathfrak{sl}_2, \quad \mathfrak{sl}_2\times\mathfrak{sl}_2\times\mathfrak{sl}_2, \quad \mathfrak{sp}_6, \quad \mathfrak{sl}_6, \quad \mathfrak{so}_{12}, \quad \mathfrak{e}_7$$

and view the series from our various perspectives.

4.6.1 Relation To the Universal Lie Algebra

This series also corresponds to a new line in Vogel's plane, namely $(\alpha,\beta,\gamma) = (-2, a, a+4)$, where $a = -2/3, 0, 1, 2, 4, 8$. Moreover, Vogel's formulas are valid for the semi-simple Lie algebra $\mathfrak{sl}_2\times\mathfrak{sl}_2\times\mathfrak{sl}_2$ despite the plane being defined only for absolutely simple objects. Unlike the three lines discovered by Vogel, this line is generic to order three among actual Lie algebras, in the sense that no Vogel space is zero except X_3'', which is zero for all actual Lie algebras. We have $\dim\mathfrak{g}(a) = 3(2a+3)(3a+4)/(a+4)$.

4.6.2 Freudenthalia

For this series there are three preferred representations from Freudenthal's perspective, corresponding to the ambient spaces for points, lines, and planes in his incidence geometries. The points representation is just $\mathfrak{g} = \mathfrak{g}(\mathbb{A},\mathbb{H})$ and has a nice model thanks to the triality construction. For a discussion of the line space, which we will denote $V_2 = V_2(a)$, see [61]. We have $\dim V_2(a) = 9(a+1)(2a+3)$. The most preferred representation is for the planes, which we denote $V = V(a)$ and has $\dim V(a) = 6a+8$. This space is the complexification of the algebra of *Zorn matrices* we encountered in Section 3.8:

$$\mathcal{Z}_3(\mathbb{A}) = \left\{ \begin{pmatrix} a & X \\ Y & b \end{pmatrix} \mid a,b\in\mathbb{R},\ X,Y\in\mathcal{J}_3(\mathbb{A}) \right\}.$$

It can be endowed with an algebra structure with multiplication

$$\begin{pmatrix} a_1 & X_1 \\ Y_1 & b_1 \end{pmatrix}\begin{pmatrix} a_2 & X_2 \\ Y_2 & b_2 \end{pmatrix} = \begin{pmatrix} a_1a_2 + \mathrm{trace}\,(X_1Y_2) & a_1X_2 + b_2X_1 + Y_1\times Y_2 \\ a_2Y_1 + b_1Y_2 + X_1\times X_2 & b_1b_2 + \mathrm{trace}\,(X_2Y_1) \end{pmatrix}.$$

Here the product $X\times Y$ is defined by the identity $\mathrm{tr}((X\times Y)Z) = \det(X,Y,Z)$, where det is the polarization of the determinant. The algebra is in fact a structurable algebra (see Section 2.5), with $V(a) = \mathcal{Z}_3(\mathbb{A}) = \langle \hat{Y}_x\rangle \subset T_xX_{\mathrm{adjoint}}(\mathbb{A},\mathbb{O})$ where $X_{\mathrm{adjoint}}(\mathbb{A},\mathbb{O})$ are the exceptional adjoint varieties. In fact, Zorn matrices give all simple structurable algebras of skew-symmetric rank equal to one, and one can show directly that the derivation algebra of $\mathcal{Z}_3(\mathbb{A})$ is $\mathfrak{g}(\mathbb{A},\mathbb{H})$.

Using this construction one can consider the automorphism group of $\mathcal{Z}_3(\mathbb{A})$ in addition to its derivation algebra, and this can be done on any field (of characteristic not equal to 2 or 3). Using this, Garibaldi constructs forms of E_7 over an arbitrary field [33].

The closed orbit $X_{\mathrm{planes}} \subset \mathbb{P}\mathcal{Z}_3(\mathbb{A})$ is a natural compactification of the Jordan algebra $\mathcal{J}_3(\mathbb{A})$, and its embedding inside $\mathbb{P}\mathcal{Z}_3(\mathbb{A})$ is given by the translates of the determinant on $\mathcal{J}_3(\mathbb{A})$. This is the *conformal compactification* considered by Faraut and Gindikin in [30]. In [61] we propose a geometric interpretation of X_{planes} as the Grassmannian of $\mathbb{O}^3$'s in $\mathbb{O}^6$ isotropic for a Hermitian symplectic form and use the notation $G_w(\mathbb{O}^3, \mathbb{O}^6)$.

4.6.3 Decomposition Formulas

The plethysms of V are extraodinarily well behaved in series. For example, here is a decomposition of S^kV into irreducible components for each k.

$$\bigoplus_{k\geq 0} t^k S^k V = (1-tV)^{-1}(1-t^2\mathfrak{g})^{-1}(1-t^3V)^{-1}(1-t^4)^{-1}(1-t^4V_2)^{-1}.$$

The right hand side in the formula is to be expanded out in geometric series and multiplication of representations is taken in terms of Cartan products $V_\lambda V_\mu = V_{\lambda+\mu}$. The case of $\mathfrak{sl}_2\times\mathfrak{sl}_2\times\mathfrak{sl}_2$ is somewhat special, since the formula above holds true only if the natural action of $\mathfrak{S}_3$ is taken into account. See [63].

Each of the factors in the rational function above can be accounted for in terms of diagram induction. In particular, $\mathfrak{g} = V_Q$, the ambient space for the variety parametrizing the G-homogeneous quadrics on X_{planes} and $V_2 = \langle \mathbb{F}_1(X_{\text{planes}})\rangle$ is the ambient space for the variety parametrizing the lines on X_{planes}.

4.6.4 Dimension Formulas

Our method for applying the Weyl dimension formula in series also works for the three distinguished representations. For example, we have

$$\dim V^{(k)} = \frac{2a+2k+2}{a+1}\,\frac{\binom{k+2a+1}{2a+1}\binom{k+\frac{3a}{2}+1}{\frac{3a}{2}+1}}{\binom{k+\frac{a}{2}+1}{\frac{a}{2}+1}},$$

where the binomial coefficients are defined by $\binom{k+x}{k} = (1+x)\cdots(k+x)/k!$ and thus are rational polynomials of degre k in x.

Note that since $V^{(k)}$ is the complement of $I_k(X_{\text{planes}})$ in S^kV^*, the above formula also gives the Hilbert functions of the varieties X_{planes} in a uniform manner.

4.6.5 Zakology

The varieties $X_{\text{planes}} \subset \mathbb{P}V$ are important for the following geometric classification problem: Given a smooth variety $X \subset \mathbb{P}V$, one defines its dual variety $X^* \subset \mathbb{P}V^*$ to be the set of hyperplanes tangent to X. Usually the degree of X^* is quite large with respect to X and Zak has proposed the problem of classifying smooth varieties whose duals are of low degree [86].

Here the duals of $X_{\text{planes}} \subset \mathbb{P}V$ are of degree four. Moreover, they are the tangential varieties of the closed orbits in the dual projective space, i.e., we have $X^*_{\text{planes}} \simeq \tau(X_{\text{planes}})$.

The first variety in the series is $X_{\text{planes}}(-2/3) = v_3(\mathbb{P}^1)$ and the equation of its dual is the classical discriminant of a cubic polynomial. Using $\mathbb{A}$-valued variables, we write the equations for the duals in a uniform fashion, see [61]. In particular, we characterize the other quartics by their restriction to the preferred subspace. The second variety in the series is $X = \text{Seg}(\mathbb{P}^1 \times \mathbb{P}^1 \times \mathbb{P}^1)$ and its dual is Cayley's *hyperdeterminant*, see [35]. Our restriction result gives a new characterization of the hyperdeterminant.

4.6.6 Orbits

The orbit structure is also uniform.

Proposition 4.4. [61] *For each of the varieties* $X_{\text{planes}} = G_w(\mathbb{A}^3, \mathbb{A}^6) \subset \mathbb{P}V$ *there are exactly four orbits, the closures of which are ordered by inclusion:*

$$G_w(\mathbb{A}^3, \mathbb{A}^6) \subset \sigma_+(G_w(\mathbb{A}^3, \mathbb{A}^6)) \subset \tau(G_w(\mathbb{A}^3, \mathbb{A}^6)) \subset \mathbb{P}V.$$

The dimensions are respectively $3a + 3, 5a + 3$ *and* $6a + 6$.

Needless to say, the singular orbit closures have uniform desingularizations by Kempf's method.

Once again, the triality model plays a unifying role for geometric and representation geometric phenomena which at first glance seem sporadic. Another striking example of this role, which is currently under investigation, concerns nilpotent orbits in exceptional Lie algebras and the associated unipotent characters of exceptional Chevalley groups [66].

References

[1] Alexander, J., Hirschowitz, A.: Polynomial interpolation in several variables. J. Algebraic Geom. **4**(2) 201–222 (1995)

[2] Allison B.N.: A class of nonassociative algebras with involution containing the class of Jordan algebras. Math. Ann. **237**(2) 133–156 (1978)

[3] Allison, B.N., Faulkner, J.R.: Nonassociative coefficient algebras for Steinberg unitary Lie algebras. J. Algebra **161**(1) 1–19 (1993)

[4] Arnol'd, V.I.: Normal forms of functions near degenerate critical points, the Weyl groups of A_k, D_k, E_k and Lagrangian singularities. Funk. Anal. **6** 3–25 (1972).

[5] Baez, J.: The octonions. Bull. Amer. Math. Soc. (N.S.) **39**(2) 145–205 (2002)

[6] Bar–Natan, D.: On the Vassiliev knot invariants. Topology **34**(2) 423–472 (1995)

[7] Barton, C.H., Sudbery, A.: Magic squares of Lie algebras. Preprint arXiv: math.RA/0001083

[8] Beauville, A.: Fano contact manifolds and nilpotent orbits. Comment. Math. Helv. **73**(4) 566–583 (1998)
[9] Bourbaki, N.: *Groupes et algèbres de Lie.* (Hermann, Paris 1968)
[10] Cartan, E.: Le principe de dualité et la théorie des groupes simples et semi-simples. Bull. Sc. math. **49** 361–374 (1925)
[11] Cartan, E.: Sur les variétés de courbure constante d'un espace euclidien ou non-euclidien. Collected Works. **III**. (Gauthier–Villars, Paris 1955) pp. 321–432.
[12] Cartan, E.: Sur des familles remarquables d'hypersurfaces isoparamétriques dans les espaces sphériques. Math. Z. **45** 335–367 (1939)
[13] Carlini, E., Chipalkatti, J.: On Waring's problem for several algebraic forms. Preprint arXiv:math.ag/0112110.
[14] Cohen, A., de Man, R.: Computational evidence for Deligne's conjecture regarding exceptional groups. C.R.A.S. **322** 427–432 (1996)
[15] Chaput, P.E.: Severi varieties. Math. Z. **240**(2) 451–459 (2002)
[16] Chaput, P.E.: Scorza varieties and Jordan algebras. Indagationes Math. **14** 169–182 (2003)
[17] Clerc, J.L.: Special prehomogeneous vector spaces associated to F_4, E_6, E_7, E_8 and simple Jordan algebras of rank 3. J. Algebra **264** (1) 98–128 (2003)
[18] Cohen, A., Cooperstein, B.: Lie incidence systems from projective varieties. Proc. Am. Math. Soc. **126** 2095–2102 (1998)
[19] Cvitanović P.: Group theory for Feynman diagrams in non-Abelian gauge theories. Phys. Rev. **D14** 1536 (1976)
[20] Cvitanović P.: Negative dimensions and E_7 symmetry. Nucl. Phys. **B188** 373 (1981)
[21] Cvitanović P.: *Group Theory.* Available at www.nbi.dk/GroupTheory/, a monograph in preparation (1984–2002).
[22] Cohen, A.M., van Leeuwen, M.A., Lisser, B.: *LiE, A Package for Lie Group Computations.* (CAN, Amsterdam 1992)
[23] Dadok, J., Harvey, R.: A triality unification of Lie algebras. Preprint.
[24] Deligne, P.: La série exceptionnelle des groupes de Lie. C.R.A.S. **322** 321–326 (1996)
[25] Deligne, P., de Man, R.: The exceptional series of Lie groups. C.R.A.S. **323** 577–582 (1996).
[26] De Concini, C., Weyman, J.: A formula with nonnegative terms for the degree of the dual variaety of homogeneous space. Proc. AMS **125:1** 1–8 (1997)
[27] Dixmier, J.: *Algèbres enveloppantes.* (Gauthier-Villars, Paris 1974)
[28] Dynkin, E.B.: Maximal subgroups of the classical groups. Amer. Math. Soc. Trans. Series **6** 245–378 (1957)
[29] Ein, L., Shepherd–Barron, N.: Some special Cremona transformations. Amer. J. Math. **111**(5) 783–800 (1989).
[30] Faraut, J., Gindikin, S.: Pseudo-Hermitian symmetric spaces of tube type. In: Gindikin, S. (ed.): *Topics in Geometry, In Memory of Joseph D'Atri.* (Birkhäuser, Basel 1996), pp. 123–154
[31] Freudenthal, H.: Oktaven, Ausnahmegruppen und Oktavengeometrie (notes from 1951). Geom. Dedicata **19** 7–63 (1985)
[32] Freudenthal, H.: Lie groups in the foundations of geometry. Advances in Math. **1** 145–190 (1964)
[33] Garibaldi, R.S.: Structurable algebras and groups of type E_6 and E_7. J. Algebra **236** 651–691 (2001)

[34] Gray, A., Wolf, J.: Homogeneous spaces defined by Lie group automorphisms. J. Diff. Geometry **2** 77–114 and 115–159 (1968)

[35] Gelfand, I.M., Kapranov, M.M., Zelevinsky, A.V.: *Discriminants, Resultants, and Multidimensional Determinants.* Mathematics: Theory and Applications. (Birkhäuser, Basel 1994)

[36] Gelfand, I.M., Zelevinsky, A.V.: Representation models for classical groups and their higher symmetries. The mathematical heritage of Elie Cartan (Lyon, 1984). Numéro Hors Série (Astérisque 1985) pp. 117–128

[37] Griffiths, P.A., Harris, J.: Algebraic geometry and local differential geometry. Ann. scient. Ec. Norm. Sup. **12** 355–432 (1979)

[38] Gross, B., Wallach, R.: On quaternionic discrete series representations and their continuations. J. reine angew. Math. **481** 73–123 (196)

[39] Harvey, F.R.: *Spinors and Calibrations.* Perspectives in Math. vol. 9 (Academic Press, 1990)

[40] Hawkins: *Emergence of the Theory of Lie Groups. An Essay in the History of Mathematics 1869–1926*, Sources and Studies in the History of Mathematics and Physical Sciences. (Springer, New York 2000)

[41] Holweck, F.: Singularities of hyperplane sections of homogeneous varieties. In preparation

[42] Hong, J.: Fano manifolds with geometric structures modeled after homogeneous contact manifolds. Internat. J. Math. **11**(9) 1203–1230 (2000)

[43] Howe, R.: Transcending classical invariant theory. J. Am. Math. Soc. **2** 535–552 (1989)

[44] Hwang, J.–M., Mok, N.: Varieties of minimal rational tangents on uniruled projective manifolds. In: Schneider, M., Siu, Y.T. (eds): *Several Complex Variables* (Berkeley, CA, 1995–1996), 351–389, Math. Sci. Res. Inst. Publ. **37**, Cambridge University Press 1999

[45] Hwang, J.–M., Mok, N.: Uniruled projective manifolds with irreducible reductive G-structures. J. reine Angew. Math. **490** 55–64 (1997)

[46] Hwang, J.–M., Mok, N.: Rigidity of irreducible Hermitian symmetric spaces of the compact type under Kähler deformation. Invent. Math. **131**(2) 393–418 (1998)

[47] Iarrobino, A., Kanev, V.: *Power Sums, Gorenstein Algebras and Determinantal Loci.* LNM **1721** (Springer, Berlin, Heidelberg 1999)

[48] Kac, V.: *Infinite Dimensional Lie Algebras.* 3rd edn. (Cambridge University Press, Cambridge 1990)

[49] Kac, V.: Some remarks on nilpotent orbits. J. Algebra, **64**(1) 190–213 (1980)

[50] Kantor, I.: Certain generalizations of Jordan algebras. (Russian) Trudy Sem. Vektor. Tenzor. Anal. **16** 407–499 (1972)

[51] Kebekus, S., Peternell, Th., Sommese, A., Wisniewski, J.: Projective contact manifolds. Invent. Math. **142** 1–15 (2000)

[52] Kebekus, S.: Lines on contact manifolds. J. reine Angew. Math. **539** 167–177 (2001)

[53] Kempf, G.: On the collapsing of homogeneous bundles. Invent. Math. **37** 229–239 (1976)

[54] Knop, F.: Ein neuer Zusammenhang zwischen einfachen Gruppen und einfachen Singularitaten. Invent. Math. **90** 579–604 (1987)

[55] Kostant, B.: The principle of triality and a distinguished unitary representation of SO(4, 4). Differential geometrical methods in theoretical physics. Proc. 16th Int. Conf., Como/Italy 1987, NATO ASI Ser., Ser. C 250, 65–108 (1988)

[56] Kostant, B.: A cubic Dirac operator and the emergence of Euler number multiplets of representations for equal rank subgroups. Duke Math. J. **100**(3) 447–501 (1999)
[57] Landsberg, J.M.: On degenerate secant and tangential varieties and local differential geometry. Duke Math. J. **85** 605–634 (1996)
[58] Landsberg, J.M.: On the infinitesimal rigidity of homogeneous varieties. Compositio Math. **118** 189–201 (1999)
[59] Landsberg, J.M., Manivel, L.: On the projective geometry of homogeneous varieties. Commentari Math. Helv. **78** 65–100 (2003)
[60] Landsberg, J.M., Manivel, L.: Classification of simple Lie algebras via projective geometry. Selecta Mathematica **8** 137–159 (2002)
[61] Landsberg, J.M., Manivel, L.: The projective geometry of Freudenthal's magic square. J. Algebra **239** 477–512 (2001)
[62] Landsberg, J.M., Manivel, L.: Triality, exceptional Lie algebras, and Deligne dimension formulas. Advances Math. **171** (1) 59–85 (2002)
[63] Landsberg, J.M., Manivel, L.: Series of Lie algebras. Preprint
[64] Landsberg, J.M., Manivel, L.: Legendrian and contact Fano varieties. In preparation
[65] Landsberg, J.M., Manivel, L., Westbury, B.: Sextonions and the dimension formulas of Cvitanović, Deligne and Vogel. In preparation
[66] Landsberg, J.M., Manivel, L., Westbury, B.: Series of nilpotent orbits. To appear in Experimental Math.
[67] Lascoux A.: Degree of the dual of a Grassmann variety. Comm. Algebra **9** 1215–1225 (1981)
[68] LeBrun, C.: Fano manifolds, contact structures, and quaternionic geometry. Internat. J. Math. **6** 419–437 (1995)
[69] LeBrun, C., Salamon, S.: Strong rigidity of positive quaternion-Kahler manifolds. Invent. Math. **118** 109–132 (1994)
[70] Loos O.: *Symmetric spaces I and II.* (Benjamin, 1969)
[71] Mukai, Sh.: Simple Lie algebra and Legendre variety. Nagoya Sūri Forum **3** 1–12 (1996) (in Japanese). Revised translation into English: http://www.math.nagoya-u.ac.jp/~mukai/warwick15.pdf (1998) pp. 1–14
[72] Onishchik, A.L., Vinberg, E.B.: *Lie Groups and Lie Algebras, III. Structure of Lie Groups and Lie Algebras.* Encyclopaedia of Mathematical Sciences 41. (Springer, Berlin, Heidelberg 1994)
[73] Proctor, I.: Odd symplectic groups. Invent. Math. **92** 307–332 (1988)
[74] Postnikov, M.: *Groupes et algèbres de Lie* (Editions Mir, Moscou 1985)
[75] Ranestad, K., Schreyer, F.–O.: Varieties of sums of powers. J. reine Angew. Math. **525** 1455–1468 (2000)
[76] Rosenfeld, B.: *Geometry of Lie Groups.* Mathematics and its Applications, vol. 393 (Kluwer 1997)
[77] Rubenthaler, H.: *Les paires duales dans les algèbres de Lie réductives.* Astérisque **219** (1994)
[78] Tevelev, E.: *Projectively Dual Varieties.* Preprint math.AG/0112028
[79] Tits J.: Algèbres alternatives, algèbres de Jordan et algèbres de Lie exceptionnelles I. Construction. Indag. Math. **28** 223–237 (1966)
[80] Tits, J.: Les groupes de Lie exceptionnels et leur interprétation géométrique. Bull. Soc. Math. Belg. **8** 48–81 (1956)
[81] Tits, J.: Groupes semi-simples complexes et géométrie projective. Séminaire Bourbaki **7** (1954/1955), exposé 112, 11 pages.

[82] Vogel, P.: Algebraic structures on modules of diagrams. Preprint 1996
[83] Vogel, P.: The universal Lie algebra. Preprint 1999
[84] Wolf, J.: Complex homogeneous contact manifolds and quaternionic symmetric spaces. Journal of Math. and Mech. **14** 1033–1047 (1965)
[85] Zak, F.: *Tangents and Secants of Algebraic Varieties.* Translations of Math. Monographs **127**, AMS 1993
[86] Zak, F.: Some properties of dual varieties and their applications in projective geometry. In: *Algebraic Geometry*, Proceedings, Chicago 1989, Springer LNM 1479, 273–280

Geometric Realization Of T-shaped Root Systems and Counterexamples To Hilbert's Fourteenth Problem

Shigeru Mukai*

Research Institute for Mathematical Sciences, Kyoto University, Kyoto 606-8502, Japan, mukai@kurims.kyoto-u.ac.jp

Abstract. Generalizing a result of Dolgachev, we realize the root system $T_{p,q,r}$ in the cohomology group of a certain rational variety of Picard number $p+q+r-1$. As an application we show that the invariant ring of a tensor product of the actions of Nagata type is infinitely generated if the Weyl group of the corresponding root system $T_{p,q,r}$ is indefinite. In this sense this article is a continuation of [4].

1 The Dynkin Diagram $T_{p,q,r}$

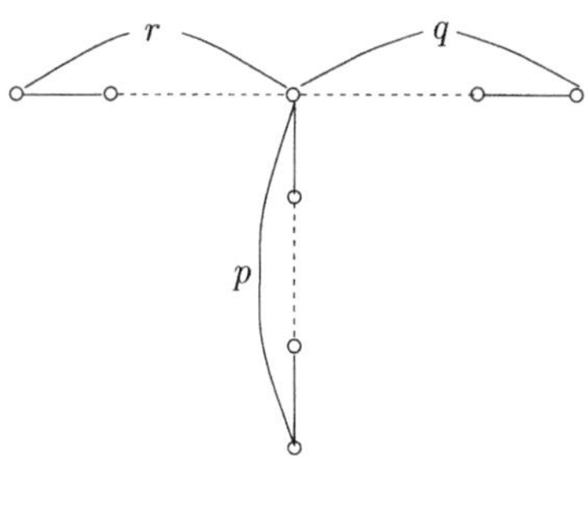

defines a lattice $L_{p,q,r}$ of rank $p+q+r-2$. The set of vertices α_i's is its basis as a free $\mathbf{Z}$-module. The bilinear form is defined to be $(\alpha_i.\alpha_j) = -2, 0$ or 1 according as $i=j$, α_i and α_j are disjoint or joined by an edge. It is known that the root system $T_{p,q,r}$ is of finite type, affine or indefinite according as $1/p+1/q+1/r$ is $>1, =1$ or <1 ([2], Ex. 4.2).

In the case $p=2$, Dolgachev, [1], realizes this root system $T_{2,q,r}$ in the cohomology group of the blow-up $\mathrm{Bl}_{q+r\,\mathrm{pts}}\,\mathbf{P}^{r-1}$ of the $(r-1)$-dimensional

* Supported in part by the JSPS Grant-in-Aid for Exploratory Research 12874002.

projective space $\mathbf{P}^{r-1}$ at $q+r$ points in general position. This is a generalization of the classical discovery that the configuration of the 27 lines on a nonsingular cubic surface $S \subset \mathbf{P}^3$, which is the blow-up of $\mathbf{P}^2$ at six points, has a symmetry of the Weyl group of $E_6 = T_{2,3,3}$ (cf. [3]). The moduli of cubic surfaces is a quotient of an open set of $\mathbf{P}^2 \times \mathbf{P}^2$ by $W(E_6)$ (cf. [1]) and the cohomology group $H^2(S, \mathbf{Z})$ has a monodromy action of $W(E_6)$.

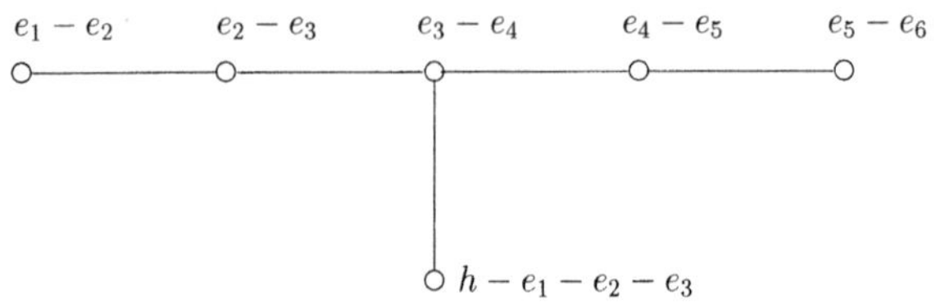

The reflections by $e_i - e_{i+1}$, $1 \le i \le 5$, generate all permutations of the 6 points which are the centers of blowing up and that by $h - e_1 - e_2 - e_3$ corresponds to the quadratic Cremona transformation

$$\mathbf{P}^2 \cdots \to \mathbf{P}^2, \qquad (x_1 : x_2 : x_3) \mapsto (1/x_1 : 1/x_2 : 1/x_3).$$

It is natural to extend the result of Dolgachev to all diagrams of T-shape. The answer is simple: just generalize $\mathbf{P}^{r-1}$ to the product $(\mathbf{P}^{r-1})^{p-1}$ of its $p-1$ copies. Let X be a blow-up of the product $(\mathbf{P}^{r-1})^{p-1}$ at $q+r$ points in general position. The second cohomology group $H^2(X, \mathbf{Z})$, or equivalently $\operatorname{Pic} X$, is a free $\mathbf{Z}$-module of rank $p+q+r-1$ and has a basis consisting of

$$h_i, \quad 1 \le i \le p-1, \quad \text{and} \quad e_j, \quad 1 \le j \le q+r, \tag{1}$$

where h_i is the pull-back of the hyperplane class on the ith factor of $(\mathbf{P}^{r-1})^{p-1}$ and e_j the class of the exceptional divisor over the jth center of blowing up. We refer (1) as the *tautological basis.*

Theorem 1. *The root system $T_{p,q,r}$ is realized in the orthogonal complement L of the anti-canonical class $c_1(X)$ in the second cohomology group $H^2(X, \mathbf{Z})$ endowed with a certain symmetric bilinear form. (See Section 3.) Moreover, for each element $w : H^2(X, \mathbf{Z}) \longrightarrow H^2(X, \mathbf{Z})$ of the Weyl group $W(T_{p,q,r})$, there is a* strong *birational map $\Psi_w : X_w \cdots \to X$ of a a blow-up X_w of $(\mathbf{P}^{r-1})^{p-1}$ at $q+r$ points such that the pull-back of a tautological basis of X_w coincides with the transformation of that of X by w.*

A birational map is called *strong* if it is an isomorphism in codimension one.

In the special case $q = 1$, X has a birational action of $W(T_{p,1,r}) = W(A_{p+r-1})$, which is the symmetric group of degree $p+r$. In fact X is a GIT quotient of the Grassmannian variety $G(p, p+r)$ by the maximal torus

$T \simeq (\mathbf{C}^*)^{p+r-1}$ of its automorphism group $G \simeq \mathrm{PGL}(p+r)$. Hence the Weyl group of G acts on X birationally. X is a compactification of the configuration space of ordered $p+r$ points on $\mathbf{P}^{r-1}$.

Remark. The isomorphism between $G(p, p+r)$ and $G(r, p+r)$ induces a strong birational map between $\mathrm{Bl}_{r+1\,\mathrm{pts}}(\mathbf{P}^{r-1})^{p-1}$ and $\mathrm{Bl}_{p+1\,\mathrm{pts}}(\mathbf{P}^{p-1})^{r-1}$ and hence that between $\mathrm{Bl}_{q+r\,\mathrm{pts}}(\mathbf{P}^{r-1})^{p-1}$ and $\mathrm{Bl}_{p+q\,\mathrm{pts}}(\mathbf{P}^{p-1})^{r-1}$. For example $\mathrm{Bl}_{q+r\,\mathrm{pts}}\mathbf{P}^{r-1}$ is strongly birationally equivalent to $\mathrm{Bl}_{2+q\,\mathrm{pts}}(\mathbf{P}^1)^{r-1}$.

2 Generalized Nagata Action

Our interest in Theorem 1 comes from Nagata's counterexample to Hilbert's fourteenth problem also. Let

$$\begin{aligned} (t_1, \ldots, t_n) \in \mathbf{C}^n \ \downarrow \mathbf{C}[x_1, \ldots, x_n, y_1, \ldots, y_n] =: S, \\ \begin{cases} x_i \mapsto x_i \\ y_i \mapsto y_i + t_i x_i \end{cases}, \quad 1 \le i < n, \end{aligned} \tag{2}$$

be the standard unipotent action of $\mathbf{C}^n$, or the additive algebraic group $\mathbf{G}_a^n$ more precisely, on the polynomial ring S of $2n$ variables and $G \subset \mathbf{C}^n$ a general linear subspace. In [5], Nagata studied the invariant ring S^G of the subaction of G. The key fact is that the ring S^G is isomorphic to the total coordinate ring

$$\begin{aligned} \mathcal{TC}(X) &:= \bigoplus_{a, b_1, \ldots, b_n \in \mathbf{Z}} H^0(X, \mathcal{O}_X(ah - b_1 e_1 - \cdots - b_n e_n)) \\ &\simeq \bigoplus_{L \in \mathrm{Pic}\, X} H^0(X, L) \end{aligned} \tag{3}$$

of the variety $X = \mathrm{Bl}_{n\,\mathrm{pts}}\mathbf{P}^{r-1}$, where r is the codimension of $G \subset \mathbf{C}^n$.

In [4], we pay attention to the support of this graded ring $\mathcal{TC}(X)$, which is the semi-group $\mathrm{Eff}\, X \subset \mathrm{Pic}\, X$ of effective divisor classes on X. A divisor $D \subset X$ is called a (-1)-divisor if there is a strong birational map $X \cdots \to X'$ such that the image of D can be contracted to a smooth point. Obviously the linear equivalence class of a (-1)-divisor is indispensable as generator of $\mathrm{Eff}\, X$.

Assume that the inequality

$$\frac{1}{2} + \frac{1}{n-r} + \frac{1}{r} \le 1. \tag{4}$$

holds. Then the Weyl group $W(T_{2,n-r,r})$ of X is infinite and infinitely many (-1)-divisors on X are obtained as its orbit. Hence $\mathrm{Eff}\, X$ and $\mathcal{TC}(X)$ are not finitely generated. This is an outline of the main argument of [4].

In order to obtain more examples, we take $p-1$ actions

$$G_i \ \downarrow \mathbf{C}[x_1, \ldots, x_n, y_1, \ldots, y_n] =: S, \qquad G_i \subset \mathbf{C}^n,\ 1 \le i \le p-1$$

of Nagata type on the same polynomial ring S and take their tensor product

$$G = \bigoplus_{i=1}^{p-1} G_i \downarrow S \otimes_{\mathbf{C}[x]} \cdots \otimes_{\mathbf{C}[x]} S =: \tilde{S} \tag{5}$$

over $\mathbf{C}[x_1, \ldots, x_n]$. $\tilde{S}$ is a polynomial ring of pn variables.

Theorem 2. *The invariant ring $\tilde{S}^G$ of the above action (5) is isomorphic to the total coordinate ring*

$$\bigoplus_{a_1,\ldots,a_{p-1},b_1,\ldots,b_n \in \mathbf{Z}} H^0(X, \mathcal{O}_X(a_1h_1 + \cdots + a_{p-1}h_{p-1} - b_1e_1 - \cdots - b_ne_n))$$

of the blow-up X of the product $\mathbf{P}^{r_1-1} \times \cdots \times \mathbf{P}^{r_{p-1}-1}$ of $p-1$ projective spaces at n points, where h_i is the pull-back of the hyperplane class of $\mathbf{P}^{r_i-1}$.

We can localize the action (5) by $x_1, \ldots, x_n$ since they are G-invariant. Then the additive group G acts on

$$\begin{aligned}\tilde{S}[x_1^{-1}, \ldots, x_n^{-1}] &= S[x_1^{-1}, \ldots, x_n^{-1}] \otimes_{\mathbf{C}[x,x^{-1}]} \cdots \otimes_{\mathbf{C}[x,x^{-1}]} S[x_1^{-1}, \ldots, x_n^{-1}] \\ &= \mathbf{C}[x_1^{\pm 1}, \ldots, x_n^{\pm 1}, y_1/x_1 \ldots, y_n/x_n] \otimes_{\mathbf{C}[x,x^{-1}]} \cdots \\ &\quad \cdots \otimes_{\mathbf{C}[x,x^{-1}]} \mathbf{C}[x_1^{\pm 1}, \ldots, x_n^{\pm 1}, y_1/x_1 \ldots, y_n/x_n].\end{aligned}$$

Since $(t_1, \ldots, t_n) \in G_i$ acts by the translation $y_j/x_j \mapsto y_j/x_j + t_j$, $1 \le j \le n$, the invariant ring $\tilde{S}[x_1^{-1}, \ldots, x_n^{-1}]^G$ is a polynomial ring of $r_1 + \cdots + r_{p-1}$ variables. The rest of the proof is similar to that of the case $p = 2$ in [4] and we omit it.

If r_i's are all the same, then we can apply Theorem 1 and obtain the following by the same reason, that is, X has infinitely many (-1)-divisors.

Theorem 3. *The invariant ring S^G of (5) is not finitely generated if $G_i \subset \mathbf{C}^n$ are general subspaces of codimension r and if the inequality*

$$\frac{1}{p} + \frac{1}{n-r} + \frac{1}{r} \le 1$$

holds.

In the case $p = 2$ there are three cases where the diagram is of affine type:

r	q	X	diagram	$\dim G = q$
3	6	$\mathrm{Bl}_{9\,\mathrm{pts}}\,\mathbf{P}^2$	$T_{2,[6],3}$	6
4	4	$\mathrm{Bl}_{8\,\mathrm{pts}}\,\mathbf{P}^3$	$T_{2,[4],4}$	4
6	3	$\mathrm{Bl}_{9\,\mathrm{pts}}\,\mathbf{P}^5$	$T_{2,[3],6}$	3

Allowing $p \ge 3$, we obtain three new ones with $p \le r$. (See Remark at the end of Section 1.)

p	r	q	X	diagram	$\dim G = (p-1)q$
3	3	3	$\mathrm{Bl}_{6\,\mathrm{pts}}\,\mathbf{P}^2 \times \mathbf{P}^2$	$T_{3,[3],3}$	6
3	6	2	$\mathrm{Bl}_{8\,\mathrm{pts}}\,\mathbf{P}^5 \times \mathbf{P}^5$	$T_{3,[2],6}$	4
4	4	2	$\mathrm{Bl}_{6\,\mathrm{pts}}\,\mathbf{P}^3 \times \mathbf{P}^3 \times \mathbf{P}^3$	$T_{4,[2],4}$	6

$T_{p,[q],r}$ is the diagram $T_{p,q,r}$ plus an extra vertex, which is defined in the next section.

3 Proof Of Theorem 1

Let X be as in the theorem. The anti-canonical class $c_1(X)$ is equal to

$$r(h_1 + \cdots + h_{p-1}) - (r-2)(e_1 + \cdots + e_{q+r}).$$

We define an integral symmetric bilinear form on $H^2(X, \mathbf{Z})$ as follows:

1. h_i and e_j are orthogonal for every $1 \le i \le p-1$ and $1 \le j \le q+r$,
2. $$(e_i.e_j) = \begin{cases} -1 & i = j, \\ 0 & \text{otherwise} \end{cases} \quad \text{and} \quad (h_i.h_j) = \begin{cases} r-2 & i=j, \\ r-1 & \text{otherwise.} \end{cases}$$

We take a new $\mathbf{Z}$-basis of $H^2(X, \mathbf{Z})$ consisting of

1. $h_1 - e_1 - \cdots - e_r$,
2. $h_i - h_{i+1}$, $1 \le i \le p-2$,
3. $e_j - e_{j+1}$, $1 \le j \le q+r-1$, and
4. e_{q+r}.

Then $H^2(X, \mathbf{Z})$ becomes a root system with the diagram on the next page which is denoted by $T_{p,[q],r}$. (See [2], Section 5.11 also for $T_{2,[q],3}$.)

The $p+q+r-2$ classes except for e_n are of length -2 and form a basis of the orthogonal complement L of $c_1(X)$ in $H^2(X, \mathbf{Z})$. Hence L is isomorphic to the root lattice $L_{p,q,r}$.

In order to show the latter half of Theorem 1, it is enough to check it for the simple reflections. This is obvious for $e_j - e_{j+1}$'s and $h_i - h_{i+1}$'s since they correspond to transpositions of a pair of centers and a pair of factors. The reflection with respect to $h_1 - e_1 - \cdots - e_r$ transforms the tautological basis (1) of $H^2(X, \mathbf{Z})$ as follows:

$$\begin{cases} h_1 \mapsto (r-1)h_1 - (r-2)\sum_{j=1}^r e_j, & \\ h_i \mapsto (r-1)h_1 + h_i - (r-1)\sum_{j=1}^r e_j, & 2 \le i \le p-1, \\ e_j \mapsto h_1 - e_1 - \cdots - \check{e_j} - \cdots - e_r, & 1 \le j \le r, \\ e_j \mapsto e_j, & r+1 \le j \le q+r. \end{cases} \tag{6}$$

Let $P = \{p_1, \ldots, p_r\}$ be a set of r distinct points on $(\mathbf{P}^{r-1})^{p-1}$. P is *non-degenerate* if ith components $p_1^{(i)}, \ldots, p_r^{(i)}$ spans $\mathbf{P}^{r-1}$ for every $1 \le i \le p-1$. If

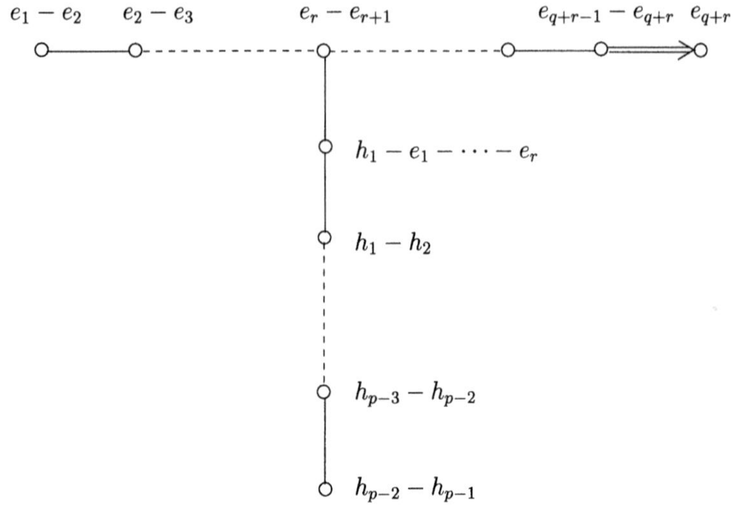

P is non-degenerate we can choose homogeneous coordinates of $\mathbf{P}^{r-1}$'s such that P is the image of the r coordinate points by the diagonal morphism $\Delta : \mathbf{P}^{r-1} \hookrightarrow (\mathbf{P}^{r-1})^{p-1}$.

Lemma. *Let $P = \{p_1, \ldots, p_r\}$ and $Q = \{q_1, \ldots, q_r\}$ be non-degenerate sets of r points of $(\mathbf{P}^{r-1})^{p-1}$ and X_P and X_Q be the blow-ups with center P and Q, respectively. Then there exists a strong birational map*

$$\Psi = \Psi_{P,Q} : X_P \cdots \longrightarrow X_Q$$

such that

$$\begin{cases} \Psi^* h_1' = (r-1)h_1 - (r-2)\sum_{j=1}^r e_j, \\ \Psi^* h_i' = (r-1)h_1 + h_i - (r-1)\sum_{j=1}^r e_j, \quad 2 \le i \le p-1, \\ \Psi^* e_j' = h_1 - e_1 - \cdots - \check{e}_j - \cdots - e_r, \quad 1 \le j \le r, \end{cases} \tag{7}$$

where $\{h_i, e_j\}$ and $\{h_i', e_j'\}$ are tautological bases of Pic X_P *and* Pic X_Q, *respectively.*

Proof. We may assume that both P and Q are the image of the coordinate points by the diagonal morphism Δ. Consider the (toric) Cremona transformation

$$\bar{\Psi} : \mathbf{P}^{r-1} \times \mathbf{P}^{r-1} \times \cdots \times \mathbf{P}^{r-1} \cdots \to \mathbf{P}^{r-1} \times \mathbf{P}^{r-1} \times \cdots \times \mathbf{P}^{r-1},$$
$$((x_1 : x_2 : \ldots : x_r), (y_1 : y_2 : \ldots : y_r), \ldots, (z_1 : z_2 : \ldots : z_r)) \mapsto$$
$$((\frac{1}{x_1} : \frac{1}{x_2} : \ldots : \frac{1}{x_r}), (\frac{y_1}{x_1} : \frac{y_2}{x_2} : \ldots : \frac{y_r}{x_r}), \ldots, (\frac{z_1}{x_1} : \frac{z_1}{x_2} : \ldots : \frac{z_1}{x_r})).$$

Its indeterminacy locus is the union $\cup_{1\le i<j\le r} H_i \cap H_j$ of the intersection of all pairs of H_i's, where $H_1, \ldots, H_r$ are the pull-backs of coordinate hyperplanes of the first factor. The map $\bar{\Psi}$ is an isomorphism off the union $\cup_{1\le i\le r} H_i$ and $\bar{\Psi}^2$ is the identity. By blowing-up, we obtain the commutative diagram:

$$\begin{array}{ccc} & \Psi & \\ X_P & \cdots\longrightarrow & X_Q \\ \downarrow & & \downarrow \\ \mathbf{P}^{r-1} & \cdots\cdots\rightarrow & \mathbf{P}^{r-1} \\ & \bar{\Psi} & \end{array}.$$

Let $(X_2, \ldots, X_r), (Y_2, \ldots, Y_r), \ldots, (Z_2, \ldots, Z_r)$ be the standard inhomogeneous coordinate of $(\mathbf{P}^{r-1})^{p-1}$ arround $p_1 = \Delta(1 : 0 : \ldots : 0)$. Then the rational map $X_P \cdots \rightarrow (\mathbf{P}^{r-1})^{p-1}$ is given by

$$E_1 \ni (X_2 : \ldots : X_r : Y_2 : \ldots : Y_r : \ldots : Z_2 : \ldots : Z_r) \mapsto$$
$$\left(\left(0 : \frac{1}{X_2} : \ldots : \frac{1}{X_r}\right), \left(1 : \frac{Y_2}{X_2} : \ldots : \frac{Y_r}{X_r}\right), \ldots, \left(1 : \frac{Z_2}{X_2} : \ldots : \frac{Z_r}{X_r}\right)\right).$$

over the exceptional divisor E_1 over p_1. Hence Ψ resrtricted to E_1 is a birational map onto (the strict transform of) the divisor H_1. Therefore, Ψ is an isomorphism in codimension one. (7) is obvious. □

References

[1] Dolgachev, I.: Weyl groups and Cremona transformations. Proc. Symp. Pure Math. **40** 283–194 (1978)

[2] Kac, V.: *Infinite Dimensional Lie Algebras*, 3rd ed. (Cambridge Univ. Press, 1990)

[3] Manin, Yu.I.: *Cubic forms: Algebra, Geometry, Arithmetic* (North-Holland Publ. Comp. 1974)

[4] Mukai, S.: Counterexample to Hilbert's fourteenth problem for three dimensional additive groups. RIMS preprint # 1343, Kyoto, 2001.

[5] Nagata, M.: On the fourteenth problem of Hilbert, Proc. Internat. Cong. Math., Edinburgh, 1958, pp. 459–462 (Cambridge Univ. Press, 1960)

Self-dual Projective Algebraic Varieties Associated With Symmetric Spaces

Vladimir L. Popov[1⋆] and Evgueni A. Tevelev[2⋆⋆]

[1] Steklov Mathematical Institute, Russian Academy of Sciences, Gubkina 8, Moscow 117966, Russia. popovvl@orc.ru
[2] Department of Mathematics, The University of Texas, Austin, Texas 78712, USA. tevelev@mail.ma.utexas.edu

Abstract. We discover a class of projective self-dual algebraic varieties. Namely, we consider actions of isotropy groups of complex symmetric spaces on the projectivized nilpotent varieties of isotropy modules. For them, we classify all orbit closures X such that $X = \check{X}$ where $\check{X}$ is the projective dual of X. We give algebraic criteria of projective self-duality for the considered orbit closures.

1 Introduction

Under different guises dual varieties of projective algebraic varieties have been considered in various branches of mathematics for over a hundred years. In fact, the dual variety is the generalization to algebraic geometry of the Legendre transform in classical mechanics, and the Biduality Theorem essentially rephrases the duality between the Lagrange and Hamilton–Jacobi approaches in classical mechanics.

Let X be an n-dimensional projective subvariety of an N-dimensional projective space $\mathbf{P}$, and let $\check{X}$ be the dual variety of X in the dual projective space $\check{\mathbf{P}}$. Since various kinds of geometrically meaningful unusual behavior of hyperplane sections are manifested more explicitly in terms of dual varieties, it makes sense to consider their natural invariants. The simplest invariant of $\check{X}$ is its dimension $\check{n}$. "Typically", $\check{n} = N - 1$, i.e., $\check{X}$ is a hypersurface. The deviation from the "typical" behavior admits a geometric interpretation: if $\check{X}$ is not a hypersurface and, say, $\operatorname{codim} \check{X} = s + 1$, then X is uniruled by s-planes.

Assume that X is a smooth variety not contained in a hyperplane. Then $n \leqslant \check{n}$, by [Z, Chapter 1]. If the extremal case $n = \check{n}$ holds and $n = \check{n} \leqslant 2N/3$, then, by [Ei1], [Ei2], such X are classified by the following list:

⋆ Partly supported by ESI, Vienna, Austria and ETH, Zürich, Switzerland.
⋆⋆ Partly supported by ESI, Vienna, Austria.

(i) hypersurfaces,
(ii) $\mathbf{P}^1 \times \mathbf{P}^{n-1}$ embedded in $\mathbf{P}^{2n-1}$ by the Segre embedding,
(iii) the Grassmannian of lines in $\mathbf{P}^4$ embedded in $\mathbf{P}^9$ by the Plücker map, or
(iv) the 10-dimensional spinor variety of 4-dimensional linear subspaces on a nonsingular 8-dimensional quadric in $\mathbf{P}^{15}$.

According to Hartshorne's famous conjecture, if $n > 2N/3$, then X is a complete intersection, and hence, by [Ei2], $\check{X}$ is a hypersurface. Therefore it is plausible that the above-stated list contains every smooth X such that $n = \check{n}$. Furthermore, in cases (ii), (iii) and (iv), the variety X is *self-dual* in the sense that, as an embedded variety, $\check{X}$ is isomorphic to X. In case (i), it is self-dual if and only if it is a quadric. Thus, modulo Hartshorne's conjecture, the above-stated list gives the complete classification of all *smooth* self-dual projective algebraic varieties. In particular it shows that there are not many of them.

In [P2], it was found a method for constructing many *nonsmooth* self-dual projective algebraic varieties. It is related to algebraic transformation groups theory: the self-dual projective algebraic varieties appearing in this construction are certain projectivized orbit closures of some linear actions of reductive algebraic groups. We recall this construction.

Let G be a connected reductive algebraic group, let V be a finite dimensional algebraic G-module and let B be a nondegenerate symmetric G-invariant bilinear form on V. We assume that the ground field is $\mathbb{C}$. For a subset S of V, put $S^{\perp} := \{v \in V \mid B(v, s) = 0 \ \forall\ s \in S\}$. We identify V and V^* by means of B and denote by $\mathbf{P}$ the associated projective space of $V = V^*$. Thereby the projective dual $\check{X}$ of a Zariski closed irreducible subset X of $\mathbf{P}$ is a Zariski closed subset of $\mathbf{P}$ as well. Let $\mathcal{N}(V)$ be the *null-cone* of V, i.e.,

$$\mathcal{N}(V) := \{v \in V \mid 0 \in \overline{G \cdot v}\},$$

where bar stands for Zariski closure in V.

Theorem 1. ([P2, Theorem 1]) *Assume that there are only finitely many G-orbits in $\mathcal{N}(V)$. Let $v \in \mathcal{N}(V)$ be a nonzero vector and let $X := \mathbf{P}(\overline{G \cdot v}) \subseteq \mathbf{P}$ be the projectivization of its orbit closure. Then the following properties are equivalent:*

(i) $X = \check{X}$.
(ii) $(\mathrm{Lie}(G) \cdot v)^{\perp} \subseteq \mathcal{N}(V)$.

Among the modules covered by this method, there are two naturally allocated classes, namely, that of the adjoint modules and that of the isotropy modules of symmetric spaces. In fact the first class is a subclass of the second one; this subclass has especially nice geometric properties and is studied in more details. Projective self-dual algebraic varieties associated with the adjoint modules by means of Theorem 1 were explicitly classified and studied in

[P2]. In particular, according to [P2], these varieties are precisely the projectivized orbit closures of nilpotent elements in the Lie algebra of G that are *distinguished* in the sense of Bala and Carter (see below Theorem 2). This purely algebraic notion plays an important role in the Bala–Carter classification of nilpotent elements, [BaCa].

For the isotropy modules of symmetric spaces, in [P2] it was introduced the notion of (-1)-*distinguished element* of a $\mathbb{Z}_2$-graded semisimple Lie algebra and shown that the projective self-dual algebraic varieties associated with such modules by means of Theorem 1 are precisely the projectivized orbit closures of (-1)-distinguished elements. Thereby classification of such varieties was reduced to the problem of classifying (-1)-distinguished elements. In [P2], it was announced that the latter problem will be addressed in a separate publication.

The goal of the present paper is to give the announced classification: here we explicitly classify (-1)-distinguished elements of all $\mathbb{Z}_2$-graded complex semisimple Lie algebras; this yields the classification of all projective self-dual algebraic varieties associated with the isotropy modules of symmetric spaces by means of the above-stated construction from [P2]. In the last section we briefly discuss some geometric properties of these varieties.

Notice that there are examples of singular projective self-dual algebraic varieties constructed in a different way. For instance, the Kummer surface in $\mathbf{P}^3$, see [GH], or the Coble quartic hypersurface in $\mathbf{P}^7$, see [Pa], are projective self-dual. A series of examples is given by means of "Pyasetskii pairing", [T] (e.g., projectivization of the cone of $n \times m$-matrices of rank $\leqslant n/2$ is projective self-dual for $n \leqslant m$, n even).

Given a projective variety, in general it may be difficult to explicitly identify its dual variety. So our classification contributes to the problem of finding projective varieties X for which $\check{X}$ can be explicitly identified. Another application concerns the problem of explicit describing the projective dual varieties of the projectivized nilpotent orbit closures in the isotropy modules of symmetric spaces: our classification yields its solution for all (-1)-distinguished orbits. To the best of our knowledge, at this writing this problem is largely open; even finding the dimensions of these projective dual varieties would be interesting (for the minimal nilpotent orbit closures in the adjoint modules, a solution was found in [KM]; cf. [Sn] for a short conceptual proof).

2 Main Reductions

Let $\mathfrak{g}$ be a semisimple complex Lie algebra, let G be the adjoint group of $\mathfrak{g}$, and let $\theta \in \operatorname{Aut} \mathfrak{g}$ be an element of order 2. We set

$$\mathfrak{k} := \{x \in \mathfrak{g} \mid \theta(x) = x\}, \quad \mathfrak{p} := \{x \in \mathfrak{g} \mid \theta(x) = -x\}. \tag{1}$$

Then $\mathfrak{k}$ and $\mathfrak{p} \neq 0$, the subalgebra $\mathfrak{k}$ of $\mathfrak{g}$ is reductive, and $\mathfrak{g} = \mathfrak{k} \oplus \mathfrak{p}$ is a $\mathbb{Z}_2$-grading of the Lie algebra $\mathfrak{g}$, cf., e.g., [OV]. Denote by G the adjoint group of $\mathfrak{g}$. Let K be the connected reductive subgroup of G with the Lie algebra $\mathfrak{k}$; it is the adjoint group of $\mathfrak{k}$.

Consider the adjoint G-module $\mathfrak{g}$. Its null-cone $\mathcal{N}(\mathfrak{g})$ is the cone of all nilpotent elements of $\mathfrak{g}$; it contains only finitely many G-orbits, [D], [K1]. The space $\mathfrak{p}$ is K-stable, and we have

$$\mathcal{N}(\mathfrak{p}) = \mathcal{N}(\mathfrak{g}) \cap \mathfrak{p}.$$

By [KR], there are only finitely many K-orbits in $\mathcal{N}(\mathfrak{p})$ and

$$2\dim_{\mathbb{C}} K \cdot x = \dim_{\mathbb{C}} G \cdot x \text{ for any } x \in \mathcal{N}(\mathfrak{p}). \tag{2}$$

The Killing form $B_{\mathfrak{g}}$ of $\mathfrak{g}$ (resp., its restriction $B_{\mathfrak{g}}|_{\mathfrak{p}}$ to $\mathfrak{p}$) is symmetric nondegenerate and G-invariant (resp., K-invariant). We identify the linear spaces $\mathfrak{g}$ and $\mathfrak{g}^*$ (resp., $\mathfrak{p}$ and $\mathfrak{p}^*$) by means of $B_{\mathfrak{g}}$ (resp., $B_{\mathfrak{g}}|_{\mathfrak{p}}$).

Recall from [BaCa], [Ca] that a nilpotent element $x \in \mathcal{N}(\mathfrak{g})$ or its G-orbit $G \cdot x$ is called *distinguished* if the centralizer $\mathfrak{z}_{\mathfrak{g}}(x)$ of x in $\mathfrak{g}$ contains no nonzero semisimple elements. According to [P2], distinguished orbits admit the following geometric characterization:

Theorem 2. ([P2, Theorem 2]) *Let $\mathcal{O}$ be a nonzero G-orbit in $\mathcal{N}(\mathfrak{g})$ and $X := \mathbf{P}(\overline{\mathcal{O}})$. The following properties are equivalent:*

(i) $X = \check{X}$,
(ii) $\mathcal{O}$ *is distinguished.*

Hence the classification of distinguished elements obtained in [BaCa] yields the classification of projective self-dual orbit closures in $\mathbf{P}(\mathcal{N}(\mathfrak{g}))$. This, in turn, helps studying geometric properties of these projective self-dual projective varieties, [P2].

There is a counterpart of Theorem 2 for the action of K on $\mathfrak{p}$. Namely, if $x \in \mathfrak{p}$, then $\mathfrak{z}_{\mathfrak{g}}(x)$ is θ-stable, hence it is a graded Lie subalgebra of $\mathfrak{g}$,

$$\mathfrak{z}_{\mathfrak{g}}(x) = \mathfrak{z}_{\mathfrak{k}}(x) \oplus \mathfrak{z}_{\mathfrak{p}}(x), \quad \text{where } \mathfrak{z}_{\mathfrak{k}}(x) := \mathfrak{z}_{\mathfrak{g}}(x) \cap \mathfrak{k}, \quad \mathfrak{z}_{\mathfrak{p}}(x) := \mathfrak{z}_{\mathfrak{g}}(x) \cap \mathfrak{p}.$$

Definition 1. ([P2]) An element $x \in \mathcal{N}(\mathfrak{p})$ and its K-orbit are called (-1)-*distinguished* if $\mathfrak{z}_{\mathfrak{p}}(x)$ contains no nonzero semisimple elements.

Remark. It is tempting to use simpler terminology and replace "(-1)-distinguished" with merely "distinguished". However this might lead to the ambiguity since, as it is shown below, there are the cases where x is (-1)-distinguished with respect to the action of K on $\mathfrak{p}$ but x is not distinguished with respect to the action of G on $\mathfrak{g}$.

Notice that since $\mathfrak{p}$ contains nonzero semisimple elements (see below the proof of Theorem 7), every (-1)-distinguished element is nonzero.

Theorem 3. ([P2, Theorem 5]) *Let $\mathcal{O}$ be a nonzero K-orbit in $\mathcal{N}(\mathfrak{p})$ and $X := \mathbf{P}(\overline{\mathcal{O}})$. The following properties are equivalent:*

(i) $X = \check{X}$.
(ii) $\mathcal{O}$ *is* (-1)-*distinguished.*

By virtue of Theorem 3, our goal is classifying (-1)-distinguished orbits.

Theorem 4. *Let $\mathcal{O}$ be a K-orbit in $\mathcal{N}(\mathfrak{p})$ and let x be an element of $\mathcal{O}$. The following properties are equivalent:*

(i) $\mathcal{O}$ *is* (-1)-*distinguished.*
(ii) *The reductive Levi factors of* $\mathfrak{z}_{\mathfrak{k}}(x)$ *and* $\mathfrak{z}_{\mathfrak{g}}(x)$ *have the same dimension.*
(iii) *The reductive Levi factors of* $\mathfrak{z}_{\mathfrak{k}}(x)$ *and* $\mathfrak{z}_{\mathfrak{g}}(x)$ *are isomorphic.*

If $\mathfrak{g} = \mathfrak{h} \oplus \mathfrak{h}$ where $\mathfrak{h}$ is a semisimple complex Lie algebra, and $\theta((y,z)) = (z,y)$, then

$$\mathfrak{k} = \{(y,y) \mid y \in \mathfrak{h}\}, \quad \mathfrak{p} = \{(y,-y) \mid y \in \mathfrak{h}\}. \tag{3}$$

If H is the adjoint group of $\mathfrak{h}$, then (3) implies that K is isomorphic to H and the K-module $\mathfrak{p}$ is isomorphic to the adjoint H-module $\mathfrak{h}$. Therefore, for $y \in \mathcal{N}(\mathfrak{h})$, the variety $\mathbf{P}(\overline{K \cdot y})$ is projective self-dual if and only if the variety $\mathbf{P}(\overline{H \cdot x})$ for $x := (y,-y) \in \mathcal{N}(\mathfrak{p})$ is projective self-dual. This is consistent with Theorems 2 and 3: the definitions clearly imply that x is (-1)-distinguished if and only y is distinguished. So Theorem 2 follows from Theorem 3.

In [N], it was made an attempt to develop an analogue of the Bala–Carter theory for nilpotent orbits in real semisimple Lie algebras. The Kostant–Sekiguchi bijection (recalled below in this section) reduces this to finding an analogue of the Bala–Carter theory for K-orbits in $\mathcal{N}(\mathfrak{p})$. Given the above discussion, we believe that if such a natural analogue exists, (-1)-distinguished elements should play a key role in it, analogous to that of distinguished elements in the original Bala–Carter theory (in [N], the so called noticed elements, different from (-1)-distinguished ones, play a central role).

Theorems 3, 4 and their proofs are valid over any algebraically closed ground field of characteristic zero. They give a key to explicit classifying (-1)-distinguished elements since there is a method (see the end of this section) for finding reductive Levi factors of $\mathfrak{z}_{\mathfrak{k}}(x)$ and $\mathfrak{z}_{\mathfrak{g}}(x)$.

However, over the complex numbers, there is another nice characterization of (-1)-distinguished orbits. It is given below in Theorem 5. For classical $\mathfrak{g}$, our approach to explicit classifying (-1)-distinguished orbits is based on this result. The very formulation of this characterization uses complex topology. However remark that, by the Lefschetz's principle, the final explicit classification of (-1)-distinguished orbits for classical $\mathfrak{g}$ given by combining Theorem 5 with Theorems 8–13 is valid over any algebraically closed ground field of characteristic zero.

Namely, according to the classical theory (cf., e.g., [OV]), there is a θ-stable real form $\mathfrak{g}_{\mathbb{R}}$ of $\mathfrak{g}$ such that

$$\mathfrak{g}_{\mathbb{R}} = \mathfrak{k}_{\mathbb{R}} \oplus \mathfrak{p}_{\mathbb{R}}, \quad \text{where} \quad \mathfrak{k}_{\mathbb{R}} := \mathfrak{g}_{\mathbb{R}} \cap \mathfrak{k}, \ \mathfrak{p}_{\mathbb{R}} := \mathfrak{g}_{\mathbb{R}} \cap \mathfrak{p}, \tag{4}$$

is a Cartan decomposition of $\mathfrak{g}_{\mathbb{R}}$. The semisimple real algebra $\mathfrak{g}_{\mathbb{R}}$ is noncompact since $\mathfrak{p} \neq 0$. Assigning $\mathfrak{g}_{\mathbb{R}}$ to θ induces well defined bijection from the set of all conjugacy classes of elements of order 2 in $\operatorname{Aut} \mathfrak{g}$ to the set of conjugacy classes of noncompact real forms of $\mathfrak{g}$. If x is an element of $\mathfrak{g}_{\mathbb{R}}$, we put

$$\mathfrak{z}_{\mathfrak{g}_{\mathbb{R}}}(x) := \mathfrak{g}_{\mathbb{R}} \cap \mathfrak{z}_{\mathfrak{g}}(x);$$

$\mathfrak{z}_{\mathfrak{g}_{\mathbb{R}}}(x)$ is the centralizer of x in $\mathfrak{g}_{\mathbb{R}}$. The identity component of the Lie group of real points of G is the adjoint group $\operatorname{Ad}(\mathfrak{g}_{\mathbb{R}})$ of $\mathfrak{g}_{\mathbb{R}}$. We set

$$\mathcal{N}(\mathfrak{g}_{\mathbb{R}}) := \mathcal{N}(\mathfrak{g}) \cap \mathfrak{g}_{\mathbb{R}}.$$

Definition 2. An element $x \in \mathcal{N}(\mathfrak{g}_{\mathbb{R}})$ is called *compact* if the reductive Levi factor of its centralizer $\mathfrak{z}_{\mathfrak{g}_{\mathbb{R}}}(x)$ is a compact Lie algebra.

Recall that there is a bijection between the sets of nonzero K-orbits in $\mathcal{N}(\mathfrak{p})$ and nonzero $\operatorname{Ad}(\mathfrak{g}_{\mathbb{R}})$-orbits in $\mathcal{N}(\mathfrak{g}_{\mathbb{R}})$, cf. [CM], [M]. Namely, let σ be the complex conjugation of $\mathfrak{g}$ defined by $\mathfrak{g}_{\mathbb{R}}$, viz.,

$$\sigma(a + ib) = a - ib, \ a, b \in \mathfrak{g}_{\mathbb{R}}.$$

Let $\{e, h, f\}$ be an $\mathfrak{sl}_2$*-triple* in $\mathfrak{g}$, i.e., an ordered triple of elements of $\mathfrak{g}$ spanning a three-dimensional simple subalgebra of $\mathfrak{g}$ and satisfying the bracket relations

$$[h, e] = 2e, \ [h, f] = -2f, \ [e, f] = h. \tag{5}$$

It is called a *complex Cayley triple* if $e, f \in \mathfrak{p}$ (hence $h \in \mathfrak{k}$) and $\sigma(e) = -f$. Given a complex Cayley triple $\{e, h, f\}$, set

$$e' := i(-h + e + f)/2, \quad h' := e - f, \quad f' := -i(h + e + f)/2. \tag{6}$$

Then $\{e', h', f'\}$ is an $\mathfrak{sl}_2$-triple in $\mathfrak{g}_{\mathbb{R}}$ such that $\theta(e') = f'$ (and hence $\theta(f') = e', \theta(h') = -h'$). The $\mathfrak{sl}_2$-triples in $\mathfrak{g}_{\mathbb{R}}$ satisfying the last property are called the *real Cayley triples.* Given $\{e', h', f'\}$, the following formulas restore $\{e, h, f\}$:

$$e := (h' + if' - ie')/2, \quad h := i(e' + f'), \quad f := (-h' + if' - ie')/2. \tag{7}$$

The map $\{e, h, f\} \mapsto \{e', h', f'\}$ is a bijection from the set of complex to the set of real Cayley triples. The triple $\{e, h, f\}$ is called the *Cayley transform* of $\{e', h', f'\}$.

Now let $\mathcal{O}$ be a nonzero K-orbit in $\mathcal{N}(\mathfrak{p})$. Then, according to [KR], there exists a complex Cayley triple $\{e, h, f\}$ in $\mathfrak{g}$ such that $e \in \mathcal{O}$. Let $\{e', h', f'\}$ be the real Cayley triple in $\mathfrak{g}_{\mathbb{R}}$ such that $\{e, h, f\}$ is its Cayley transform. Let $\mathcal{O}' = \operatorname{Ad}(\mathfrak{g}_{\mathbb{R}}) e'$. Then the map assigning $\mathcal{O}'$ to $\mathcal{O}$ is well defined and establishes a bijection, called the *Kostant–Sekiguchi bijection*, between the set of nonzero K-orbits in $\mathcal{N}(\mathfrak{p})$ and the set of nonzero $\operatorname{Ad}(\mathfrak{g}_{\mathbb{R}})$-orbits in $\mathcal{N}(\mathfrak{g}_{\mathbb{R}})$. We say that the orbits $\mathcal{O}$ and $\mathcal{O}'$ *correspond* one another via the Kostant–Sekiguchi bijection. One can show, [KR], cf. [CM], that

$$\begin{aligned} G \cdot e' &= G \cdot e, \\ \dim_{\mathbb{R}} \operatorname{Ad}(\mathfrak{g}_{\mathbb{R}}) \cdot e' &= \dim_{\mathbb{C}} G \cdot e'. \end{aligned} \tag{8}$$

Theorem 5. *Let $\mathcal{O}$ be a nonzero K-orbit in $\mathcal{N}(\mathfrak{p})$ and let x be an element of the* $\mathrm{Ad}(\mathfrak{g}_{\mathbb{R}})$*-orbit in $\mathcal{N}(\mathfrak{g}_{\mathbb{R}})$ corresponding to $\mathcal{O}$ via the Kostant–Sekiguchi bijection. The following properties are equivalent:*

(i) *$\mathcal{O}$ is (-1)-distinguished.*
(ii) *x is compact.*

Next result, Theorem 6, reduces studying projective self-dual varieties associated with symmetric spaces to the case of simple Lie algebras $\mathfrak{g}$. Namely, since the Lie algebra $\mathfrak{g}$ is semisimple, it is the direct sum of all its simple ideals. As this set of ideals is θ-stable, $\mathfrak{g}$ is the direct sum of its θ-stable ideals, $\mathfrak{g} = \mathfrak{g}_1 \oplus \ldots \oplus \mathfrak{g}_d$, where each $\mathfrak{g}_l$ is either

(a) simple or
(b) direct sum of two isomorphic simple ideals permuted by θ.

Let G_l, $\mathfrak{k}_l$, $\mathfrak{p}_l$, K_l and $\mathcal{N}(\mathfrak{p}_l)$ have the same meaning for $\mathfrak{g}_l$ with respect to $\theta_l := \theta|_{\mathfrak{g}_l}$ as resp., G, $\mathfrak{k}$, $\mathfrak{p}$, K and $\mathcal{N}(\mathfrak{p})$ have for $\mathfrak{g}$ with respect to θ. Then $G = G_1 \times \ldots \times G_d$, $\mathfrak{k} = \mathfrak{k}_1 \oplus \ldots \oplus \mathfrak{k}_d$, $\mathfrak{p} = \mathfrak{p}_1 \oplus \ldots \oplus \mathfrak{p}_d$, $K = K_1 \times \ldots \times K_d$ and $\mathcal{N}(\mathfrak{p}) = \mathcal{N}(\mathfrak{p}_1) \times \ldots \times \mathcal{N}(\mathfrak{p}_d)$.

In Theorem 6, we use the following notation. Let $X_1, \ldots, X_s$ be the closed subvarieties of a projective space $\mathbf{P}$, and let $s \geqslant 2$. Consider the variety

$$\overline{\{(x_1, \ldots, x_s, y) \in X_1 \times \ldots \times X_s \times \mathbf{P} \mid \dim\langle x_1, \ldots, x_s\rangle = s-1,\ y \in \langle x_1, \ldots, x_s\rangle\}}, \tag{9}$$

where bar denotes Zariski closure in $X_1 \times \ldots \times X_s \times \mathbf{P}$ and $\langle S\rangle$ denotes the linear span of S in $\mathbf{P}$. Consider the projection of variety (9) to $\mathbf{P}$. Its image is denoted by

$$\mathrm{Join}(X_1, \ldots, X_s) \tag{10}$$

and called the *join* of $X_1, \ldots, X_s$. If $s = 2$, then (10) is the usual join of X_1 and X_2, cf. [Ha]. If $s > 2$, then (10) is the usual join of $\mathrm{Join}(X_1, \ldots, X_{s-1})$ and X_s.

Theorem 6. *Let $x = x_1 + \ldots + x_d$ where $x_l \in \mathcal{N}(\mathfrak{p}_l)$, $l = 1, \ldots, d$. Consider in $\mathbf{P}(\mathfrak{g})$ the projective subvarieties $X := \mathbf{P}(\overline{K \cdot x})$ and $X_l := \mathbf{P}(\overline{K_l \cdot x_l})$. Then*

$$X = \mathrm{Join}(X_1, \ldots, X_d)$$

and the following properties are equivalent:

(i) *$X = \check{X}$.*
(ii) *$X_l = \check{X}_l$ for all l.*

Remark. This is a specific geometric property of the considered varieties. In general setting, projective self-duality of $\mathrm{Join}(Z_1, \ldots, Z_m)$ is not equivalent to projective self-duality of all $Z_1, \ldots, Z_m$.

If $\mathfrak{g}_l$ is of type (b), viz., $\mathfrak{g}_l$ is isomorphic to the direct sum of algebras $\mathfrak{s} \oplus \mathfrak{s}$, where $\mathfrak{s}$ is a simple algebra, and θ_l acts by $\theta_l((x, y)) = (y, x)$, then, as explained above (see (3)), classifying (-1)-distinguished orbits in $\mathcal{N}(\mathfrak{p}_l)$ amounts to classifying distinguished orbits in $\mathcal{N}(\mathfrak{s})$. Since the last classification is known, [BaCa], [Ca], this and Theorem 6 reduce classifying (-1)-distinguished orbits in $\mathfrak{g}$ to the case where $\mathfrak{g}$ is a *simple* Lie algebra.

In this case, for explicit classifying (-1)-distinguished orbits, one can apply Theorem 4 since there are algorithms, [Ka], [Vi], yielding, in principle, a classification of K-orbits in $\mathcal{N}(\mathfrak{p})$ and a description of the reductive Levi factors of $\mathfrak{z}_\mathfrak{k}(x)$ and $\mathfrak{z}_\mathfrak{g}(x)$ for $x \in \mathcal{N}(\mathfrak{p})$. Moreover, there is a computer program, [L], implementing these algorithms and yielding, for concrete pairs $(\mathfrak{g}, \theta)$, the explicit representatives x of K-orbits in $\mathcal{N}(\mathfrak{p})$ and the reductive Levi factors of $\mathfrak{z}_\mathfrak{k}(x)$ and $\mathfrak{z}_\mathfrak{g}(x)$. There is also another way for finding representatives and dimensions of K-orbits in $\mathcal{N}(\mathfrak{p})$: they may be obtained using the general algorithm of finding Hesselink strata for linear reductive group actions given in [P3], since in our case these strata coincide with K-orbits, see [P3, Proposition 4].

Fortunately, for exceptional simple $\mathfrak{g}$, explicit finding of the aforementioned classification and description already has been performed by D. Ž. Đoković in [D3], [D4], cf. [CM]. In these papers, the answers are given in terms of the so called *characteristics* in the sense of Dynkin (in Section 5, we recall this classical approach, [D], [K1], [KR], cf. [CM] and [Vi]). Combining either of Theorems 4 and 5 with these results, we obtain, for all exceptional simple $\mathfrak{g}$ and all θ, the explicit classification of (-1)-distinguished K-orbits in $\mathcal{N}(\mathfrak{p})$ in terms of their characteristics (see Theorem 14).

For classical $\mathfrak{g}$, we use another approach and obtain the classification of (-1)-distinguished K-orbits in $\mathcal{N}(\mathfrak{p})$ by means of the Kostant–Sekiguchi bijection, Theorem 5 and elementary representation theory of $\mathfrak{sl}_2$ (see Theorems 8–13).

Theorems 4, 5, 6 are proved in Section 3. For simple Lie algebras $\mathfrak{g}$, the classifications of (-1)-distinguished K-orbits in $\mathcal{N}(\mathfrak{p})$ are obtained in the next two sections: in Section 4, we consider the case of classical $\mathfrak{g}$, and in Section 5, that of exceptional simple $\mathfrak{g}$. In Section 6, we briefly discuss some geometric properties of the constructed self-dual projective algebraic varieties.

3 Proofs Of Theorems 4–6

Proof of Theorem 4. Theorem 4 immediately follows from Theorem 7 proved below.

Let k be a field of characteristic 0. If $\mathfrak{c}$ is a finite dimensional algebraic Lie algebra over k, we denote by $\operatorname{rad}\mathfrak{c}$ (resp., $\operatorname{rad}_u\mathfrak{c}$) the radical (resp., the unipotent radical, i.e., the maximal ideal whose elements are nilpotent) of $\mathfrak{c}$. We call $\mathfrak{c}/\operatorname{rad}\mathfrak{c}$ (resp., $\mathfrak{c}/\operatorname{rad}_u\mathfrak{c}$) the *Levi factor* (resp., the *reductive Levi factor*) of $\mathfrak{c}$. According to [Ch], there is a semisimple (resp., reductive) subalgebra $\mathfrak{l}$ in $\mathfrak{c}$ such that $\mathfrak{c}$ is the semidirect sum of $\mathfrak{l}$ and $\operatorname{rad}\mathfrak{c}$ (resp., $\operatorname{rad}_u\mathfrak{c}$). Every such $\mathfrak{l}$ is called *Levi subalgebra* (resp., *reductive Levi subalgebra*) of $\mathfrak{c}$.

Theorem 7. *Assume that k is algebraically closed and let $\mathfrak{a}$ be a finite dimensional algebraic Lie algebra over k. Let $\theta \in \operatorname{Aut}\mathfrak{a}$ be an element of order 2. For any θ-stable linear subspace $\mathfrak{l}$ of $\mathfrak{a}$, set*

$$\mathfrak{l}^{\pm}=\{x\in\mathfrak{l}\mid\theta(x)=\pm x\}.$$

The following properties are equivalent:

(i) $\mathfrak{a}^-$ *contains no nonzero semisimple elements.*
(ii) $\mathfrak{a}^- = (\operatorname{rad}_u\mathfrak{a})^-$.
(iii) $\dim\mathfrak{a}^- = \dim(\operatorname{rad}_u\mathfrak{a})^-$.
(iv) $\mathfrak{a}^- \subseteq \operatorname{rad}_u\mathfrak{a}$.
(v) *The reductive Levi factors of $\mathfrak{a}^+$ and $\mathfrak{a}$ have the same dimension.*
(vi) *The reductive Levi factors of $\mathfrak{a}^+$ and $\mathfrak{a}$ are isomorphic.*
(vii) *The set of all reductive Levi subalgebras of $\mathfrak{a}^+$ coincides with the set of all θ-stable reductive Levi subalgebras of $\mathfrak{a}$.*

Proof. It is known that $\mathfrak{a}$ contains a θ-stable Levi subalgebra $\mathfrak{s}$ (e. g., see [KN, Appendix 9]), and $\operatorname{rad}\mathfrak{a}$ contains a θ-stable maximal torus $\mathfrak{t}$, see [St]. Thus we have the following θ-stable direct sum decompositions of vector spaces:

$$\left.\begin{array}{c}\mathfrak{a}=\mathfrak{s}\oplus\operatorname{rad}\mathfrak{a},\quad \operatorname{rad}\mathfrak{a}=\mathfrak{t}\oplus\operatorname{rad}_u\mathfrak{a},\\ \mathfrak{s}=\mathfrak{s}^+\oplus\mathfrak{s}^-,\quad \mathfrak{t}=\mathfrak{t}^+\oplus\mathfrak{t}^-,\quad \operatorname{rad}_u\mathfrak{a}=(\operatorname{rad}_u\mathfrak{a})^+\oplus(\operatorname{rad}_u\mathfrak{a})^-,\\ \mathfrak{a}^{\pm}=\mathfrak{s}^{\pm}\oplus\mathfrak{t}^{\pm}\oplus(\operatorname{rad}_u\mathfrak{a})^{\pm}.\end{array}\right\}\tag{11}$$

(i) $\Rightarrow$ (ii): Assume that (i) holds. Then, by (11), we have $\mathfrak{t}^- = 0$. If $\mathfrak{s}^- \neq 0$, then $\theta|_{\mathfrak{s}} \in \operatorname{Aut}\mathfrak{a}$ is an element of order 2. By [Vu], this implies that $\mathfrak{s}^-$ contains a nonzero θ-stable algebraic torus. By (11), this contradicts (i). Thus $\mathfrak{s}^- = 0$. Whence (ii) by (11).

(ii) $\Rightarrow$ (i) and (iv) $\Rightarrow$ (i): This is because all elements of $\operatorname{rad}_u\mathfrak{a}$ are nilpotent.

(ii) $\Rightarrow$ (iii) and (ii) $\Rightarrow$ (iv): This is evident.

(iii) $\Rightarrow$ (ii): This follows from (11).

(ii) $\Rightarrow$ (v): It follows from (11) that (ii) is equivalent to

$$\mathfrak{s} \oplus \mathfrak{t} \subseteq \mathfrak{a}^+. \tag{12}$$

Assume that (12) holds. Let $\pi : \mathfrak{a} \to \mathfrak{a}/\mathrm{rad}_u\mathfrak{a}$ be the natural homomorphism. By (11), (12), we have $\pi(\mathfrak{a}^+) = \mathfrak{a}/\mathrm{rad}_u\mathfrak{a}$. Since the algebra $\mathfrak{a}/\mathrm{rad}_u\mathfrak{a}$ is reductive, this implies

$$\begin{aligned} \dim \mathfrak{a}/\mathrm{rad}_u\mathfrak{a} &\leqslant \text{dimension of reductive Levi subalgebras of } \mathfrak{a}^+ \\ &= \dim \mathfrak{a}^+/\mathrm{rad}_u(\mathfrak{a}^+). \end{aligned} \tag{13}$$

On the other hand, since reductive Levi subalgebras of $\mathfrak{a}$ and $\mathfrak{a}^+$ are precisely their maximal reductive subalgebras, cf., e.g., [OV] , the inclusion $\mathfrak{a}^+ \subseteq \mathfrak{a}$ implies

$$\begin{aligned} \dim \mathfrak{a}/\mathrm{rad}_u\mathfrak{a} &= \text{dimension of reductive Levi subalgebras of } \mathfrak{a} \\ &\geqslant \text{dimension of reductive Levi subalgebras of } \mathfrak{a}^+. \end{aligned} \tag{14}$$

Now (v) follows from (13), (14).

(v) $\Rightarrow$ (ii): Assume that (v) holds. Let $\mathfrak{l}$ be a reductive Levi subalgebra of $\mathfrak{a}^+$. Since reductive Levi subalgebras of $\mathfrak{a}$ are precisely its maximal reductive subalgebras, it follows from (v) that $\mathfrak{l}$ is a reductive Levi subalgebra of $\mathfrak{a}$ as well. Let $\mathfrak{s}$ and $\mathfrak{t}$ be resp. the derived subalgebra and the center of $\mathfrak{l}$; so we have $\mathfrak{l} = \mathfrak{s} \oplus \mathfrak{t}$. Then $\mathfrak{s}$ is a θ-stable Levi subalgebra of $\mathfrak{a}$, and $\mathfrak{t}$ is a θ-stable maximal torus of $\mathrm{rad}\,\mathfrak{a}$. Hence, as above, (ii) is equivalent to the inclusion (12). As the latter holds by the very construction of $\mathfrak{l}$, we conclude that (ii) holds as well.

(vii) $\Rightarrow$ (v): This is clear.

(i) $\Rightarrow$ (vii): Let $\mathfrak{l}$ be a θ-stable reductive Levi subalgebra of $\mathfrak{a}$. If $\mathfrak{l}^- \neq 0$, then $\theta|_{\mathfrak{l}} \in \mathrm{Aut}\,\mathfrak{l}$ is an element of order 2. Hence, by [Vu], there is a nonzero θ-stable algebraic torus in $\mathfrak{l}^-$. This contradicts (i). Whence $\mathfrak{l}^- = 0$, i.e., $\mathfrak{l}$ lies in $\mathfrak{a}^+$.

(vi) $\Rightarrow$ (v) and (vii) $\Rightarrow$ (vi): This is clear. □

Proof of Theorem 5. According to [KR], there is a complex Cayley triple $\{e, h, f\}$ in $\mathfrak{g}$ such that $\mathcal{O} = K \cdot e$. Let $\{e', h', f'\}$ be the real Cayley triple in $\mathfrak{g}_\mathbb{R}$ whose Cayley transform is $\{e, h, f\}$. Since $\mathcal{O}' = \mathrm{Ad}(\mathfrak{g}_\mathbb{R}) \cdot e'$, we may, and will, assume that $x = e'$.

Let $\mathfrak{s}$ (resp., $\mathfrak{s}_\mathbb{R}$) be the simple three-dimensional subalgebra of $\mathfrak{g}$ (resp., $\mathfrak{g}_\mathbb{R}$) spanned by $\{e, h, f\}$ (resp., $\{e', h', f'\}$),

$$\mathfrak{s} = \mathbb{C}e + \mathbb{C}h + \mathbb{C}f, \quad \mathfrak{s}_\mathbb{R} = \mathbb{R}e' + \mathbb{R}h' + \mathbb{R}f'. \tag{15}$$

Denote by $\mathfrak{z}_\mathfrak{g}(\mathfrak{s})$ (resp., $\mathfrak{z}_{\mathfrak{g}_\mathbb{R}}(\mathfrak{s}_\mathbb{R})$) the centralizer of $\mathfrak{s}$ (resp., $\mathfrak{s}_\mathbb{R}$) in $\mathfrak{g}$ (resp., $\mathfrak{g}_\mathbb{R}$).

It follows from (6), (7), (15) that $\mathfrak{s}_\mathbb{R}$ is a real form of $\mathfrak{s}$. Since $\mathfrak{g}_\mathbb{R}$ is a real form of $\mathfrak{g}$, this yields that $\mathfrak{z}_{\mathfrak{g}_\mathbb{R}}(\mathfrak{s}_\mathbb{R})$ is a real form of $\mathfrak{z}_\mathfrak{g}(\mathfrak{s})$. The definitions of complex and real Cayley triples imply that $\mathfrak{s}$ and $\mathfrak{s}_\mathbb{R}$ are θ-stable subalgebras. Therefore $\mathfrak{z}_\mathfrak{g}(\mathfrak{s})$ and $\mathfrak{z}_{\mathfrak{g}_\mathbb{R}}(\mathfrak{s}_\mathbb{R})$ are θ-stable as well. Whence

$$\begin{aligned}\mathfrak{z}_{\mathfrak{g}}(\mathfrak{s}) &= (\mathfrak{z}_{\mathfrak{g}}(\mathfrak{s}) \cap \mathfrak{k}) \oplus (\mathfrak{z}_{\mathfrak{g}}(\mathfrak{s}) \cap \mathfrak{p}),\\ \mathfrak{z}_{\mathfrak{g}_{\mathbb{R}}}(\mathfrak{s}_{\mathbb{R}}) &= (\mathfrak{z}_{\mathfrak{g}_{\mathbb{R}}}(\mathfrak{s}_{\mathbb{R}}) \cap \mathfrak{k}_{\mathbb{R}}) \oplus (\mathfrak{z}_{\mathfrak{g}_{\mathbb{R}}}(\mathfrak{s}_{\mathbb{R}}) \cap \mathfrak{p}_{\mathbb{R}}),\end{aligned} \tag{16}$$

$$\mathfrak{z}_{\mathfrak{g}}(\mathfrak{s}) \cap \mathfrak{k} = \mathbb{C}(\mathfrak{z}_{\mathfrak{g}_{\mathbb{R}}}(\mathfrak{s}_{\mathbb{R}}) \cap \mathfrak{k}_{\mathbb{R}}), \quad \mathfrak{z}_{\mathfrak{g}}(\mathfrak{s}) \cap \mathfrak{p} = \mathbb{C}(\mathfrak{z}_{\mathfrak{g}_{\mathbb{R}}}(\mathfrak{s}_{\mathbb{R}}) \cap \mathfrak{p}_{\mathbb{R}}). \tag{17}$$

Now take into account that $\mathfrak{z}_{\mathfrak{g}}(\mathfrak{s})$ (resp., $\mathfrak{z}_{\mathfrak{g}_{\mathbb{R}}}(\mathfrak{s}_{\mathbb{R}})$) is the reductive Levi subalgebra of $\mathfrak{z}_{\mathfrak{g}}(e)$ (resp., $\mathfrak{z}_{\mathfrak{g}_{\mathbb{R}}}(e')$): regarding $\mathfrak{z}_{\mathfrak{g}}(\mathfrak{s})$, see, e.g., [SpSt]; for $\mathfrak{z}_{\mathfrak{g}_{\mathbb{R}}}(\mathfrak{s}_{\mathbb{R}})$, the arguments are the same. Hence, by Theorem 7, the orbit $\mathcal{O}$ is (-1)-distinguished if and only if $\mathfrak{z}_{\mathfrak{g}}(\mathfrak{s}) \cap \mathfrak{p} = 0$. By (17), the latter condition is equivalent to $\mathfrak{z}_{\mathfrak{g}_{\mathbb{R}}}(\mathfrak{s}_{\mathbb{R}}) \cap \mathfrak{p}_{\mathbb{R}} = 0$. On the other hand, since (4) is the Cartan decomposition of $\mathfrak{g}_{\mathbb{R}}$, it follows from (16) that $\mathfrak{z}_{\mathfrak{g}_{\mathbb{R}}}(\mathfrak{s}_{\mathbb{R}}) \cap \mathfrak{p}_{\mathbb{R}} = 0$ if and only if the Lie algebra $\mathfrak{z}_{\mathfrak{g}_{\mathbb{R}}}(\mathfrak{s}_{\mathbb{R}})$ is compact. This completes the proof of Theorem 5. □

Proof of Theorem 6. Clearly the following proposition implies the first statement.

Proposition 1. ([P2, Proposition 2]) *Let H_t be an algebraic group and let L_t be a finite dimensional algebraic H_t-module, $t = 1, 2$. Let $v_t \in L_s$ be a nonzero vector such that the orbit $H_t \cdot v_t$ is stable with respect to scalar multiplications. Put $H := H_1 \times H_2$, $L := L_1 \oplus L_2$. Identify L_t with the linear subspace of L and set $v := v_1 + v_2$. Then*

$$\mathbf{P}(\overline{H \cdot v}) = \mathrm{Join}(\mathbf{P}(\overline{H_1 \cdot v_1}), \mathbf{P}(\overline{H_2 \cdot v_2})).$$

Regarding the second statement, notice that the centralizers $\mathfrak{z}_{\mathfrak{g}_l}(x_l)$ and $\mathfrak{z}_{\mathfrak{g}}(x)$ are resp. θ_l- and θ-stable and $\mathfrak{z}_{\mathfrak{p}}(x) = \mathfrak{z}_{\mathfrak{p}_1}(x_1) \oplus \ldots \oplus \mathfrak{z}_{\mathfrak{p}_d}(x_d)$. Now the claim follows from this decomposition, Definition 1 and Theorem 3. □

4 Classification For Classical $\mathfrak{g}$

If $\mathfrak{g}$ is classical, our approach to classifying (-1)-distinguished nilpotent K-orbits in $\mathfrak{p}$ is based on the Kostant–Sekiguchi bijection and Theorem 5. Namely, for every noncompact real form $\mathfrak{g}_{\mathbb{R}}$ of $\mathfrak{g}$, we will find in $\mathcal{N}(\mathfrak{g}_{\mathbb{R}})$ all $\mathrm{Ad}(\mathfrak{g}_{\mathbb{R}})$-orbits $\mathcal{O}$ whose elements are compact. If this is done, then for the complexification of a Cartan decomposition of $\mathfrak{g}_{\mathbb{R}}$, the K-orbits in $\mathcal{N}(\mathfrak{p})$ corresponding to all such $\mathcal{O}$ via the Kostant–Sekiguchi bijection are precisely all (-1)-distinguished orbits.

First we recall the classification of real forms $\mathfrak{g}_{\mathbb{R}}$ of classical complex Lie algebras $\mathfrak{g}$. Let D be either real numbers $\mathbb{R}$, complex numbers $\mathbb{C}$, or quaternions $\mathbb{H}$. If $a \in D$, we denote by $\overline{a}$ the conjugate of a. Let V be a finite dimensional vector space over D (left D-module). Denote by $\mathrm{GL}_D(V)$ the Lie group of all D-linear automorphisms of V. Its Lie algebra $\mathfrak{gl}_D(V)$ is identified with the real Lie algebra of all D-linear endomorphisms of V. The derived algebra of $\mathfrak{gl}_D(V)$ is denoted by $\mathfrak{sl}_D(V)$.

Let a map $\Phi : V \times V \to D$ be D-linear with respect to the first argument. It is called

- *symmetric bilinear form on* V if $\Phi(x,y) = \Phi(y,x)$ for all $x, y \in V$,
- *skew-symmetric bilinear form on* V if $\Phi(x,y) = -\Phi(y,x)$ for all $x, y \in V$,
- *Hermitian form on* V if $\Phi(x,y) = \overline{\Phi(y,x)}$ for all $x, y \in V$,
- *skew-Hermitian form on* V if $\Phi(x,y) = -\overline{\Phi(y,x)}$ for all $x, y \in V$.

By "form on V" we always mean form of one of these four types. A form Φ on V is called *nondegenerate* if, for any $x \in V$, there is $y \in V$ such that $\Phi(x,y) \neq 0$. If Φ_s, $s = 1, 2$, is a form on a finite dimensional vector space V_s over D, then Φ_1 and Φ_2 are called *equivalent forms* if there exists an isometry $\psi : V_1 \to V_2$ with respect to Φ_1 and Φ_2, i.e., an isomorphism of vector spaces over D such that $\Phi_2(\psi(x), \psi(y)) = \Phi_1(x,y)$ for all $x, y \in V_1$.

The automorphism group of a form Φ on V,

$$\mathrm{GL}_D^{\Phi}(V) := \{g \in \mathrm{GL}_D(V) \mid \Phi(g \cdot x, g \cdot y) = \Phi(x,y) \text{ for all } x, y \in V\},$$

is a Lie subgroup of $\mathrm{GL}_D(V)$ whose Lie algebra is

$$\mathfrak{gl}_D^{\Phi}(V) := \{A \in \mathfrak{gl}_D(V) \mid \Phi(A \cdot x, y) + \Phi(x, A \cdot y) = 0 \text{ for all } x, y \in V\}.$$

For $\Phi = 0$ (zero form), we have $\mathrm{GL}_D^{\Phi}(V) = \mathrm{GL}_D(V)$ and $\mathfrak{gl}_D^{\Phi}(V) = \mathfrak{gl}_D(V)$.

Let $n := \dim_D V$ and let $(e) := \{e_1, \dots, e_n\}$ be a basis of V over D. Identifying D-linear endomorphisms of V with their matrices with respect to (e), we identify $\mathfrak{gl}_D(V)$, $\mathfrak{sl}_D(V)$ and $\mathfrak{gl}_D^{\Phi}(V)$ with the corresponding matrix real Lie algebras. For $D = \mathbb{R}, \mathbb{H}$, the real Lie algebras $\mathfrak{gl}_D(V)$ and $\mathfrak{sl}_D(V)$ are denoted resp. by $\mathfrak{gl}_n(D)$ and $\mathfrak{sl}_n(D)$, and for $D = \mathbb{C}$, by $\mathfrak{gl}_n(\mathbb{C})_{\mathbb{R}}$ and $\mathfrak{sl}_n(\mathbb{C})_{\mathbb{R}}$. The last algebras are endowed with the natural structures of complex Lie algebras that are denoted resp. by $\mathfrak{gl}_n(\mathbb{C})$ and $\mathfrak{sl}_n(\mathbb{C})$. If $D = \mathbb{H}$, then V is a $2n$-dimensional vector space over the subfield $\mathbb{C}$ of $\mathbb{H}$, and elements of $\mathfrak{gl}_{\mathbb{H}}(V)$ are its $\mathbb{C}$-linear endomorphisms. Identifying them with their matrices with respect to the basis $e_1, \dots, e_n, je_1, \dots, je_n$, we identify $\mathfrak{gl}_n(\mathbb{H})$ (resp., $\mathfrak{sl}_n(\mathbb{H})$) with the corresponding Lie subalgebra of $\mathfrak{gl}_{2n}(\mathbb{C})_{\mathbb{R}}$ (resp., $\mathfrak{sl}_{2n}(\mathbb{C})_{\mathbb{R}}$).

If Φ is a form on V, then the matrix (Φ_{st}) of Φ with respect to (e) defines Φ by

$$\Phi\left(\textstyle\sum_{s=1}^n x_s e_s, \sum_{t=1}^n y_t e_t\right) = \begin{cases} \sum_{s,t=1}^n x_s \Phi_{st} y_t & \text{if } \Phi \text{ is symmetric or skew-symmetric,} \\ \sum_{s,t=1}^n x_s \Phi_{st} \overline{y_t} & \text{if } \Phi \text{ is Hermitian or skew-Hermitian.} \end{cases}$$

The map $\Phi \mapsto (\Phi_{st})$ is a bijection between all symmetric (resp., skew-symmetric, Hermitian, skew-Hermitian) forms on V and all symmetric (resp., skew-symmetric, Hermitian, skew-Hermitian) $n \times n$-matrices over D. Φ is nondegenerate if and only if (Φ_{st}) is nonsingular.

We denote by I_d the unit matrix of size d and put $\mathrm{I}_{p,q} := \mathrm{diag}(1, \dots, 1, -1, \dots, -1)$ where p (resp., q) is the number of 1's (resp., -1's).

Now we summarize the results from linear algebra about classification of forms and fix some notation and terminology. In the sequel, $\mathbb{N}$ denotes the set of all nonnegative integers.

Symmetric Bilinear Forms

$D = \mathbb{R}$. There are exactly $n+1$ equivalence classes of nondegenerate symmetric bilinear forms on V. They are represented by the forms Φ with $(\Phi_{st}) = \mathrm{I}_{p,q}$, $p + q = n$, $p = 0, \ldots, n$. If $(\Phi_{st}) = \mathrm{I}_{p,q}$, then $\operatorname{sgn} \Phi := p - q$ is called the *signature of* Φ, and the real Lie algebra $\mathfrak{gl}_D^{\Phi}(V)$ is denoted by $\mathfrak{so}_{p,q}$.

$D = \mathbb{C}$. There is exactly one equivalence class of nondegenerate symmetric bilinear forms on V. It is represented by the form Φ with $(\Phi_{st}) = \mathrm{I}_n$. The corresponding real Lie algebra $\mathfrak{gl}_D^{\Phi}(V)$ is denoted by $\mathfrak{so}_n(\mathbb{C})_{\mathbb{R}}$. It has a natural structure of complex Lie algebra that is denoted by $\mathfrak{so}_n(\mathbb{C})$.

$D = \mathbb{H}$. There are no nonzero symmetric bilinear forms on V.

Skew-symmetric Bilinear Forms

$D = \mathbb{R}$ and $\mathbb{C}$. Nondegenerate skew-symmetric bilinear forms on V exist if and only if n is even. In this case, there is exactly one equivalence class of such forms. It is represented by the form Φ with $(\Phi_{st}) = \begin{bmatrix} 0 & \mathrm{I}_{n/2} \\ -\mathrm{I}_{n/2} & 0 \end{bmatrix}$. Corresponding real Lie algebra $\mathfrak{gl}_D^{\Phi}(V)$ is denoted by $\mathfrak{sp}_n(\mathbb{R})$ for $D = \mathbb{R}$, and by $\mathfrak{sp}_n(\mathbb{C})_{\mathbb{R}}$ for $D = \mathbb{C}$. The algebra $\mathfrak{sp}_n(\mathbb{C})_{\mathbb{R}}$ has a natural structure of complex Lie algebra that is denoted by $\mathfrak{sp}_n(\mathbb{C})$.

$D = \mathbb{H}$. There are no nonzero skew-symmetric bilinear forms on V.

Hermitian Forms

$D = \mathbb{R}$. Hermitian forms on V coincide with symmetric bilinear forms.

$D = \mathbb{C}$ and $\mathbb{H}$. There are exactly $n+1$ equivalence classes of nondegenerate Hermitian forms on V. They are represented by the forms Φ with $(\Phi_{st}) = \mathrm{I}_{p,q}$, $p+q = n$, $p = 0, \ldots, n$. If $(\Phi_{st}) = \mathrm{I}_{p,q}$, then $\operatorname{sgn} \Phi := p-q$ is called the *signature* of Φ, and the real Lie algebra $\mathfrak{gl}_D^{\Phi}(V)$ is denoted by $\mathfrak{su}_{p,q}$ for $D = \mathbb{C}$, and by $\mathfrak{sp}_{p,q}$ for $D = \mathbb{H}$.

Skew-Hermitian Forms

$D = \mathbb{R}$. Skew-Hermitian forms on V coincide with skew-symmetric forms.

$D = \mathbb{C}$. The map $\Phi \mapsto i\Phi$ is a bijection between all nondegenerate Hermitian forms on V and all nondegenerate skew-Hermitian forms on V.

$D = \mathbb{H}$. There is exactly one equivalence class of nondegenerate skew-Hermitian forms on V. It is represented by the form Φ with $(\Phi_{st}) = j\mathrm{I}_n$. The corresponding real Lie algebra $\mathfrak{gl}_D^{\Phi}(V)$ is denoted by $\mathfrak{u}_n^*(\mathbb{H})$.

Some of the above-defined Lie algebras are isomorphic to each other. Obviously $\mathfrak{su}_{p,q} = \mathfrak{su}_{q,p}$, $\mathfrak{so}_{p,q} = \mathfrak{so}_{q,p}$, $\mathfrak{sp}_{p,q} = \mathfrak{sp}_{q,p}$ and we also have

$$\begin{gathered}
\mathfrak{so}_2 = \mathfrak{u}_1^*(\mathbb{H}),\ \mathfrak{so}_3(\mathbb{C}) \simeq \mathfrak{sl}_2(\mathbb{C}) = \mathfrak{sp}_2(\mathbb{C}),\ \mathfrak{so}_3 \simeq \mathfrak{su}_2 = \mathfrak{sp}_{1,0} = \mathfrak{sl}_1(\mathbb{H}),\\
\mathfrak{so}_{1,2} \simeq \mathfrak{su}_{1,1} \simeq \mathfrak{sl}_2(\mathbb{R}) = \mathfrak{sp}_2(\mathbb{R}),\ \mathfrak{so}_4(\mathbb{C}) \simeq \mathfrak{sl}_2(\mathbb{C}) \oplus \mathfrak{sl}_2(\mathbb{C}),\\
\mathfrak{so}_4 \simeq \mathfrak{su}_2 \oplus \mathfrak{su}_2, \mathfrak{so}_{1,3} \simeq \mathfrak{sl}_2(\mathbb{C})_{\mathbb{R}}, \mathfrak{so}_{2,2} \simeq \mathfrak{sl}_2(\mathbb{R}) \oplus \mathfrak{sl}_2(\mathbb{R}),\\
\mathfrak{u}_2^*(\mathbb{H}) \simeq \mathfrak{su}_2 \oplus \mathfrak{sl}_2(\mathbb{R}), \mathfrak{so}_5(\mathbb{C}) \simeq \mathfrak{sp}_4(\mathbb{C}), \mathfrak{so}_5 \simeq \mathfrak{sp}_2, \mathfrak{so}_{1,4} \simeq \mathfrak{sp}_{1,1},\\
\mathfrak{so}_{2,3} \simeq \mathfrak{sp}_4(\mathbb{R}),\ \mathfrak{so}_6(\mathbb{C}) \simeq \mathfrak{sl}_4(\mathbb{C}),\ \mathfrak{so}_6 \simeq \mathfrak{su}_4,\ \mathfrak{so}_{1,5} \simeq \mathfrak{sl}_2(\mathbb{H}),\\
\mathfrak{so}_{2,4} \simeq \mathfrak{su}_{2,2},\ \mathfrak{so}_{3,3} \simeq \mathfrak{sl}_4(\mathbb{R}),\ \mathfrak{u}_3(\mathbb{H}) \simeq \mathfrak{su}_{1,3},\ \mathfrak{u}_4(\mathbb{H}) \simeq \mathfrak{so}_{2,6}.
\end{gathered} \tag{18}$$

By definition, *classical complex Lie algebras* $\mathfrak{g}$ are the algebras $\mathfrak{sl}_n(\mathbb{C})$, $\mathfrak{so}_n(\mathbb{C})$ and $\mathfrak{sp}_n(\mathbb{C})$. According to E. Cartan's classification (cf., e.g., [OV]), up to isomorphism, all their real forms $\mathfrak{g}_{\mathbb{R}}$ are listed in Table 1.

Table 1.

$\mathfrak{g}$	$\mathfrak{g}_{\mathbb{R}}$	compact $\mathfrak{g}_{\mathbb{R}}$
$\mathfrak{sl}_n(\mathbb{C}),\ n \geqslant 2$	$\mathfrak{sl}_n(\mathbb{R})$, $\mathfrak{sl}_l(\mathbb{H}),\ n = 2l$, $\mathfrak{su}_{n-q,q},\ q = 0, 1, \ldots, [n/2]$	$\mathfrak{su}_n := \mathfrak{su}_{n,0}$
$\mathfrak{so}_n(\mathbb{C}),\ n = 3$ or $n \geqslant 5$	$\mathfrak{so}_{n-q,q},\ q = 0, 1, \ldots, [n/2]$, $\mathfrak{u}_l^*(\mathbb{H}),\ n = 2l$	$\mathfrak{so}_n := \mathfrak{so}_{n,0}$
$\mathfrak{sp}_n(\mathbb{C}),\ n = 2l \geqslant 2$	$\mathfrak{sp}_n(\mathbb{R})$, $\mathfrak{sp}_{l-q,q},\ q = 0, 1, \ldots, [l/2]$	$\mathfrak{sp}_l := \mathfrak{sp}_{l,0}$

There are the isomorphisms between some of the $\mathfrak{g}_{\mathbb{R}}$'s in Table 1 given by (18).

Now we restate the classification of nilpotent $\mathrm{Ad}(\mathfrak{g}_{\mathbb{R}})$-orbits in all real forms $\mathfrak{g}_{\mathbb{R}}$ of classical complex Lie algebras $\mathfrak{g}$ (cf. [BoCu], [CM], [SpSt], [M], [W]) in the form adapted to our goal of classifying compact nilpotent elements in $\mathfrak{g}_{\mathbb{R}}$.

If a subalgebra of $\mathfrak{g}_{\mathbb{R}}$ is isomorphic to $\mathfrak{sl}_2(\mathbb{R})$, we call it an $\mathfrak{sl}_2(\mathbb{R})$-*subalgebra.* We use the following basic facts, cf. [CM], [KR], [M], [SpSt]:

(F1) For any nonzero nilpotent element $x \in \mathfrak{g}_{\mathbb{R}}$, there is an $\mathfrak{sl}_2(\mathbb{R})$-subalgebra of $\mathfrak{g}_{\mathbb{R}}$ containing x.

(F2) If $\mathfrak{a}_1$ and $\mathfrak{a}_2$ are the $\mathfrak{sl}_2(\mathbb{R})$-subalgebras of $\mathfrak{g}_{\mathbb{R}}$, and a nonzero nilpotent $\mathrm{Ad}(\mathfrak{g}_{\mathbb{R}})$-orbit intersects both $\mathfrak{a}_1$ and $\mathfrak{a}_2$, then $\mathfrak{a}_2 = g \cdot \mathfrak{a}_1$ for some $g \in \mathrm{Ad}(\mathfrak{g}_{\mathbb{R}})$.

(F3) There are exactly two nonzero nilpotent $\mathrm{Ad}(\mathfrak{sl}_2(\mathbb{R}))$-orbits in $\mathfrak{sl}_2(\mathbb{R})$. Scalar multiplication by -1 maps one of them to the other.

(F4) Any finite dimensional $\mathfrak{sl}_2(\mathbb{R})$-$D$-module is completely reducible. For any integer $d \in \mathbb{N}$, there is a unique, up to isomorphism, d-dimensional simple $\mathfrak{sl}_2(\mathbb{R})$-$D$-module S_d. The module S_1 is trivial. Let T_m be the direct sum of m copies of S_1, and $T_0 := 0$.

(F5) If $\mathfrak{a}$ is an $\mathfrak{sl}_2(\mathbb{R})$-subalgebra of $\mathfrak{g}_{\mathbb{R}}$ and $x \in \mathfrak{a}$ is a nonzero nilpotent element, then $\mathfrak{z}_{\mathfrak{g}_{\mathbb{R}}}(\mathfrak{a})$ is the reductive Levi subalgebra of $\mathfrak{z}_{\mathfrak{g}_{\mathbb{R}}}(x)$.

(F6) If there is a nonzero $\mathfrak{sl}_2(\mathbb{R})$-invariant form on S_d of a given type (symmetric, skew-symmetric, Hermitian or skew-Hermitian), it is nondegenerate and unique up to proportionality. We fix such a form. Table 2 contains information about its existence and the notation for the fixed forms.

Table 2.

D	d	symmetric	sgn	skew–symmetric	Hermitian	sgn	skew–Hermitian
$\mathbb{R}$	even			Δ_d^{ss}			
	odd	Δ_d^s	1				
$\mathbb{C}$	even			Δ_d^{ss}	Δ_d^H	0	$i\Delta_d^H$
	odd	Δ_d^s				1	
$\mathbb{H}$	even						Δ_d^{sH}
	odd				Δ_d^H	1	

Partitions, $\mathfrak{sl}_2(\mathbb{R})$-$D$-modules, and $\mathfrak{sl}_2(\mathbb{R})$-invariant Forms

We call any vector

$$\boldsymbol{m} = (m_1, \dots, m_p) \in \mathbb{N}^p, \quad \text{where } m_p \neq 0, \tag{19}$$

a *partition* of the number

$$|\boldsymbol{m}| := \textstyle\sum_d dm_d$$

(this is nontraditional usage of the term "partition" but it is convenient for our purposes). If $p = 1$, then $\boldsymbol{m}$ is called a *trivial* partition.

The *Young diagram* of $\boldsymbol{m}$ is a left-justified array $\mathrm{Y}(\boldsymbol{m})$ of empty boxes with p boxes in each of the first m_p rows, $p-1$ boxes in each of the next m_{p-1} rows, and so on. The partition

$$\check{\boldsymbol{m}} = \{\check{m}_1, \dots, \check{m}_q\}$$

is called the *transpose partition* to $\boldsymbol{m}$ if $\mathrm{Y}(\check{\boldsymbol{m}})$ is the transpose of $\mathrm{Y}(\boldsymbol{m})$, i.e., the rows of $\mathrm{Y}(\check{\boldsymbol{m}})$ are the columns of $\mathrm{Y}(\boldsymbol{m})$ from left to right. We have $|\check{\boldsymbol{m}}| = |\boldsymbol{m}|$.

Let $\boldsymbol{m}$ be a nontrivial partition. It is called a *symmetric* (resp., *skew-symmetric*) *partition* of $|\boldsymbol{m}|$ if m_d in (19) is even for every even (resp., odd) d.

Let $\underline{\boldsymbol{m}}$ be a sequence obtained from $\boldsymbol{m}$ by replacing m_d in (19) for every d (resp., every odd d, every even d) with a pair $(p_d, q_d) \in \mathbb{N}^2$ such that $p_d + q_d = m_d$. Such $\underline{\boldsymbol{m}}$ is called a *fine* (resp., *fine Hermitian*, *fine skew-Hermitian*) *partition* of $|\boldsymbol{m}|$ *associated* with $\boldsymbol{m}$. If $\underline{\boldsymbol{m}}$ is fine or fine Hermitian,

$$\operatorname{sgn} \underline{\boldsymbol{m}} := \textstyle\sum_{d \text{ odd}} (p_d - q_d) \tag{20}$$

is called the *signature* of $\underline{m}$. If $\underline{m}$ is fine Hermitian (resp., fine skew-Hermitian) and m is symmetric (resp., skew-symmetric), then $\underline{m}$ is called a *fine symmetric* (resp., *fine skew-symmetric*) *partition* of $|m|$.

Any partition (19) defines the $|m|$-dimensional $\mathfrak{sl}_2(\mathbb{R})$-$D$-module

$$V_{m} := \bigoplus_{d\geqslant 1}(T_{m_d}\otimes S_d) \tag{21}$$

(tensor product is taken over D, with respect to the canonical D-bimodule structure on D-vector spaces). By (F4), any nonzero finite dimensional $\mathfrak{sl}_2(\mathbb{R})$-$D$-module is isomorphic to V_{m} for a unique m. It is trivial if and only if m is trivial.

Identifying every $\varphi \in \mathfrak{gl}_D(T_{m_d})$ with the transformation of V_{m} acting as $\varphi \otimes \mathrm{id}$ on the summand $T_{m_d}\otimes S_d$ and as 0 on the other summands in the right-hand side of (21), we identify $\oplus_{d\geqslant 1}\,\mathfrak{gl}_D(T_{m_d})$ with the subalgebra of $\mathfrak{gl}_D(V_{m})$.

Two $\mathfrak{sl}_2(\mathbb{R})$-invariant forms on V_{m} are called $\mathfrak{sl}_2(\mathbb{R})$-*equivalent* if there is an $\mathfrak{sl}_2(\mathbb{R})$-equivariant isometry $V_{m}\to V_{m}$ with respect to them. To describe the $\mathfrak{sl}_2(\mathbb{R})$-equivalency classes of $\mathfrak{sl}_2(\mathbb{R})$-invariant forms on V_{m}, we fix, for every positive integer r and pair $(p,q)\in\mathbb{N}^2$ with $p+q=r$, a nondegenerate form on T_r whose notation, type and signature (if applicable) are specified in Table 3.

Table 3.

D	symmetric	sgn	skew–symmetric	Hermitian	sgn	skew–Hermitian
$\mathbb{R}$	$\Theta^s_{p,q}$	$p-q$	Θ^{ss}_r, r even			
$\mathbb{C}$	Θ^s_r		Θ^{ss}_r, r even	$\Theta^H_{p,q}$	$p-q$	
$\mathbb{H}$				$\Theta^H_{p,q}$	$p-q$	Θ^{sH}_r

According to the above discussion, any nondegenerate form on T_r is equivalent to a unique form from Table 3.

If Ψ is a nondegenerate $\mathfrak{sl}_2(\mathbb{R})$-invariant form on V_{m}, then $\Psi|_d$, its restriction to the summand $T_{m_d}\otimes S_d$ in (21), is a nondegenerate $\mathfrak{sl}_2(\mathbb{R})$-invariant form of the same type (i.e., symmetric, skew-symmetric, Hermitian or skew-Hermitian) as Ψ, and different summands in (21) are orthogonal with respect to Ψ. If Ψ' is another nondegenerate $\mathfrak{sl}_2(\mathbb{R})$-invariant form on V_{m}, then Ψ and Ψ' are $\mathfrak{sl}_2(\mathbb{R})$-equivalent if and only if $\Psi|_d$ and $\Psi'|_d$ are $\mathfrak{sl}_2(\mathbb{R})$-equivalent for every d. If, for every d, a nondegenerate $\mathfrak{sl}_2(\mathbb{R})$-invariant form Φ_d on $T_{m_d}\otimes S_d$ is fixed, and all Φ_d's are of the same type, then there is a nondegenerate $\mathfrak{sl}_2(\mathbb{R})$-invariant form Ψ on V_{m} such that $\Psi|_d=\Phi_d$ for all d. This reduces describing the $\mathfrak{sl}_2(\mathbb{R})$-equivalency classes of nondegenerate $\mathfrak{sl}_2(\mathbb{R})$-invariant forms on V_{m} to describing that on $T_{m_d}\otimes S_d$ for all d. For any positive integers r and d,

Table 4 describes all, up to $\mathfrak{sl}_2(\mathbb{R})$-equivalency, nondegenerate $\mathfrak{sl}_2(\mathbb{R})$-invariant forms on $T_r \otimes S_d$ (in these tables, $p+q=r$).

Table 4.

D	d	symmetric	sgn	skew−symmetric	Hermitian	sgn	skew−Hermitian
$\mathbb{R}$	even	$\Theta_r^{ss} \otimes \Delta_d^{ss}$	0	$\Theta_{p,q}^{s} \otimes \Delta_d^{ss}$			
	odd	$\Theta_{p,q}^{s} \otimes \Delta_d^{s}$	$p-q$	$\Theta_r^{ss} \otimes \Delta_d^{s}$			
$\mathbb{C}$	even	$\Theta_r^{ss} \otimes \Delta_d^{ss}$		$\Theta_r^{s} \otimes \Delta_d^{ss}$	$\Theta_{p,q}^{H} \otimes \Delta_d^{H}$	0	
	odd	$\Theta_r^{s} \otimes \Delta_d^{s}$		$\Theta_r^{ss} \otimes \Delta_d^{s}$		$p-q$	
$\mathbb{H}$					$\Theta_r^{sH} \otimes \Delta_d^{sH}$	0	$\Theta_{p,q}^{H} \otimes \Delta_d^{sH}$
					$\Theta_{p,q}^{H} \otimes \Delta_d^{H}$	$p-q$	$\Theta_r^{sH} \otimes \Delta_d^{H}$

Returning back to the n-dimensional vector space V over D, fix a form Φ (not necessarily nondegenerate) on V. Let $\boldsymbol{m}$ be a nontrivial partition of n and let

$$\alpha_{\boldsymbol{m}} : \mathfrak{sl}_2(\mathbb{R}) \hookrightarrow \mathfrak{gl}_D(V_{\boldsymbol{m}}) \tag{22}$$

be the injection defining the $\mathfrak{sl}_2(\mathbb{R})$-$D$-module structure on $V_{\boldsymbol{m}}$. Let Ψ be an $\mathfrak{sl}_2(\mathbb{R})$-invariant form on $V_{\boldsymbol{m}}$. If Ψ and Φ are equivalent and $\iota : V_{\boldsymbol{m}} \to V$ is an isometry with respect to Ψ and Φ, then the image of the homomorphism

$$\mathfrak{sl}_2(\mathbb{R}) \to \mathfrak{gl}_D(V), \quad \varphi \mapsto \iota \circ \alpha_{\boldsymbol{m}}(\varphi) \circ \iota^{-1},$$

is an $\mathfrak{sl}_2(\mathbb{R})$-subalgebra of $\mathfrak{gl}_D^{\Phi}(V)$. The above discussion shows that any $\mathfrak{sl}_2(\mathbb{R})$-subalgebra of $\mathfrak{gl}_D^{\Phi}(V)$ is obtained in this way from some pair $\{\boldsymbol{m}, \Psi\}$. The $\mathfrak{sl}_2(\mathbb{R})$-subalgebras of $\mathfrak{gl}_D^{\Phi}(V)$ obtained in this way from the pairs $\{\boldsymbol{m}, \Psi\}$, $\{\boldsymbol{m}', \Psi'\}$ are $\mathrm{GL}_D^{\Phi}(V)$-conjugate if and only if $\boldsymbol{m} = \boldsymbol{m}'$, and Ψ and Ψ' are $\mathfrak{sl}_2(\mathbb{R})$-equivalent. This yields a bijection between the union of $\mathfrak{sl}_2(\mathbb{R})$-equivalency classes of $\mathfrak{sl}_2(\mathbb{R})$-invariant forms equivalent to Φ on $V_{\boldsymbol{m}}$, where $\boldsymbol{m}$ ranges over all nontrivial partitions of n, and the set of all $\mathrm{GL}_D^{\Phi}(V)$-conjugacy classes of $\mathfrak{sl}_2(\mathbb{R})$-subalgebras in $\mathfrak{gl}_D^{\Phi}(V)$. *Denote this bijection by* $\star$. If Ψ is either zero or nondegenerate, considering $\star$ leads to the classification of nilpotent $\mathrm{Ad}(\mathfrak{gl}_D^{\Phi}(V))$-orbits in $\mathfrak{gl}_D^{\Phi}(V)$ described below. In each case, we give the formulas for the orbit dimensions (one obtains them using (8) and [CM], [M]); then (2), (8) yield the dimensions of the corresponding K-orbits in $\mathcal{N}(\mathfrak{p})$.

• Nilpotent Orbits In $\mathfrak{sl}_D(V)$

Take $\Phi = 0$. Then $\mathrm{GL}_D^{\Phi}(V) = \mathrm{GL}_D(V)$ and $\mathfrak{gl}_D^{\Phi}(V) = \mathfrak{gl}_D(V)$. Any $\mathfrak{sl}_2(\mathbb{R})$-subalgebra of $\mathfrak{gl}_D(V)$ is contained in $\mathfrak{sl}_D(V)$. For any nontrivial partition $\boldsymbol{m}$ of n, consider the $\mathrm{GL}_D^{\Phi}(V)$-conjugacy class of $\mathfrak{sl}_2(\mathbb{R})$-subalgebras in $\mathfrak{sl}_D(V)$

corresponding under $\star$ to the $\mathfrak{sl}_2(\mathbb{R})$-equivalency class of zero form on $V_{\boldsymbol{m}}$. Then by (F1)–(F3), [KR], the subset of $\mathfrak{gl}_D^{\Phi}(V)$ that is the union of all $\mathfrak{sl}_2(\mathbb{R})$-subalgebras from this class contains a unique nonzero nilpotent $\mathrm{GL}_D(V)$-orbit $\mathcal{O}_{\boldsymbol{m}}$.

$\mathrm{Ad}(\mathfrak{sl}_D(V))$-orbits in $\mathcal{N}(\mathfrak{sl}_D(V))$:

> $\mathcal{O}_{\boldsymbol{m}}$ is an $\mathrm{Ad}(\mathfrak{sl}_D(V))$-orbit, except that $\mathcal{O}_{\boldsymbol{m}}$ is the union of two $\mathrm{Ad}(\mathfrak{sl}_D(V))$-orbits $\mathcal{O}^1_{\boldsymbol{m}}$ and $\mathcal{O}^2_{\boldsymbol{m}}$ (that are the connected components of $\mathcal{O}_{\boldsymbol{m}}$) if $D = \mathbb{R}$ and $m_d = 0$ for every odd d. Such $\mathrm{Ad}(\mathfrak{sl}_D(V))$-orbits, taken over all partitions $\boldsymbol{m}$ of n, are pairwise different and exhaust all nonzero $\mathrm{Ad}(\mathfrak{sl}_D(V))$-orbits in $\mathcal{N}(\mathfrak{sl}_D(V))$.

Orbit dimensions:

$$n^2 - \textstyle\sum_d d^2 \check{m}_d = \begin{cases} \dim_{\mathbb{R}} \mathcal{O}_{\boldsymbol{m}} & \text{for } D = \mathbb{R}, \\ \dim_{\mathbb{C}} \mathcal{O}_{\boldsymbol{m}} & \text{for } D = \mathbb{C}, \\ \dim_{\mathbb{R}} \mathcal{O}_{\boldsymbol{m}}/4 & \text{for } D = \mathbb{H}. \end{cases}$$

• Nilpotent Orbits In $\mathfrak{gl}_D^{\Phi}(V)$ for $D = \mathbb{R}$ and Nondegenerate Symmetric Φ

Let $\underline{\boldsymbol{m}}$ be a fine symmetric partition of n associated with (19). Let $\Psi_{\underline{\boldsymbol{m}}}$ be the $\mathfrak{sl}_2(\mathbb{R})$-invariant form on $V_{\boldsymbol{m}}$ such that for all d,

$$\Psi_{\underline{\boldsymbol{m}}}|_d = \begin{cases} \Theta^{s}_{p_d,q_d} \otimes \Delta^{s}_d & \text{if } d \text{ is odd}, \\ \Theta^{ss}_{m_d} \otimes \Delta^{ss}_d & \text{if } d \text{ is even}. \end{cases} \tag{23}$$

Then $\Psi_{\underline{\boldsymbol{m}}}$ is equivalent to Φ if and only if

$$\operatorname{sgn} \underline{\boldsymbol{m}} = \operatorname{sgn} \Phi. \tag{24}$$

If (24) holds, consider the $\mathrm{GL}_D^{\Phi}(V)$-conjugacy class of $\mathfrak{sl}_2(\mathbb{R})$-subalgebras in $\mathfrak{gl}_D^{\Phi}(V)$ corresponding under $\star$ to the $\mathfrak{sl}_2(\mathbb{R})$-equivalency class of $\Psi_{\underline{\boldsymbol{m}}}$. Then by (F1)–(F3), [KR], the union of all $\mathfrak{sl}_2(\mathbb{R})$-subalgebras from this class contains a unique nilpotent $\mathrm{GL}_D^{\Phi}(V)$-orbit $\mathcal{O}_{\underline{\boldsymbol{m}}}$.

$\mathrm{Ad}(\mathfrak{gl}_D^{\Phi}(V))$-orbits in $\mathcal{N}(\mathfrak{gl}_D^{\Phi}(V))$:

> If (24) holds, $\mathcal{O}_{\underline{\boldsymbol{m}}}$ is an $\mathrm{Ad}(\mathfrak{gl}_D^{\Phi}(V))$-orbit, except the following cases. If $m_d = 0$ for all odd d, then $\mathcal{O}_{\underline{\boldsymbol{m}}}$ is the union of four $\mathrm{Ad}(\mathfrak{gl}_D^{\Phi}(V))$-orbits $\mathcal{O}^1_{\underline{\boldsymbol{m}}}$, $\mathcal{O}^2_{\underline{\boldsymbol{m}}}$, $\mathcal{O}^3_{\underline{\boldsymbol{m}}}$, $\mathcal{O}^4_{\underline{\boldsymbol{m}}}$ that are the connected components of $\mathcal{O}_{\underline{\boldsymbol{m}}}$. If there is an odd d such that $m_d \neq 0$ and
>
> $$\text{either} \quad \begin{cases} p_d = 0 & \text{for all } d \equiv 1 \bmod 4, \\ q_d = 0 & \text{for all } d \equiv 3 \bmod 4 \end{cases}$$
>
> $$\text{or} \quad \begin{cases} p_d = 0 & \text{for all } d \equiv 3 \bmod 4, \\ q_d = 0 & \text{for all } d \equiv 1 \bmod 4, \end{cases}$$

then $\mathcal{O}_{\underline{\boldsymbol{m}}}$ is the union of two $\mathrm{Ad}(\mathfrak{gl}_D^{\Phi}(V))$-orbits $\mathcal{O}^1_{\underline{\boldsymbol{m}}}$, $\mathcal{O}^2_{\underline{\boldsymbol{m}}}$ that are the connected components of $\mathcal{O}_{\underline{\boldsymbol{m}}}$. Such $\mathrm{Ad}(\mathfrak{gl}_D^{\Phi}(V))$-orbits, taken over all fine symmetric partitions $\underline{\boldsymbol{m}}$ of n satisfying (24), are pairwise different and exhaust all nonzero $\mathrm{Ad}(\mathfrak{gl}_D^{\Phi}(V))$-orbits in $\mathcal{N}(\mathfrak{gl}_D^{\Phi}(V))$.

Orbit dimensions:

$$\dim_{\mathbb{R}} \mathcal{O}_{\underline{\boldsymbol{m}}} = \big(n^2 - n - \textstyle\sum_d d^2 \check{m}_d + \sum_{d \text{ odd}} m_d\big)/2.$$

• Nilpotent Orbits In $\mathfrak{gl}_D^{\Phi}(V)$ for $D = \mathbb{C}$ and Nondegenerate Symmetric Φ

Let $\boldsymbol{m}$ be a symmetric partition of n. Let $\Psi_{\boldsymbol{m}}$ be the $\mathfrak{sl}_2(\mathbb{R})$-invariant form on $V_{\boldsymbol{m}}$ such that for all d,

$$\Psi_{\boldsymbol{m}}|_d = \begin{cases} \Theta^s_{m_d} \otimes \Delta^s_d & \text{if } d \text{ is odd,} \\ \Theta^{ss}_{m_d} \otimes \Delta^{ss}_d & \text{if } d \text{ is even.} \end{cases}$$

Then $\Psi_{\boldsymbol{m}}$ is equivalent to Φ. Consider the $\mathrm{GL}_D^{\Phi}(V)$-conjugacy class of $\mathfrak{sl}_2(\mathbb{R})$-subalgebras in $\mathfrak{gl}_D^{\Phi}(V)$ corresponding under $\star$ to the $\mathfrak{sl}_2(\mathbb{R})$-equivalency class of $\Psi_{\boldsymbol{m}}$. Then by (F1)–(F3), [KR], the union of all $\mathfrak{sl}_2(\mathbb{R})$-subalgebras from this class contains a unique nilpotent $\mathrm{GL}_D^{\Phi}(V)$-orbit $\mathcal{O}_{\boldsymbol{m}}$.

$\mathrm{Ad}(\mathfrak{gl}_D^{\Phi}(V))$-orbits in $\mathcal{N}(\mathfrak{gl}_D^{\Phi}(V))$:

$\mathcal{O}_{\boldsymbol{m}}$ is an $\mathrm{Ad}(\mathfrak{gl}_D^{\Phi}(V))$-orbit, except that if $m_d = 0$ for all odd d, then $\mathcal{O}_{\boldsymbol{m}}$ is the union of two $\mathrm{Ad}(\mathfrak{gl}_D^{\Phi}(V))$-orbits $\mathcal{O}^1_{\boldsymbol{m}}$ and $\mathcal{O}^2_{\boldsymbol{m}}$ that are the connected components of $\mathcal{O}_{\boldsymbol{m}}$. Such $\mathrm{Ad}(\mathfrak{gl}_D^{\Phi}(V))$-orbits, taken over all symmetric partitions $\boldsymbol{m}$ of n, are pairwise different and exhaust all nonzero $\mathrm{Ad}(\mathfrak{gl}_D^{\Phi}(V))$-orbits in $\mathcal{N}(\mathfrak{gl}_D^{\Phi}(V))$.

Orbit dimensions:

$$\dim_{\mathbb{C}} \mathcal{O}_{\boldsymbol{m}} = \big(n^2 - n - \textstyle\sum_d d^2 \check{m}_d + \sum_{d \text{ odd}} m_d\big)/2.$$

• Nilpotent Orbits In $\mathfrak{gl}_D^{\Phi}(V)$ for $D = \mathbb{R}$ and Nondegenerate Skew-symmetric Φ

Let $\underline{\boldsymbol{m}}$ be a fine skew-symmetric partition of n associated with (19). Let $\Psi_{\underline{\boldsymbol{m}}}$ be the $\mathfrak{sl}_2(\mathbb{R})$-invariant form on $V_{\boldsymbol{m}}$ such that for all d,

$$\Psi_{\underline{\boldsymbol{m}}}|_d = \begin{cases} \Theta^s_{p_d, q_d} \otimes \Delta^{ss}_d & \text{if } d \text{ is even,} \\ \Theta^{ss}_{m_d} \otimes \Delta^s_d & \text{if } d \text{ is odd.} \end{cases} \tag{25}$$

Then $\Psi_{\underline{\boldsymbol{m}}}$ is equivalent to Φ. Consider the $\mathrm{GL}_D^\Phi(V)$-conjugacy class of $\mathfrak{sl}_2(\mathbb{R})$-subalgebras in $\mathfrak{gl}_D^\Phi(V)$ corresponding under $\star$ to the $\mathfrak{sl}_2(\mathbb{R})$-equivalency class of $\Psi_{\underline{\boldsymbol{m}}}$. Then by (F1)–(F3), [KR], the union of all $\mathfrak{sl}_2(\mathbb{R})$-subalgebras from this class contains a unique nilpotent $\mathrm{GL}_D^\Phi(V)$-orbit $\mathcal{O}_{\underline{\boldsymbol{m}}}$.

$\mathrm{Ad}(\mathfrak{gl}_D^\Phi(V))$-orbits in $\mathcal{N}(\mathfrak{gl}_D^\Phi(V))$:

> $\mathcal{O}_{\underline{\boldsymbol{m}}}$ is an $\mathrm{Ad}(\mathfrak{gl}_D^\Phi(V))$-orbit. Such $\mathrm{Ad}(\mathfrak{gl}_D^\Phi(V))$-orbits, taken over all fine skew-symmetric partitions $\underline{\boldsymbol{m}}$ of n, are pairwise different and exhaust all nonzero $\mathrm{Ad}(\mathfrak{gl}_D^\Phi(V))$-orbits in $\mathcal{N}(\mathfrak{gl}_D^\Phi(V))$.

Orbit dimensions:

$$\dim_{\mathbb{R}} \mathcal{O}_{\underline{\boldsymbol{m}}} = \left(n^2 + n - \textstyle\sum_d d^2 \check{m}_d - \sum_{d \text{ odd}} m_d\right)/2.$$

• Nilpotent Orbits In $\mathfrak{gl}_D^\Phi(V)$ for $D = \mathbb{C}$ and Nondegenerate Skew-symmetric Φ

Let $\boldsymbol{m}$ be a skew-symmetric partition of n. Let $\Psi_{\boldsymbol{m}}$ be the $\mathfrak{sl}_2(\mathbb{R})$-invariant form on $V_{\boldsymbol{m}}$ such that for all d,

$$\Psi_{\underline{\boldsymbol{m}}}|_d = \begin{cases} \Theta^s_{m_d} \otimes \Delta^{ss}_d & \text{if } d \text{ is even,} \\ \Theta^{ss}_{m_d} \otimes \Delta^s_d & \text{if } d \text{ is odd.} \end{cases}$$

Then $\Psi_{\underline{\boldsymbol{m}}}$ is equivalent to Φ. Consider the $\mathrm{GL}_D^\Phi(V)$-conjugacy class of $\mathfrak{sl}_2(\mathbb{R})$-subalgebras in $\mathfrak{gl}_D^\Phi(V)$ corresponding under $\star$ to the $\mathfrak{sl}_2(\mathbb{R})$-equivalency class of $\Psi_{\boldsymbol{m}}$. Then by (F1)–(F3), [KR], the union of all $\mathfrak{sl}_2(\mathbb{R})$-subalgebras from this class contains a unique nilpotent $\mathrm{GL}_D^\Phi(V)$-orbit $\mathcal{O}_{\boldsymbol{m}}$.

$\mathrm{Ad}(\mathfrak{gl}_D^\Phi(V))$-orbits in $\mathcal{N}(\mathfrak{gl}_D^\Phi(V))$:

> $\mathcal{O}_{\boldsymbol{m}}$ is an $\mathrm{Ad}(\mathfrak{gl}_D^\Phi(V))$-orbit. Such $\mathrm{Ad}(\mathfrak{gl}_D^\Phi(V))$-orbits, taken over all skew-symmetric partitions $\boldsymbol{m}$ of n, are pairwise different and exhaust all nonzero $\mathrm{Ad}(\mathfrak{gl}_D^\Phi(V))$-orbits in $\mathcal{N}(\mathfrak{gl}_D^\Phi(V))$.

Orbit dimensions:

$$\dim_{\mathbb{C}} \mathcal{O}_{\boldsymbol{m}} = \left(n^2 + n - \textstyle\sum_d d^2 \check{m}_d - \sum_{d \text{ odd}} m_d\right)/2.$$

• Nilpotent Orbits In $\mathfrak{gl}_D^\Phi(V)$ for $D = \mathbb{C}$ and Nondegenerate Hermitian Φ

Let $\underline{\boldsymbol{m}}$ be a fine partition of n associated with (19). Let $\Psi_{\underline{\boldsymbol{m}}}$ be the $\mathfrak{sl}_2(\mathbb{R})$-invariant form on $V_{\boldsymbol{m}}$ such that for all d,

$$\Psi_{\underline{\boldsymbol{m}}}|_d = \Theta^H_{p_d, q_d} \otimes \Delta^H_d. \tag{26}$$

Then $\Psi_{\boldsymbol{m}}$ is equivalent to Φ if and only if (24) holds. In the last case, consider the $\mathrm{GL}_D^\Phi(V)$-conjugacy class of $\mathfrak{sl}_2(\mathbb{R})$-subalgebras in $\mathfrak{gl}_D^\Phi(V)$ corresponding under $\star$ to the $\mathfrak{sl}_2(\mathbb{R})$-equivalency class of $\Psi_{\underline{m}}$. Then by (F1)–(F3), [KR], the union of all $\mathfrak{sl}_2(\mathbb{R})$-subalgebras from this class contains a unique nilpotent $\mathrm{GL}_D^{\Phi}(V)$-orbit $\mathcal{O}_{\underline{m}}$.

$\mathrm{Ad}(\mathfrak{gl}_D^{\Phi}(V))$-orbits in $\mathcal{N}(\mathfrak{gl}_D^{\Phi}(V))$:

If (24) holds, $\mathcal{O}_{\underline{m}}$ is an $\mathrm{Ad}(\mathfrak{gl}_D^{\Phi}(V))$-orbit. Such $\mathrm{Ad}(\mathfrak{gl}_D^{\Phi}(V))$-orbits, taken over all fine partitions $\underline{m}$ of n satisfying (24), are pairwise different and exhaust all nonzero $\mathrm{Ad}(\mathfrak{gl}_D^{\Phi}(V))$-orbits in $\mathcal{N}(\mathfrak{gl}_D^{\Phi}(V))$.

Orbit dimensions:

$$\dim_{\mathbb{R}} \mathcal{O}_{\underline{m}} = n^2 - \textstyle\sum_d d^2 \check{m}_d.$$

• Nilpotent Orbits In $\mathfrak{gl}_D^{\Phi}(V)$ for $D = \mathbb{H}$ and Nondegenerate Hermitian Φ

Let $\underline{m}$ be a fine Hermitian partition of n associated with (19). Let $\Psi_{\underline{m}}$ be the $\mathfrak{sl}_2(\mathbb{R})$-invariant form on $V_{\underline{m}}$ such that for all d,

$$\Psi_{\underline{m}}|_d = \begin{cases} \Theta_{m_d}^{sH} \otimes \Delta_d^{sH} & \text{if } d \text{ is even,} \\ \Theta_{p_d,q_d}^{H} \otimes \Delta_d^{H} & \text{if } d \text{ is odd.} \end{cases} \tag{27}$$

Then $\Psi_{\underline{m}}$ is equivalent to Φ if and only if (24) holds. In the last case, consider the $\mathrm{GL}_D^{\Phi}(V)$-conjugacy class of $\mathfrak{sl}_2(\mathbb{R})$-subalgebras in $\mathfrak{gl}_D^{\Phi}(V)$ corresponding under $\star$ to the $\mathfrak{sl}_2(\mathbb{R})$-equivalency class of $\Psi_{\underline{m}}$. Then by (F1)–(F3), [KR], the union of all $\mathfrak{sl}_2(\mathbb{R})$-subalgebras from this class contains a unique nilpotent $\mathrm{GL}_D^{\Phi}(V)$-orbit $\mathcal{O}_{\underline{m}}$.

$\mathrm{Ad}(\mathfrak{gl}_D^{\Phi}(V))$-orbits in $\mathcal{N}(\mathfrak{gl}_D^{\Phi}(V))$:

If (24) holds, $\mathcal{O}_{\underline{m}}$ is an $\mathrm{Ad}(\mathfrak{gl}_D^{\Phi}(V))$-orbit. Such $\mathrm{Ad}(\mathfrak{gl}_D^{\Phi}(V))$-orbits, taken over all fine Hermitian partitions $\underline{m}$ of n satisfying (24), are pairwise different and exhaust all nonzero $\mathrm{Ad}(\mathfrak{gl}_D^{\Phi}(V))$-orbits in $\mathcal{N}(\mathfrak{gl}_D^{\Phi}(V))$.

Orbit dimensions:

$$\dim_{\mathbb{R}} \mathcal{O}_{\underline{m}} = 2n^2 + n - 2\textstyle\sum_d d^2 \check{m}_d - \sum_{d \text{ odd}} m_d.$$

• Nilpotent Orbits In $\mathfrak{gl}_D^{\Phi}(V)$ for $D = \mathbb{H}$ and Skew-Hermitian Φ

Let $\underline{m}$ be a fine skew-Hermitian partition of n associated with (19). Let $\Psi_{\underline{m}}$ be the $\mathfrak{sl}_2(\mathbb{R})$-invariant form on $V_{\underline{m}}$ such that for all d,

$$\Psi_{\underline{m}}|_d = \begin{cases} \Theta_{p_d,q_d}^{H} \otimes \Delta_d^{sH} & \text{if } d \text{ is even,} \\ \Theta_{m_d}^{sH} \otimes \Delta_d^{H} & \text{if } d \text{ is odd.} \end{cases} \tag{28}$$

Then $\Psi_{\underline{\boldsymbol{m}}}$ is equivalent to Φ. Consider the $\mathrm{GL}_D^\Phi(V)$-conjugacy class of $\mathfrak{sl}_2(\mathbb{R})$-subalgebras in $\mathfrak{gl}_D^\Phi(V)$ corresponding under $\star$ to the $\mathfrak{sl}_2(\mathbb{R})$-equivalency class of $\Psi_{\underline{\boldsymbol{m}}}$. Then by (F1)–(F3), [KR], the union of all $\mathfrak{sl}_2(\mathbb{R})$-subalgebras from this class contains a unique nilpotent $\mathrm{GL}_D^\Phi(V)$-orbit $\mathcal{O}_{\underline{\boldsymbol{m}}}$.

$\mathrm{Ad}(\mathfrak{gl}_D^\Phi(V))$-orbits in $\mathcal{N}(\mathfrak{gl}_D^\Phi(V))$:

> $\mathcal{O}_{\underline{\boldsymbol{m}}}$ is an $\mathrm{Ad}(\mathfrak{gl}_D^\Phi(V))$-orbit. Such $\mathrm{Ad}(\mathfrak{gl}_D^\Phi(V))$-orbits, taken over all fine skew-Hermitian partitions $\underline{\boldsymbol{m}}$ of n, are pairwise different and exhaust all $\mathrm{Ad}(\mathfrak{gl}_D^\Phi(V))$-orbits in $\mathcal{N}(\mathfrak{gl}_D^\Phi(V))$.

Orbit dimensions:

$$\dim_{\mathbb{R}} \mathcal{O}_{\underline{\boldsymbol{m}}} = 2n^2 - n - 2\sum_d d^2 \check{m}_d + \sum_{d \text{ odd}} m_d.$$

Now we are ready to classify compact nilpotent elements in the real forms of complex classical Lie algebras. Recall that $n = \dim_D V$.

• Compact Nilpotent Elements In $\mathfrak{sl}_n(\mathbb{R})$ and $\mathfrak{sl}_n(\mathbb{H})$

Theorem 8. *Let $\mathcal{O}$ be the $\mathrm{Ad}(\mathfrak{sl}_D(V))$-orbit of a nonzero element $x \in \mathcal{N}(\mathfrak{sl}_D(V))$ where D is $\mathbb{R}$ or $\mathbb{H}$. Let $\mathfrak{a}$ be an $\mathfrak{sl}_2(\mathbb{R})$-subalgebra of $\mathfrak{sl}_D(V)$ containing x. The following properties are equivalent:*

(i) *x is compact.*
(ii) *$\mathfrak{a}$-module V is simple.*
(iii) *If $D = \mathbb{R}$, then $\mathcal{O} = \mathcal{O}_{(0,\dots,0,1)}$ for odd n, and $\mathcal{O}^1_{(0,\dots,0,1)}$ or $\mathcal{O}^2_{(0,\dots,0,1)}$ for even n. If $D = \mathbb{H}$, then $\mathcal{O} = \mathcal{O}_{(0,\dots,0,1)}$.*

Proof. According to the previous discussion, we may, and will, assume that $V = V_{\boldsymbol{m}}$ and $\mathfrak{a} = \alpha_{\boldsymbol{m}}(\mathfrak{sl}_2(\mathbb{R}))$ (see (22)) for some nontrivial partition $\boldsymbol{m}$ of n. The Double Centralizer Theorem implies that

$$\mathfrak{z}_{\mathfrak{gl}_D(V)}(\mathfrak{a}) = \bigoplus_{d \geqslant 1} \mathfrak{gl}_D(T_{m_d}). \tag{29}$$

Since $\mathfrak{gl}_D(T_{m_d})$ (resp., $\mathfrak{sl}_D(T_{m_d})$) is compact if and only if $m_d = 0$ (resp., 0 or 1), the claim follows from (29) and (F5). □

• Compact Nilpotent Elements In $\mathfrak{so}_{n-q,q}$

Theorem 9. *Let $\mathcal{O}$ be the $\mathrm{Ad}(\mathfrak{gl}_{\mathbb{R}}^\Phi(V))$-orbit of a nonzero element $x \in \mathcal{N}(\mathfrak{gl}_{\mathbb{R}}^\Phi(V))$ for a nondegenerate symmetric form Φ. The following properties are equivalent:*

(i) *x is compact.*
(ii) *$\mathcal{O} = \mathcal{O}^1_{\underline{\boldsymbol{m}}}$ or $\mathcal{O}^2_{\underline{\boldsymbol{m}}}$, where $\underline{\boldsymbol{m}}$ is a fine symmetric partition of n such that (24) holds, $p_d q_d = 0$ for all odd d, and $m_d = 0$ for all even d.*

Proof. We may, and will, assume that $V = V_{\boldsymbol{m}}$, Φ is $\Psi_{\underline{\boldsymbol{m}}}$ defined by (23), and $x \in \mathfrak{a} := \alpha_{\boldsymbol{m}}(\mathfrak{sl}_2(\mathbb{R}))$ for some fine symmetric partition $\underline{\boldsymbol{m}}$ of n such that (24) holds. Then (29) holds, whence by (23)

$$\mathfrak{z}_{\mathfrak{gl}_{\mathbb{R}}^{\Phi}(V)}(\mathfrak{a}) = \bigoplus_{d \geqslant 1} \mathfrak{gl}_{\mathbb{R}}^{\Theta_d}(T_{m_d}) \text{ where } \Theta_d = \begin{cases} \Theta^{s}_{p_d, q_d} & \text{if } d \text{ is odd,} \\ \Theta^{ss}_{m_d} & \text{if } d \text{ is even.} \end{cases} \tag{30}$$

Since $\mathfrak{gl}_{\mathbb{R}}^{\Theta_d}(T_{m_d})$ for Θ_d given by (30) is compact if and only if $p_d q_d = 0$ for odd d and $m_d = 0$ for even d (see Table 1 and (18)), the claim follows from (30) and (F5). □

• Compact Nilpotent Elements In $\mathfrak{sp}_n(\mathbb{R})$

Theorem 10. *Let $\mathcal{O}$ be the $\mathrm{Ad}(\mathfrak{gl}_{\mathbb{R}}^{\Phi}(V))$-orbit of a nonzero element $x \in \mathcal{N}(\mathfrak{gl}_{\mathbb{R}}^{\Phi}(V))$ for a nondegenerate skew-symmetric form Φ. The following properties are equivalent:*

(i) *x is compact.*
(ii) *$\mathcal{O} = \mathcal{O}_{\underline{\boldsymbol{m}}}$ where $\underline{\boldsymbol{m}}$ is a fine skew-symmetric partition of n such that $p_d q_d = 0$ for all even d, and $m_d = 0$ for all odd d.*

Proof. The arguments are similar to that in the proof of Theorem 9 with (23) replaced with (25). □

• Compact Nilpotent Elements In $\mathfrak{su}_{n-q,q}$

Theorem 11. *Let $\mathcal{O}$ be the $\mathrm{Ad}(\mathfrak{gl}_{\mathbb{C}}^{\Phi}(V))$-orbit of a nonzero element $x \in \mathcal{N}(\mathfrak{gl}_{\mathbb{C}}^{\Phi}(V))$ for a nondegenerate Hermitian form Φ. The following properties are equivalent:*

(i) *x is compact.*
(ii) *$\mathcal{O} = \mathcal{O}_{\underline{\boldsymbol{m}}}$ where $\underline{\boldsymbol{m}}$ is a fine partition of n such that (24) holds and $p_d q_d = 0$ for all d.*

Proof. By Table 1, the algebra $\mathfrak{gl}_{\mathbb{C}}^{\Psi}(W)$ for $\Psi = \Theta^{H}_{p_d, q_d}$ and $W = T_{m_d}$ is compact if and only if $p_d q_d = 0$. Now one completes the proof using the arguments similar to that in the proof of Theorem 9 with (23) replaced with (26). □

• Compact Nilpotent Elements In $\mathfrak{sp}_{n/2-q,q}$

Theorem 12. *Let $\mathcal{O}$ be the $\mathrm{Ad}(\mathfrak{gl}_{\mathbb{H}}^{\Phi}(V))$-orbit of a nonzero element $x \in \mathcal{N}(\mathfrak{gl}_{\mathbb{H}}^{\Phi}(V))$ for a nondegenerate Hermitian form Φ. The following properties are equivalent:*

(i) *x is compact.*

(ii) $\mathcal{O} = \mathcal{O}_{\underline{\mathbf{m}}}$ *where* $\underline{\mathbf{m}}$ *is a fine Hermitian partition of* n *such that* (24) *holds,* $p_d q_d = 0$ *for all odd* d*, and* $m_d = 0$ *or* 1 *for all even* d*.*

Proof. It follows from Table 1 and (18) that the algebra $\mathfrak{gl}_{\mathbb{H}}^{\Psi}(W)$ for $\Psi = \Theta_{p_d,q_d}^{H}$ (resp., $\Theta_{m_d}^{sH}$) and $W = T_{m_d}$ is compact if and only if $p_d q_d = 0$ (resp., $m_d = 0$ or 1). Now one completes the proof using the arguments similar to that in the proof of Theorem 9 with (23) replaced with (27). □

• Compact Nilpotent Elements In $\mathfrak{u}_n^*(\mathbb{H})$

Theorem 13. *Let* $\mathcal{O}$ *be the* $\mathrm{Ad}(\mathfrak{gl}_{\mathbb{H}}^{\Phi}(V))$*-orbit of a nonzero element* $x \in \mathcal{N}(\mathfrak{gl}_{\mathbb{H}}^{\Phi}(V))$ *for a nondegenerate skew-Hermitian form* Φ*. The following properties are equivalent:*

(i) x *is compact.*
(ii) $\mathcal{O} = \mathcal{O}_{\underline{\mathbf{m}}}$ *where* $\underline{\mathbf{m}}$ *is a fine skew-Hermitian partition of* n *such that* $p_d q_d = 0$ *for all even* d*, and* $m_d = 0$ *or* 1 *for all odd* d*.*

Proof. The arguments are similar to that in the proof of Theorem 12 with (27) replaced with (28). □

5 Classification For Exceptional Simple $\mathfrak{g}$

In this section, we assume that $\mathfrak{g}$ is an exceptional simple complex Lie algebra and $\mathfrak{g}_{\mathbb{R}}$ is its noncompact real form. According to E. Cartan's classification (cf., e.g., [OV]), up to isomorphism there are, in total, twelve possibilities listed below in Table 5. Its second column gives both types of E. Cartan's notation for $\mathfrak{g}_{\mathbb{R}}$; the notation $\mathtt{X}_{s(t)}$ means that $\mathtt{X}_s$ is the type of $\mathfrak{g}$ and t is the signature of the Killing form of $\mathfrak{g}_{\mathbb{R}}$, i.e., $t = \dim_{\mathbb{C}} \mathfrak{p} - \dim_{\mathbb{C}} \mathfrak{k}$.

Recall the classical approach to classifying nilpotent orbits by means of their characteristics and weighted Dynkin diagrams, cf. [CM], [M]. Fix a θ-stable Cartan subalgebra $\mathfrak{t}$ of $\mathfrak{g}$ such that $\mathfrak{t} \cap \mathfrak{k}$ is a Cartan subalgebra of $\mathfrak{k}$. Denote

$$l := \dim_{\mathbb{C}} \mathfrak{t}, \quad s := \dim_{\mathbb{C}}(\mathfrak{t} \cap \mathfrak{k}).$$

Let $\Delta_{(\mathfrak{g},\mathfrak{t})}$ and $\Delta_{(\mathfrak{k},\mathfrak{t}\cap\mathfrak{k})}$ be the root systems of resp. $(\mathfrak{g}, \mathfrak{t})$ and $(\mathfrak{k}, \mathfrak{t} \cap \mathfrak{k})$. Given a nonzero element $e \in \mathcal{N}(\mathfrak{g})$ (resp., $e \in \mathcal{N}(\mathfrak{p})$), there are the elements $h, f \in \mathfrak{g}$ (resp., $h \in \mathfrak{k}$, $f \in \mathfrak{p}$) such that $\{e, h, f\}$ is an $\mathfrak{sl}_2$-triple. The intersection of $\mathfrak{t}$ (resp., $\mathfrak{t}\cap\mathfrak{k}$) with the orbit $\mathcal{O} = G\cdot e$ (resp., $K\cdot e$) contains a unique element h_0 lying in a fixed Weyl chamber of $\Delta_{(\mathfrak{g},\mathfrak{t})}$ (resp., $\Delta_{(\mathfrak{k},\mathfrak{t}\cap\mathfrak{k})}$). The mapping $\mathcal{O} \mapsto h_0$ is a well defined injection of the set of all nonzero G-orbits in $\mathcal{N}(\mathfrak{g})$ (resp., K-orbits in $\mathcal{N}(\mathfrak{p})$) into $\mathfrak{t}$ (resp., $\mathfrak{t} \cap \mathfrak{k}$). The image h_0 of the orbit $\mathcal{O}$ under this injection is called the *characteristic* of $\mathcal{O}$ (and e). Thus the G-orbits in $\mathcal{N}(\mathfrak{g})$ (resp., K-orbits in $\mathcal{N}(\mathfrak{p})$) are defined by their characteristics. In turn, the characteristics are defined by the numerical data, namely, the system of

Table 5.

$\mathfrak{g}$	$\mathfrak{g}_{\mathbb{R}}$	type of $\mathfrak{k}_{\mathbb{R}}$	$\dim_{\mathbb{C}} \mathfrak{k}$	$\dim_{\mathbb{C}} \mathfrak{p}$
E_6	$EI=E_{6(6)}$	$\mathfrak{sp}_4$	36	42
E_6	$EII=E_{6(2)}$	$\mathfrak{su}_2 \oplus \mathfrak{su}_6$	38	40
E_6	$EIII=E_{6(-14)}$	$\mathfrak{so}_{10} \oplus \mathbb{R}$	46	32
E_6	$EIV=E_{6(-26)}$	$F_{4(-52)}$	52	26
E_7	$EV=E_{7(7)}$	$\mathfrak{su}_8$	63	70
E_7	$EVI=E_{7(-5)}$	$\mathfrak{su}_2 \oplus \mathfrak{so}_{12}$	69	64
E_7	$EVII=E_{7(-25)}$	$E_{6(-78)} \oplus \mathbb{R}$	79	54
E_8	$EVIII=E_{8(8)}$	$\mathfrak{so}_{16}$	120	128
E_8	$EIX=E_{8(-24)}$	$\mathfrak{su}_2 \oplus E_{7(-133)}$	136	112
F_4	$FI=F_{4(4)}$	$\mathfrak{su}_2 \oplus \mathfrak{sp}_3$	24	28
F_4	$FII=F_{4(-20)}$	$\mathfrak{so}_9$	36	16
G_2	$GI=G_{2(2)}$	$\mathfrak{so}_3 \oplus \mathfrak{so}_3$	6	8

values $\{\beta_j(h_0)\}$ where $\{\beta_j\}$ is a fixed basis of $\mathfrak{t}^*$ (resp., $(\mathfrak{t}\cap\mathfrak{k})^*$). In practice, $\{\beta_j\}$ is always chosen to be a base of some root system. Then assigning the integer $\beta_j(h_0)$ to each node β_j of the Dynkin diagram of this base yields the *weighted Dynkin diagram* of the orbit $\mathcal{O}$, denoted by $\operatorname{Dyn}\mathcal{O}$. It uniquely defines $\mathcal{O}$. This choice of the base $\{\beta_j\}$ is the following.

For $\mathfrak{t}^*$, since $\mathfrak{g}$ is semisimple, it is natural to take $\{\beta_j\}$ to be the base $\alpha_1,\ldots,\alpha_l$ of $\Delta_{(\mathfrak{g},\mathfrak{t})}$ defining the fixed Weyl chamber that is used in the definition of characteristics. Below the Dynkin diagrams of G-orbits in $\mathcal{N}(\mathfrak{g})$ are considered with respect to this base $\alpha_1,\ldots,\alpha_l$.

To describe $\{\beta_j\}$ for $(\mathfrak{t}\cap\mathfrak{k})^*$, denote by α_0 the lowest root of $\Delta_{(\mathfrak{g},\mathfrak{t})}$. The extended Dynkin diagrams of $\Delta_{(\mathfrak{g},\mathfrak{t})}$ indicating the location of the α_i's are given in Table 6.

The real form $\mathfrak{g}_{\mathbb{R}}$ of $\mathfrak{g}$ is of inner type (i.e., $s=l$, $\mathfrak{t}\subset\mathfrak{k}$, and thereby $\Delta_{(\mathfrak{k},\mathfrak{t}\cap\mathfrak{k})}\subset\Delta_{(\mathfrak{g},\mathfrak{t})}$) if and only if $\mathfrak{g}_{\mathbb{R}}\neq E_{6(6)}, E_{6(-26)}$. If $\mathfrak{g}_{\mathbb{R}}$ is of inner type, then $\mathfrak{k}$ is semisimple if and only if $\mathfrak{g}_{\mathbb{R}}\neq E_{6(-14)}, E_{7(-25)}$; in the last two cases the center of $\mathfrak{k}$ is one-dimensional. If $\mathfrak{g}_{\mathbb{R}}$ is of inner type, then the set $\{\alpha_0,\ldots,\alpha_l\}$ contains a unique base $\beta_1,\ldots,\beta_r$ of $\Delta_{(\mathfrak{k},\mathfrak{t}\cap\mathfrak{k})}$. Moreover, α_0 is one of the β_i's $\iff$ $\mathfrak{k}$ is semisimple $\iff r=l$. If $\mathfrak{g}_{\mathbb{R}}$ is of inner type and $\mathfrak{k}$ is not semisimple, then $r=l-1$ and $\beta_j\in\{\alpha_1,\ldots,\alpha_l\}$ for all $j=1,\ldots,r$; in this case we set $\beta_l:=\alpha_l$.

Thus in all cases where $\mathfrak{g}_{\mathbb{R}}$ is of inner type, we have a basis $\{\beta_j\}$ of $(\mathfrak{t}\cap\mathfrak{k})^*$ that is the system of simple roots of some root system. The location of the β_j's is given in Table 6.

If $\mathfrak{g}_{\mathbb{R}}$ is of outer type, then $\mathfrak{k}$ is semisimple, so we may, and will, take $\{\beta_j\}$ to be a base of $\Delta_{(\mathfrak{k},\mathfrak{t}\cap\mathfrak{k})}$. We take the following base. If $\mathfrak{g}_{\mathbb{R}}=E_{6(-26)}$, then

$$\beta_4=\alpha_1|_{\mathfrak{t}\cap\mathfrak{k}}=\alpha_6|_{\mathfrak{t}\cap\mathfrak{k}},\quad \beta_3=\alpha_3|_{\mathfrak{t}\cap\mathfrak{k}}=\alpha_5|_{\mathfrak{t}\cap\mathfrak{k}},\quad \beta_2=\alpha_4|_{\mathfrak{t}\cap\mathfrak{k}},\quad \beta_1=\alpha_2|_{\mathfrak{t}\cap\mathfrak{k}},$$

Table 6.

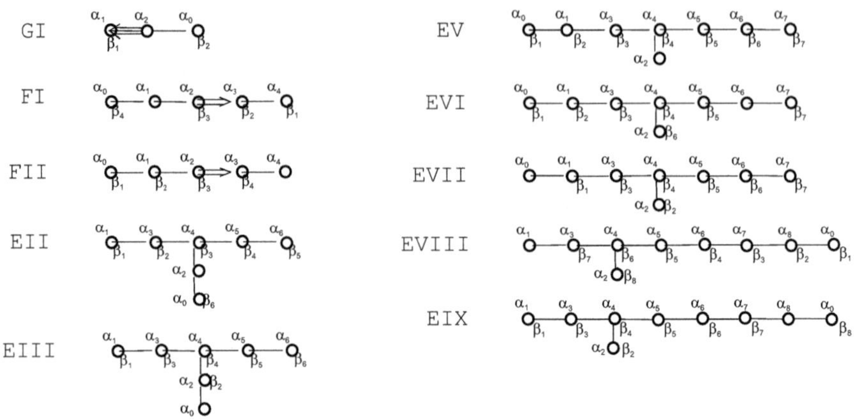

with the Dynkin diagram

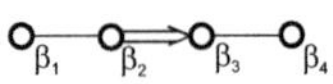

and if $\mathfrak{g}_{\mathbb{R}} = \mathbf{E}_{6(6)}$, then

$$\beta_1 = -2\beta_2 - 3\beta_3 - 2\beta_4 - \alpha_2|_{\mathfrak{t}\cap\mathfrak{k}},$$
$$\beta_2 = \alpha_1|_{\mathfrak{t}\cap\mathfrak{k}} = \alpha_6|_{\mathfrak{t}\cap\mathfrak{k}}, \quad \beta_3 = \alpha_3|_{\mathfrak{t}\cap\mathfrak{k}} = \alpha_5|_{\mathfrak{t}\cap\mathfrak{k}}, \quad \beta_4 = \alpha_4|_{\mathfrak{t}\cap\mathfrak{k}}$$

with the Dynkin diagram

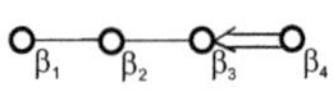

Below the weighted Dynkin diagrams of K-orbits in $\mathcal{N}(\mathfrak{g})$ are considered with respect to the described base $\{\beta_j\}$.

In [D3], [D4] one finds:

(D1) the weighted Dynkin diagrams of all nonzero orbits $G \cdot x$ and $K \cdot x$ for $x \in \mathcal{N}(\mathfrak{p})$,

(D2) the type and dimension of the reductive Levi factor of $\mathfrak{z}_{\mathfrak{k}}(x)$ and $\dim_{\mathbb{C}} \mathfrak{z}_{\mathfrak{k}}(x)$,

(D3) the type of the reductive Levi factor of $\mathfrak{z}_{\mathfrak{g}_{\mathbb{R}}}(y)$ for an element y of the $\mathrm{Ad}(\mathfrak{g}_{\mathbb{R}})$-orbit in $\mathcal{N}(\mathfrak{g}_{\mathbb{R}})$ corresponding to $K \cdot x$ via the Kostant–Sekiguchi bijection.

Given the weighted Dynkin diagram of $G \cdot x$, one finds the type of reductive Levi factor of $\mathfrak{z}_{\mathfrak{g}}(x)$ in [El], [Ca]. So using (D1), one can apply Theorem 4 for explicit classifying (-1)-distinguished K-orbits in $\mathcal{N}(\mathfrak{p})$. On the other hand, given (D3), Theorem 5 can be applied for this purpose as well. So following either of these ways, one obtains, for every exceptional simple $\mathfrak{g}$, the explicit classification of (-1)-distinguished K-orbits in $\mathcal{N}(\mathfrak{p})$ in terms of their weighted Dynkin diagrams. The final result is the following.

Theorem 14. *For all exceptional simple Lie algebras* $\mathfrak{g}$ *and all conjugacy classes of elements* $\theta \in \operatorname{Aut}\mathfrak{g}$ *of order* 2, *all* (-1)*-distinguished* K*-orbits in* $\mathcal{N}(\mathfrak{p})$ *are listed in Tables* 7–18 *at the end of this paper.*

Tables 7–18 contain further information as summarized below.

(T1) The conjugacy class of θ is defined by specifying the type of noncompact real form $\mathfrak{g}_{\mathbb{R}}$ of $\mathfrak{g}$ canonically corresponding to this class.
(T2) Column 2 gives the weights $\{\beta_j(h)\}$ (listed in the order of increasing of j) of the weighted Dynkin diagram $\operatorname{Dyn} K \cdot x$ of the (-1)-distinguished K-orbit $K \cdot x$ in $\mathcal{N}(\mathfrak{p})$, where h is the characteristic of $K \cdot x$.
(T3) Column 3 gives the weights $\{\alpha_j(H)\}$ (listed in the order of increasing of j) of the weighted Dynkin diagram $\operatorname{Dyn} G \cdot x$ of the G-orbit $G \cdot x$ in $\mathcal{N}(\mathfrak{g})$, where H is the characteristic of $G \cdot x$.
(T4) Column 4 gives $\dim_{\mathbb{C}} K \cdot x$, and hence, by (2), also $\dim_{\mathbb{C}} G \cdot x$.
(T5) Column 5 gives the number of K-orbits (not necessarily (-1)-distinguished) in $G \cdot x \cap \mathfrak{p}$. One of them is $K \cdot x$. Since this number is equal to the number of K-orbits $K \cdot x'$ in $\mathcal{N}(\mathfrak{p})$ such that $\operatorname{Dyn} G \cdot x' = \operatorname{Dyn} G \cdot x$, one finds it using the tables in [D3], [D4].
(T6) Columns 6 gives the type of the reductive Levi factor of $\mathfrak{z}_{\mathfrak{k}}(x)$. By T_m is denoted the Lie algebra of an m-dimensional torus.
(T7) Column 7 gives the complex dimension of the unipotent radical of $\mathfrak{z}_{\mathfrak{k}}(x)$.

Remark 1. For $\mathrm{E}_{6(6)}$ and $\mathrm{E}_{6(-26)}$, our numeration of $\{\alpha_j\}$ is the same as in [D3]; it differs from that in [D4]. For $\mathrm{E}_{6(-26)}$, our numeration of $\{\beta_j\}$ differs from that in [D3].

Remark 2. One finds in [D5], [D6], [D7] the explicit classification of all Cayley triples in $\mathfrak{g}$. This yields the explicit representatives of all K-orbits in $\mathcal{N}(\mathfrak{p})$ and $\operatorname{Ad}(\mathfrak{g}_{\mathbb{R}})$-orbits in $\mathcal{N}(\mathfrak{g}_{\mathbb{R}})$.

6 Geometric Properties

Geometric properties of the varieties $\mathbf{P}(\overline{\mathcal{O}})$ where $\mathcal{O}$ is a nilpotent K-orbit in $\mathcal{N}(\mathfrak{p})$ have been studies by several authors. Thereby their results provide some information on the geometry of the projective self-dual varieties that we associated with symmetric spaces. However these results are less complete than in the adjoint case (where a rather detailed information about singular loci and normality of the projective self-dual varieties $\mathbf{P}(\overline{\mathcal{O}})$ is available, see [P2]). Some phenomena valid for the projective self-dual $\mathbf{P}(\overline{\mathcal{O}})$'s in the adjoint case fail in general. Below we briefly summarize some facts about the geometry of the projective self-dual varieties that we associated with symmetric spaces.

Intersections of G-orbits In $\mathbf{P}(\mathcal{N}(\mathfrak{g}))$ With the Linear Subspace $\mathbf{P}(\mathfrak{p})$

Let $\mathcal{O}$ be a nonzero G-orbit in $\mathcal{N}(\mathfrak{g})$, and $X = \mathbf{P}(\overline{\mathcal{O}})$. If $X \cap \mathbf{P}(\mathfrak{p}) \neq \varnothing$, then, by (2), all irreducible components $Y_1, \ldots Y_s$ of the variety $X \cap \mathbf{P}(\mathfrak{p})$ have dimension $\frac{1}{2} \dim X$, and every Y_j is the closure of a K-orbit in $\mathbf{P}(\mathcal{N}(\mathfrak{p}))$. Theorems 2, 3 and Definition 1 imply that

$$X \text{ is self-dual} \implies \text{all } Y_1, \ldots, Y_s \text{ are self-dual.}$$

There are many examples showing that the converse in not true. For instance, one deduces from Table 8 that $\mathfrak{g} = \mathrm{E}_6$, $\mathfrak{g}_\mathbb{R} = \mathrm{E}_{6(2)}$ and $\operatorname{Dyn} \mathcal{O} = 111011, 121011$ or 220202 are such cases. There are also many examples where some of the Y_j's are self-dual and some are not: for instance, this is so if $\mathfrak{g} = \mathrm{E}_6$, $\mathfrak{g}_\mathbb{R} = \mathrm{E}_{6(2)}$ and $\operatorname{Dyn} \mathcal{O} = 020000, 001010, 000200, 020200$ or 220002. Finally, there are many instances where all Y_j are not self-dual (to obtain them, e.g., compare Tables 7–18 with tables in [D3], [D4]).

Affine (-1)-distinguished Orbits

Recall that by Matsushima's criterion, an orbit of a reductive algebraic group acting on an affine algebraic variety is affine if and only if the stabilizer of a point of this orbit is reductive, cf., e.g. [PV, Theorem 4.17]. Therefore distinguished G-orbits in $\mathcal{N}(\mathfrak{g})$ are never affine. However in general there exist affine (-1)-distinguished K-orbits in $\mathcal{N}(\mathfrak{p})$. For instance, if $\mathfrak{g}$ is exceptional simple, we immediately obtain their classification from Tables 7–18: these are precisely the orbits for which the number in the last column is 0.

If $K \cdot x$ is affine, then each irreducible component of the boundary $\mathbf{P}(\overline{K \cdot x}) \setminus \mathbf{P}(K \cdot x)$ has codimension 1 in $\mathbf{P}(\overline{K \cdot x})$, cf. [P1, Lemma 3]. Therefore if a point of the open K-orbit of such an irreducible component lies in the singular locus of $\mathbf{P}(\overline{K \cdot x})$, then $\mathbf{P}(\overline{K \cdot x})$ is not normal.

Orbit Closure Ordering and Orbit Decomposition Of the Orbit Boundary $\mathbf{P}(\overline{K \cdot x}) \setminus \mathbf{P}(K \cdot x)$

The closure ordering on the set of K-orbits in $\mathcal{N}(\mathfrak{p})$ (resp., $\operatorname{Ad}(\mathfrak{g}_\mathbb{R})$-orbits in $\mathcal{N}(\mathfrak{g}_\mathbb{R})$) is defined by the condition that $\mathcal{O}_1 > \mathcal{O}_2$ if and only if $\mathcal{O}_2$ is contained in the closure of $\mathcal{O}_1$ and $\mathcal{O}_1 \neq \mathcal{O}_2$. According to [BS], the Kostant–Sekiguchi bijection preserves the closure ordering. Clearly its describing can be reduced to the case of simple $\mathfrak{g}$. In this case, the explicit description of the closure ordering is obtained in [D1], [D2], [D8], [D9], [D10], [D11], [D12], [D13] (see also [O], [Se]).

This information and Theorems 8–14 yield the orbit decomposition of every irreducible component of the orbit boundary $\mathbf{P}(\overline{K\cdot x})\setminus\mathbf{P}(K\cdot x)$ for any (-1)-distinguished K-orbit $K\cdot x$. In particular, this gives the dimensions of these components and their intersection configurations.

Singular Locus Of a Self-dual Variety $\mathbf{P}(\overline{K\cdot x})$

Apart from a rather detailed information on the singular loci of the self-dual projectivized nilpotent orbit closures in the adjoint case (see [P2]), in general case only a partial information is available. In [O], for $(\mathfrak{g},\mathfrak{k})=(\mathfrak{sl}_n(\mathbb{C}),\mathfrak{so}_n(\mathbb{C}))$ and $(\mathfrak{sl}_{2n}(\mathbb{C}),\mathfrak{sp}_n(\mathbb{C}))$, the normality of the nilpotent orbit closures $\overline{K\cdot x}$ (hence that of $\mathbf{P}(\overline{K\cdot x})$) is studied. In particular, it is shown there that for $(\mathfrak{g},\mathfrak{k})=(\mathfrak{sl}_{2n}(\mathbb{C}),\mathfrak{sp}_n(\mathbb{C}))$, normality holds for any $\overline{K\cdot x}$, but if $(\mathfrak{g},\mathfrak{k})=(\mathfrak{sl}_n(\mathbb{C}),\mathfrak{so}_n(\mathbb{C}))$, this is not the case. See also [Se], [O], [SeSh], where the local equations of the generic singularities of some orbit closures in $\mathbf{P}(\mathcal{N}(\mathfrak{p}))$ are found. By [P3, Proposition 4], Hesselink's desingularization of the closures of Hesselink strata, [He], cf. [PV], yields a desingularization of any orbit closure $\mathbf{P}(\overline{K\cdot x})$ in $\mathbf{P}(\mathcal{N}(\mathfrak{p}))$ (in [R], another approach to desingularization of such orbit closured is considered).

Table 7. (-1)-distinguished K-orbits $K\cdot x$ in $\mathcal{N}(\mathfrak{p})$ for $\mathfrak{g}_{\mathbb{R}}=\mathrm{E}_{6(6)}$

No.	Dyn $K\cdot x$	Dyn $G\cdot x$	$\dim_{\mathbb{C}} K\cdot x$	$\sharp(G\cdot x\cap\mathfrak{p})/K$	$\mathfrak{z}_{\mathfrak{k}}(x)/\mathrm{rad}_u\mathfrak{z}_{\mathfrak{k}}(x)$	$\dim_{\mathbb{C}}\mathrm{rad}_u\mathfrak{z}_{\mathfrak{k}}(x)$
1	2222	202222	35	1	0	1
2	2202	220002	33	2	0	3
3	0220	220002	33	2	0	3
4	4224	222222	36	1	0	0

Table 8. (-1)-distinguished K-orbits $K\cdot x$ in $\mathcal{N}(\mathfrak{p})$ for $\mathfrak{g}_{\mathbb{R}}=\mathrm{E}_{6(2)}$

No.	Dyn $K\cdot x$	Dyn $G\cdot x$	$\dim_{\mathbb{C}} K\cdot x$	$\sharp(G\cdot x\cap\mathfrak{p})/K$	$\mathfrak{z}_{\mathfrak{k}}(x)/\mathrm{rad}_u\mathfrak{z}_{\mathfrak{k}}(x)$	$\dim_{\mathbb{C}}\mathrm{rad}_u\mathfrak{z}_{\mathfrak{k}}(x)$
1	000004	020000	21	3	$2\mathrm{A}_2$	1
2	301000	001010	25	3	$\mathrm{A}_1+\mathrm{T}_1$	9
3	001030	001010	25	3	$\mathrm{A}_1+\mathrm{T}_1$	9
4	004000	000200	29	3	T_2	7
5	020204	000200	29	3	T_2	7
6	004008	020200	30	2	A_2	0
7	400044	220002	30	2	$\mathrm{A}_1+\mathrm{T}_1$	4
8	121131	111011	31	2	T_1	6
9	311211	111011	31	2	T_1	6
10	313104	121011	32	2	T_1	5
11	013134	121011	32	2	T_1	5

Table 8. (*continued*) (-1)-distinguished K-orbits $K \cdot x$ in $\mathcal{N}(\mathfrak{p})$ for $\mathfrak{g}_{\mathbb{R}} = \mathrm{E}_{6(2)}$

No.	Dyn $K \cdot x$	Dyn $G \cdot x$	$\dim_{\mathbb{C}} K \cdot x$	$\sharp(G \cdot x \cap \mathfrak{p})/K$	$\mathfrak{z}_{\mathfrak{k}}(x)/\mathrm{rad}_u\mathfrak{z}_{\mathfrak{k}}(x)$	$\dim_{\mathbb{C}} \mathrm{rad}_u\mathfrak{z}_{\mathfrak{k}}(x)$
12	222222	200202	33	2	0	5
13	040404	200202	33	2	0	5
14	224224	220202	34	2	T_1	3
15	404048	220202	34	2	T_1	3
16	440444	222022	35	1	0	3
17	444448	222222	36	1	0	2

Table 9. (-1)-distinguished K-orbits $K \cdot x$ in $\mathcal{N}(\mathfrak{p})$ for $\mathfrak{g}_{\mathbb{R}} = \mathrm{E}_{6(-26)}$

No.	Dyn $K \cdot x$	Dyn $G \cdot x$	$\dim_{\mathbb{C}} K \cdot x$	$\sharp(G \cdot x \cap \mathfrak{p})/K$	$\mathfrak{z}_{\mathfrak{k}}(x)/\mathrm{rad}_u\mathfrak{z}_{\mathfrak{k}}(x)$	$\dim_{\mathbb{C}} \mathrm{rad}_u\mathfrak{z}_{\mathfrak{k}}(x)$
1	0001	100001	16	1	B_3	15
2	0002	200002	24	1	G_2	14

Table 10. (-1)-distinguished K-orbits $K \cdot x$ in $\mathcal{N}(\mathfrak{p})$ for $\mathfrak{g}_{\mathbb{R}} = \mathrm{E}_{6(-14)}$

No.	Dyn $K \cdot x$	Dyn $G \cdot x$	$\dim_{\mathbb{C}} K \cdot x$	$\sharp(G \cdot x \cap \mathfrak{p})/K$	$\mathfrak{z}_{\mathfrak{k}}(x)/\mathrm{rad}_u\mathfrak{z}_{\mathfrak{k}}(x)$	$\dim_{\mathbb{C}} \mathrm{rad}_u\mathfrak{z}_{\mathfrak{k}}(x)$
1	100001	100001	16	3	$\mathrm{B}_3 + \mathrm{T}_1$	8
2	10000−2	100001	16	3	$\mathrm{B}_3 + \mathrm{T}_1$	8
3	10101−2	110001	23	2	$\mathrm{A}_2 + \mathrm{T}_1$	14
4	11100−3	110001	23	2	$\mathrm{A}_2 + \mathrm{T}_1$	14
5	40000−2	200002	24	1	G_2	8
6	03001−2	120001	26	2	$\mathrm{B}_2 + \mathrm{T}_1$	9
7	01003−6	120001	26	2	$\mathrm{B}_2 + \mathrm{T}_1$	9
8	02202−6	220002	30	1	$\mathrm{A}_1 + \mathrm{T}_1$	12

Table 11. (-1)-distinguished K-orbits $K \cdot x$ in $\mathcal{N}(\mathfrak{p})$ for $\mathfrak{g}_{\mathbb{R}} = \mathrm{E}_{7(-5)}$

No.	Dyn $K \cdot x$	Dyn $G \cdot x$	$\dim_{\mathbb{C}} K \cdot x$	$\sharp(G \cdot x \cap \mathfrak{p})/K$	$\mathfrak{z}_{\mathfrak{k}}(x)/\mathrm{rad}_u\mathfrak{z}_{\mathfrak{k}}(x)$	$\dim_{\mathbb{C}} \mathrm{rad}_u\mathfrak{z}_{\mathfrak{k}}(x)$
1	0000004	2000000	33	3	A_5	1
2	4000000	0000020	42	2	$\mathrm{G}_2 + \mathrm{A}_1$	10
3	0000400	0020000	47	3	$3\mathrm{A}_1$	13
4	0002004	0020000	47	3	$3\mathrm{A}_1$	13
5	0000408	2020000	48	2	C_3	0
6	2010112	0001010	49	1	$\mathrm{A}_1 + \mathrm{T}_1$	16
7	0400004	2000020	50	2	$\mathrm{A}_2 + \mathrm{T}_1$	10
8	1111101	1001010	52	1	T_2	15
9	2010314	2001010	53	1	$\mathrm{A}_1 + \mathrm{T}_1$	12

Table 11. (*continued*) (-1)-distinguished K-orbits $K \cdot x$ in $\mathcal{N}(\mathfrak{p})$ for $\mathfrak{g}_{\mathbb{R}} = \mathtt{E}_{7(-5)}$

No.	Dyn $K \cdot x$	Dyn $G \cdot x$	$\dim_{\mathbb{C}} K \cdot x$	$\sharp(G \cdot x \cap \mathfrak{p})/K$	$\mathfrak{z}_{\mathfrak{k}}(x)/\mathrm{rad}_u\mathfrak{z}_{\mathfrak{k}}(x)$	$\dim_{\mathbb{C}} \mathrm{rad}_u\mathfrak{z}_{\mathfrak{k}}(x)$
10	0040000	0002000	53	1	$\mathtt{A}_1$	13
11	0202202	0020020	55	2	$\mathtt{A}_1$	11
12	0004004	0020020	55	2	$\mathtt{A}_1$	11
13	0202404	2020020	56	2	$2\mathtt{A}_1$	7
14	0400408	2020020	56	2	$2\mathtt{A}_1$	7
15	4004000	0002020	57	1	$\mathtt{A}_1$	9
16	0404004	2002020	59	1	$\mathtt{T}_1$	9
17	0404408	2022020	60	1	$\mathtt{A}_1$	6

Table 12. (-1)-distinguished K-orbits $K \cdot x$ in $\mathcal{N}(\mathfrak{p})$ for $\mathfrak{g}_{\mathbb{R}} = \mathtt{E}_{7(7)}$

No.	Dyn $K \cdot x$	Dyn $G \cdot x$	$\dim_{\mathbb{C}} K \cdot x$	$\sharp(G \cdot x \cap \mathfrak{p})/K$	$\mathfrak{z}_{\mathfrak{k}}(x)/\mathrm{rad}_u\mathfrak{z}_{\mathfrak{k}}(x)$	$\dim_{\mathbb{C}} \mathrm{rad}_u\mathfrak{z}_{\mathfrak{k}}(x)$
1	4000000	0200000	42	4	$\mathtt{G}_2$	7
2	0000004	0200000	42	4	$\mathtt{G}_2$	7
3	0040000	0000200	50	4	$\mathtt{A}_1$	10
4	0000400	0000200	50	4	$\mathtt{A}_1$	10
5	3101021	1001010	52	3	$\mathtt{T}_2$	9
6	1201013	1001010	52	3	$\mathtt{T}_2$	9
7	4004000	2000200	54	4	$\mathtt{A}_1$	6
8	0004004	2000200	54	4	$\mathtt{A}_1$	6
9	2220202	0002002	56	4	0	7
10	2020222	0002002	56	4	0	7
11	0400400	0002002	56	4	0	7
12	0040040	0002002	56	4	0	7
13	2222202	2002002	58	4	0	5
14	2022222	2002002	58	4	0	5
15	4004040	2002002	58	4	0	5
16	0404004	2002002	58	4	0	5
17	4220224	2002020	59	2	$\mathtt{T}_1$	3
18	2422222	2002022	60	4	0	3
19	2222242	2002022	60	4	0	3
20	4404040	2002022	60	4	0	3
21	0404044	2002022	60	4	0	3
22	4404404	2220202	61	2	0	2
23	4044044	2220202	61	2	0	2
24	4444044	2220222	62	2	0	1
25	4404444	2220222	62	2	0	1
26	8444444	2222222	63	2	0	0
27	4444448	2222222	63	2	0	0

Table 13. (-1)-distinguished K-orbits $K \cdot x$ in $\mathcal{N}(\mathfrak{p})$ for $\mathfrak{g}_{\mathbb{R}} = \mathrm{E}_{7(-25)}$

No.	Dyn $K \cdot x$	Dyn $G \cdot x$	$\dim_{\mathbb{C}} K \cdot x$	$\sharp(G \cdot x \cap \mathfrak{p})/K$	$\mathfrak{z}_{\mathfrak{k}}(x)/\mathrm{rad}_u \mathfrak{z}_{\mathfrak{k}}(x)$	$\dim_{\mathbb{C}} \mathrm{rad}_u \mathfrak{z}_{\mathfrak{k}}(x)$
1	0000002	0000002	27	4	F_4	0
2	000000−2	0000002	27	4	F_4	0
3	010010−2	1000010	38	2	$\mathrm{A}_3 + \mathrm{T}_1$	25
4	011000−3	1000010	38	2	$\mathrm{A}_3 + \mathrm{T}_1$	25
5	200002−2	2000002	43	4	B_3	15
6	400000−2	2000002	43	4	B_3	15
7	000004−6	2000002	43	4	B_3	15
8	200002−6	2000002	43	4	B_3	15
9	220002−6	2000020	50	1	$\mathrm{A}_2 + \mathrm{T}_1$	20
10	400004−6	2000022	51	2	G_2	14
11	400004−10	2000022	51	2	G_2	14

Table 14. (-1)-distinguished K-orbits $K \cdot x$ in $\mathcal{N}(\mathfrak{p})$ for $\mathfrak{g}_{\mathbb{R}} = \mathrm{E}_{8(8)}$

No.	Dyn $K \cdot x$	Dyn $G \cdot x$	$\dim_{\mathbb{C}} K \cdot x$	$\sharp(G \cdot x \cap \mathfrak{p})/K$	$\mathfrak{z}_{\mathfrak{k}}(x)/\mathrm{rad}_u \mathfrak{z}_{\mathfrak{k}}(x)$	$\dim_{\mathbb{C}} \mathrm{rad}_u \mathfrak{z}_{\mathfrak{k}}(x)$
1	40000000	20000000	78	3	$2\mathrm{G}_2$	14
2	00000004	02000000	92	3	A_2	20
3	21010100	00010001	96	3	$\mathrm{A}_1 + \mathrm{T}_1$	20
4	00400000	00000200	97	2	$2\mathrm{A}_1$	17
5	40000040	02000002	99	3	A_2	13
6	20200200	00002000	104	3	0	16
7	00004000	00002000	104	3	0	16
8	02002002	00002000	104	3	0	16
9	40040000	20000200	105	2	$2\mathrm{A}_1$	9
10	02002022	00002002	107	3	T_1	12
11	00400040	00002002	107	3	T_1	12
12	31010211	10010101	108	2	T_1	11
13	13111101	10010102	109	2	T_1	10
14	20202022	00020002	110	2	0	10
15	04004000	00020002	110	2	0	10
16	02022022	20002002	111	3	T_1	8
17	40040040	20002002	111	3	T_1	8
18	00400400	00020020	112	2	0	8
19	22202022	00020020	112	2	0	8
20	22202042	00020022	113	2	0	7
21	04004040	00020022	113	2	0	7
22	22222022	20020020	114	2	0	6
23	40040400	20020020	114	2	0	6

Table 14. (*continued*) (-1)-distinguished K-orbits $K \cdot x$ in $\mathcal{N}(\mathfrak{p})$ for $\mathfrak{g}_{\mathbb{R}} = \mathrm{E}_{8(8)}$

No.	Dyn $K \cdot x$	Dyn $G \cdot x$	$\dim_{\mathbb{C}} K \cdot x$	$\sharp(G \cdot x \cap \mathfrak{p})/K$	$\mathfrak{z}_{\mathfrak{k}}(x)/\mathrm{rad}_u \mathfrak{z}_{\mathfrak{k}}(x)$	$\dim_{\mathbb{C}} \mathrm{rad}_u \mathfrak{z}_{\mathfrak{k}}(x)$
24	22222042	20020022	115	2	0	5
25	04040044	20020022	115	2	0	5
26	22222222	20020202	116	2	0	4
27	44040400	20020202	116	2	0	4
28	24222242	20020222	117	2	0	3
29	44040440	20020222	117	2	0	3
30	44044044	22202022	118	1	0	2
31	44440444	22202222	119	1	0	1
32	84444444	22222222	120	1	0	0

Table 15. (-1)-distinguished K-orbits $K \cdot x$ in $\mathcal{N}(\mathfrak{p})$ for $\mathfrak{g}_{\mathbb{R}} = \mathrm{E}_{8(-24)}$

No.	Dyn $K \cdot x$	Dyn $G \cdot x$	$\dim_{\mathbb{C}} K \cdot x$	$\sharp(G \cdot x \cap \mathfrak{p})/K$	$\mathfrak{z}_{\mathfrak{k}}(x)/\mathrm{rad}_u \mathfrak{z}_{\mathfrak{k}}(x)$	$\dim_{\mathbb{C}} \mathrm{rad}_u \mathfrak{z}_{\mathfrak{k}}(x)$
1	00000004	00000002	57	3	E_6	1
2	00000204	00000020	83	3	D_4	25
3	00000040	00000020	83	3	D_4	25
4	00000048	00000022	84	2	F_4	0
5	01100012	10000100	89	1	$\mathrm{B}_2 + \mathrm{T}_1$	36
6	40000004	20000002	90	2	A_4	22
7	10100111	10000101	94	1	$\mathrm{A}_2 + \mathrm{T}_1$	33
8	01100034	10000102	95	1	A_3	26
9	00020000	00000200	97	1	$2\mathrm{A}_1$	33
10	20000222	20000020	99	2	G_2	23
11	00000404	20000020	99	2	G_2	23
12	20000244	20000022	100	2	B_3	15
13	40000048	20000022	100	2	B_3	15
14	00020200	20000200	105	1	$2\mathrm{A}_1$	25
15	40000404	20000202	107	1	A_2	21
16	40000448	20000222	108	1	G_2	14

Table 16. (-1)-distinguished K-orbits $K \cdot x$ in $\mathcal{N}(\mathfrak{p})$ for $\mathfrak{g}_{\mathbb{R}} = \mathsf{F}_{4(4)}$

No.	Dyn $K \cdot x$	Dyn $G \cdot x$	$\dim_{\mathbb{C}} K \cdot x$	$\sharp(G \cdot x \cap \mathfrak{p})/K$	$\mathfrak{z}_{\mathfrak{k}}(x)/\mathrm{rad}_u \mathfrak{z}_{\mathfrak{k}}(x)$	$\dim_{\mathbb{C}} \mathrm{rad}_u \mathfrak{z}_{\mathfrak{k}}(x)$
1	0004	2000	15	3	A_2	1
2	0040	0200	20	3	0	4
3	0204	0200	20	3	0	4
4	2022	0200	20	3	0	4
5	2200	0048	21	2	A_1	0
6	0404	0202	22	2	0	2
7	2222	0202	22	2	0	2
8	2244	2202	23	2	0	1
9	4048	2202	23	2	0	1
10	4448	2222	24	1	0	0

Table 17. (-1)-distinguished K-orbits $K \cdot x$ in $\mathcal{N}(\mathfrak{p})$ for $\mathfrak{g}_{\mathbb{R}} = \mathsf{F}_{4(-20)}$

No.	Dyn $K \cdot x$	Dyn $G \cdot x$	$\dim_{\mathbb{C}} K \cdot x$	$\sharp(G \cdot x \cap \mathfrak{p})/K$	$\mathfrak{z}_{\mathfrak{k}}(x)/\mathrm{rad}_u \mathfrak{z}_{\mathfrak{k}}(x)$	$\dim_{\mathbb{C}} \mathrm{rad}_u \mathfrak{z}_{\mathfrak{k}}(x)$
1	0001	0001	11	1	A_3	10
2	4000	0002	15	1	G_2	7

Table 18. (-1)-distinguished K-orbits $K \cdot x$ in $\mathcal{N}(\mathfrak{p})$ for $\mathfrak{g}_{\mathbb{R}} = \mathsf{G}_{2(2)}$

No.	Dyn $K \cdot x$	Dyn $G \cdot x$	$\dim_{\mathbb{C}} K \cdot x$	$\sharp(G \cdot x \cap \mathfrak{p})/K$	$\mathfrak{z}_{\mathfrak{k}}(x)/\mathrm{rad}_u \mathfrak{z}_{\mathfrak{k}}(x)$	$\dim_{\mathbb{C}} \mathrm{rad}_u \mathfrak{z}_{\mathfrak{k}}(x)$
1	22	02	5	2	0	1
2	04	02	5	2	0	1
3	48	22	6	1	0	0

References

[BaCa] Bala, P., Carter, R.W.: Classes of unipotent elements in simple algebraic groups, I and II. Math. Proc. Camb. Phil. Soc. **79** 401–425 (1976) and **80** 1–18 (1976)

[BS] Barbash, D., Sepanski, M.S.: Closure ordering and the Kostant–Sekiguchi correspondence. Proc. AMS **126** 311–317 (1998)

[BoCu] Bourgoyne, N., Cushman, R.: Conjugacy classes in linear groups. J. Algebra **44** 339–362 (1977)

[Ca] Carter, R.W.: *Finite Groups of Lie Type* (John Wiley and Sons, Chichester, New York, Brisbane, Toronto, Singapore 1985)

[Ch] Chevalley, C.: *Théorie des Groupes de Lie, Tome II* (Hermann, Paris 1951)

[CM] Collingwood, D.H., McGovern, W.M.: *Nilpotent Orbits in Semisimple Lie Algebras* (Van Nostrand Reinhold, New York 1992)

[D1] Đoković, D.Ž.: Closures of conjugacy classes in classical real linear Lie groups. In: *Algebra, Carbondale* 1980, Lect. Notes Math., Vol. 848, (Springer, Berlin, Heidelberg 1981) pp. 63–83

[D2] Đoković, D.Ž.: Closures of conjugacy classes in classical real linear Lie groups, II. Trans. AMS **270**(1) 217–252 (1982)

[D3] Đoković, D.Ž.: Classification of nilpotent elements in simple exceptional real Lie algebras of inner type and description of their centralizers. J. Algebra **112** 503–524 (1988)

[D4] Đoković, D.Ž.: Classification of nilpotent elements in simple real Lie algebras $E_{6(6)}$ and $E_{6(-26)}$ and description of their centralizers. J. Algebra **116** 196–207 (1988)

[D5] Đoković, D.Ž.: Explicit Cayley triples in real forms of G_2, F_4, and E_6. Pacif. J. Math. **184**(2) 231–255 (1998)

[D6] Đoković, D.Ž.: Explicit Cayley triples in real forms of E_7. Pacif. J. Math. **191**(1) 1–23 (1999)

[D7] Đoković, D.Ž.: Explicit Cayley triples in real forms of E_8. Pacif. J. Math. **194**(1) 57–82 (2000)

[D8] Đoković, D.Ž.: The closure diagrams for nilpotent orbits of real forms of F_4 and G_2. J. Lie Theory **10** 491–510 (2000)

[D9] Đoković, D.Ž.: The closure diagrams for nilpotent orbits of real forms of E_6. J. Lie Theory **11** 381–413 (2001)

[D10] Đoković, D.Ž.: The closure diagrams for nilpotent orbits of the real forms EVI and EVII of E_7. Representation Theory **5** 17–42 (2001)

[D11] Đoković, D.Ž.: The closure diagrams for nilpotent orbits of the split real form of E_7. Representation Theory **5** 284–316 (2001)

[D12] Đoković, D.Ž.: The closure diagrams for nilpotent orbits of the real form EIX of E_8. Asian J. Math. **5**(3) 561–584 (2001)

[D13] Đoković, D.Ž.: The closure diagram for nilpotent orbits of the split real form of E_8. Preprint, 2003.

[D] Дынкин, Е.Б.: Полупростые подалгебры полупростых алгебр Ли. Мат. сб. **30** (**72**)(2) 349–462 (1952). English transl.: Dynkin, E.B.: Semisimple subalgebras of semisimple Lie algebras. Amer. Math. Soc. Transl. (2) **6** 111–245 (1957)

[Ei1] Ein, L.: Varieties with small dual varieties I. Invent. Math. **86** 63–74 (1986)

[Ei2] Ein, L.: Varieties with small dual varieties, II. Duke Math. J.,**52**(4) 895–907 (1985)

[El] Elashvili, A.G.: The centralizers of nilpotent elements in semisimple Lie algebras. Trudy Tbil. Mat. Inst. Razmadze Acad. Nauk Gruz. SSR **46** 109–132 (1975) (Russian)

[GH] Griffiths, P., Harris, J.: *Principles of Algebraic Geometry* (Wiley–Interscience, London, New York 1979)

[Ha] Harris, J.: *Algebraic Geometry.* Graduate Texts in Math., Vol. 133 (Springer, Berlin, Heidelberg 1992)

[He] Hesselink, W.H.: Desingularizations of varieties of nullforms. Invent. math. **55** 141–163 (1979)

[Ka] Kawanaka, N.: Orbits and sttabilizers of nilpotent elements of graded semisimple Lie algebra. J. Fac. Sci. Univ. Tokyo, Sect. IA, Math. **34** 573–597 (1987)

[KN] Kobayashi, S., Nomizu, K.: *Foundations of Differential Geometry.* vol. 2 (Interscience Publ. 1969)

[K1] Kostant, B.: The principal three-dimensional subgroup and the Betti numbers of a complex simple Lie group. Amer. J. Math. **81** 973–1032 (1959)

[K2] Kostant, B.: Lie group representations on polynomial rings. Amer. J. Math. **86** 327–402 (1963)

[KM] Knop, F., Menzel, G.: Duale Varietäten von Fahnenvarietäten. Comm. Math. Helv. **62** 38–61 (1987)

[KR] Kostant, B., Rallis, S.: Orbits and representations associated with symmetric spaces. Amer. J. Math. **93** 753–809 (1971)

[L] Littelmann, P.: *An effective method to classify nilpotent orbits.* Progress in Math., vol. 143 (Birkhäuser Basel, 1996) pp. 255–269.

[M] McGovern, W.M.: The adjoint representation and the adjoint action. In: *Encycl. Math. Sci.*, vol. 131, Subseries "*Invariant Theory and Algebraic Transformation Groups*", Vol. II (Springer, Berlin, Heidelberg 2002) pp. 159–238

[N] Noël, A.G.: Nilpotent orbits and theta-stable parabolic subalgebras. Represent. Theory, An Electronic J. of the AMS **2** 1–32 (1998)

[O] Ohta, T.: The singularities of the closures of nilpotent orbits in certain symmetric pairs. Tôhoku Math. J. **38** 441–468 (1986)

[OV] Onishchik, A.L., Vinberg, E.B.: *Lie Groups and Algebraic Groups* (Springer, Berlin, Heidelberg, New York 1990)

[Pa] Pauly, C.: Self-duality of Coble's quartic hypersurface and applications. Michigan Math. J. **50**(3) 551–574 (2002)

[P1] Попов, В.Л.: О стабильности действия алгебраической группы на алгебраическом многообразии. Изв. АН СССР, сер. мат. **36** 371–385 (1972). Engl. transl.: Popov, V.L.: On the stability of the action of an algebraic group on an affine variety. Math. USSR Izv. **6**(2) 367–379 (1972)

[P2] Popov, V.L.: Self-dual algebraic varieties and nilpotent orbits. In: Proc. Internat. Colloq. "*Algebra, Arithmetic and Geometry, Mumbai,* 2000", Tata Inst. Fund. Research. (Narosa Publ. House 2001) pp. 509–533

[P3] Попов, В.Л.: Конус нуль-форм Гильберта. Труды Матем. инст. им. В. А. Стеклова **241** 192–209 (2003). Engl. transl.: Popov, V.L.: The cone of Hilbert nullforms. Proc. of the Steklov Inst. of Math. **241** 177–194 (2003)

[P4] Popov, V.L.: Projective duality and principal nilpotent elements of symmetric pairs. To appear in AMS Transl., Ser. 2 (2004)

[PV] Винберг, Э.Б., Попов, В.Л.: Теория инвариантов. Итоги науки и техники. Соврем. пробл. матем. Фунд. направл. **55** М., ВИНИТИ, 1989, 137–314. Engl. transl.: Popov, V.L., Vinberg, E.B.: Invariant Theory. In: *Algebraic Geometry*, IV. Encycl. Math. Sci., Vol. 55 (Springer, Berlin, Heidelberg 1994) pp. 123–284

[R] Reeder, M.: Desingularisations of some unstable orbit closures. Pacif. J. Math. **167**(2) 327–343 (1995)

[Se] Sekiguchi, J.: The nilpotent subvariety of the vector space associated to a symmetric pair. Publ. Res. Inst. Math. Sci. **20** 155–212 (1984)

[Sn] Snow, D.: The nef value and defect of homogeneous line bundles. Trans. AMS **340** 227–241 (1993)

[SeSh] Sekiguchi, J., Shimizu, Y.: Simple singularities and infinitesimally symmetric spaces. Publ. Jap. Acad. Ser. A **57** 42–46 (1981)

[SpSt] Springer, T.A., Steinberg, R.: Conjugacy classes in algebraic groups. In: *Seminar on Algebraic Groups and Related Finite Groups*, Lect. Notes Math., Vol. 131 (Springer, Berlin, Heidelberg 1969) pp. 167–266

[St] Steinberg, R.: *Endomorphisms of Linear Algebraic Groups.* Memoirs of AMS, No. 80, 1968

[T] Tevelev, E.: *Projective Duality and Homogeneous Spaces.* Preprint 2002. http://www.ma.utexas.edu/users/tevelev/research

[Vi] Винберг, Э.Б.: Классификация однородных нильпотентных элементов полупростой градуированной алгебры Ли. Труды сем. по вект. и тенз. анализу, вып. XIX, 155–177 (1979), МГУ. Engl. transl.: Vinberg, E.B.: A classification of homogeneous nilpotent elements of a semisimple graded Lie algebra. Sel. Math. Sov. **6** 15–35 (1987)

[Vu] Vust, Th.: Opération de groupes réductifs dans un type de cônes presque homogènes. Bull. Soc. Math. France **102** 317–334 (1974)

[W] Wall, G.E.: On the conjugacy classes in the unitary, symplectic and orthogonal groups. J. Austral. Math. Soc. **3** 1–62 (1963)

[Z] Zak, F.L.: *Tangents and Secants of Algebraic Varieties.* Translations of Mathematical Monographs, Vol. 127, AMS, 1993

The Role Of Exotic Affine Spaces In the Classification Of Homogeneous Affine Varieties

Dennis Snow*

Department of Mathematics, University of Notre Dame, Notre Dame, IN 46556–4618, USA. snow.1@nd.edu

Let G be a connected linear algebraic group over $\mathbb{C}$ and let H a closed algebraic subgroup. A fundamental problem in the study of homogeneous spaces is to describe, characterize, or classify those quotients G/H that are affine varieties. While cohomological characterizations of affine G/H are possible, there is still no general group-theoretic conditions that imply G/H is affine. In this article, we survey some of the known results about this problem and suggest a way of classifying affine G/H by means of its internal geometric structure as a fiber bundle.

Cohomological characterizations of affine G/H provide useful vanishing theorems and related information if one already knows G/H is affine. Such characterizations cannot be realistically applied to prove that a given homogeneous space G/H is affine. Ideally, one would like to have easily verified group-theoretic conditions on G and H that imply G/H is affine. Very few positive results are known in this direction, the most notable of which is Matsushima's Theorem for reductive groups. For general linear algebraic groups there is a natural generalization of Matsushima's Theorem that provides a necessary condition for G/H to be affine. While this criterion is also sufficient for some special situations, it is not sufficient in general.

In the absence of general group-theoretic conditions for G/H to be affine, it is worthwhile to understand the underlying geometric structure of an affine homogeneous space G/H. Such a space is always isomorphic to a fiber bundle over an orbit of a maximal reductive subgroup of G. The fiber is a smooth affine variety diffeomorphic to an affine space $\mathbb{C}^n$. Here several interesting phenomena seem possible: either the fiber is truly an "exotic" affine space or is in fact isomorphic to $\mathbb{C}^n$. If exotic structures occur, they would also provide counter-examples to the Cancellation Problem for affine spaces. So far, no such exotic examples are known. If such structures are impossible, then an affine homogeneous space G/H would always have the simple description of

* Supported by the Erwin Schrödinger Institute, Vienna, Austria

a homogeneous vector bundle. In this case, one can change groups, $G/H = \hat{G}/\hat{H}$, where $\hat{G}$ and $\hat{H}$ are easily classified, giving an indirect group-theoretic characterization for G/H to be affine.

1 Cohomological Characterizations

Recall that a subgroup $H \subset G$ is called *observable* if every finite dimensional rational H-module can be embedded as an H-submodule of a finite dimensional rational G-module. This is equivalent to the condition that for any rational H-module V, the induced module $V\big|^G = \{s : G \to V \mid s(gh^{-1}) = h \cdot s(g), \forall h \in H, \forall g \in G\}$ surjects onto V under the evaluation map $s \to s(1)$. It is well-known that H is observable in G if and only if G/H is quasi-affine [2]. The subgroup H is call *strongly observable* if, given any rational H module V, V is an H-submodule of a rational G-module W such that $V^H = W^G$. Finally, H is called an *exact* subgroup of G if induction from rational H-modules to rational G-modules preserves short exact sequences.

Theorem 1. [4, 13] *The following are equivalent:*

1. *G/H is affine.*
2. *H is a strongly observable subgroup of G.*
3. *H is an exact subgroup of G.*
4. *$H^1(R_u(H), \mathcal{O}(G)) = 0$ (or, equivalently, $H^1(G/R_u(H), \mathcal{O}) = 0$) where $R_u(H)$ is the unipotent radical of H.*

Such characterizations of affineness are basically "cohomological" in nature. They are primarily used when one already knows that G/H is affine. Verifying the properties themselves may be more difficult than directly proving that G/H is affine.

2 Group-theoretic Conditions

There is a practical need for easily verified conditions on the groups G and H that guarantee the quotient G/H is affine. We shall now investigate some of the known results in this direction.

2.1 Unipotent and Solvable Groups

If G is a unipotent linear algebraic group, then $G/H \cong \mathbb{C}^n$ for any algebraic subgroup H. More generally, if G is a solvable linear algebraic group, then $G/H \cong \mathbb{C}^n \times (\mathbb{C}^*)^m$. The corresponding statements for complex Lie groups are not automatically true. For example if $G = \mathbb{C}^* \times \mathbb{C}^*$ and $H = \{(e^z, e^{iz}) \mid z \in \mathbb{C}\}$, then G/H is a compact complex torus. Nevertheless, some generalizations are possible, see [13].

2.2 Reductive Groups

After these relatively simple cases, the best known group-theoretic criterion for G/H to be affine goes back to Matsushima [8]:

If G is reductive then G/H is affine if and only if H is reductive.

Matsushima's original theorem assumes G is a reductive complex Lie group and characterizes when G/H is Stein. However, a reductive complex Lie group G is in fact biholomorphically isomorphic to an algebraic group [6] and G/H is affine if it is Stein [1]. Matsushima's theorem has been generalized to reductive algebraic groups over algebraically closed fields of positive characteristic, see [11, 3].

2.3 General Linear Algebraic Groups

Any connected linear algebraic group G has a decomposition into a semi-direct product, $G = M \cdot R_u(G)$, where M is a maximal reductive subgroup M and $R_u(G)$ is the unipotent radical of G. A closed algebraic subgroup H has a similar decomposition, $H = L \cdot R_u(H)$ where L is a maximal reductive subgroup of H (not necessarily connected). Since the maximal reductive subgroups of G are conjugate, we may assume $L \subset M$. The group L is not important in determining whether G/H is affine:

G/H is affine if and only if $G/R_u(H)$ is affine.

This follows from the fact that L is reductive and $G/R_u(H) \to G/H$ is a principal L-bundle, see [11]. We therefore focus our attention on $R_u(H)$ and its location in G.

If $R_u(H) \subset R_u(G)$, then, of course, $G/R_u(H) \cong M \times R_u(G)/R_u(H)$ is affine and so G/H is affine. However, $R_u(H) \subset R_u(G)$ is not a necessary condition for G/H to be affine. For example, let G be the semi-direct product $\mathrm{SL}(2,\mathbb{C}) \cdot U$ where U is the standard 2-dimensional representation of $\mathrm{SL}(2,\mathbb{C})$, and let

$$H = \left\{ \begin{bmatrix} 1 & 0 \\ t & 1 \end{bmatrix} \times (0,t) \mid t \in \mathbb{C} \right\}.$$

Then $G/H \cong \mathrm{SL}(2,\mathbb{C}) \times \mathbb{C}$.

A necessary condition for G/H to be affine is not hard to discover, see [4, 13].

Lemma 1. *If G/H is affine then the intersection of $R_u(H)$ with any reductive subgroup of G is trivial.*

Proof. Let M be a maximal reductive subgroup of G. We must show that $R_u(H) \cap M^g = 1$ for all $g \in G$. Since G/H affine, so is $G/R_u(H)$, and thus any M-orbit of minimal dimension in $G/R_u(H)$, being automatically closed, is affine, see [10]. By Matsushima's Theorem, the isotropy subgroup in M of such an orbit is reductive. However, it also is of the form $M \cap R_u(H)^g$ for some $g \in G$, and, being a subgroup of a the unipotent algebraic group $R_u(H)^g$, is unipotent. Therefore, $M \cap R_u(H)^g = 1$ and any M-orbit of minimal dimension is isomorphic to M. Hence, all M-orbits have the same dimension. Therefore *every* M orbit in $G/G_u(H)$ is isomorphic to M and $M \cap R_u(H)^g = 1$ for all $g \in G$. □

The above lemma also holds when G/H is Stein [13]. Notice that the lemma is a natural generalization of Matsushima's Theorem in one direction: if G is reductive and G/H is affine, then the lemma implies $R_u(H) = 1$ and H is reductive. It is natural to explore whether the converse to Lemma 1 holds. For convenience, let us say that H satisfies *Matsushima's criterion* in G if the intersection of $R_u(H)$ with any reductive subgroup of G is trivial, or equivalently, if $R_u(H) \cap M^g = 1$ for all $g \in G$.

If H satisfies Matsushima's Criterion, then, as we have seen in the proof of the lemma, a maximal reductive subgroup M of G acts *freely* on $G/R_u(H)$. It is easy to see that this condition is equivalent to $R_u(H)$ acting freely on $M\backslash G$ from the right. Since $M\backslash G \cong R_u(G) \cong \mathbb{C}^n$, we are led to considering free actions of unipotent groups on affine space. In fact, in this situation, the actions are also triangular [14]. However, free unipotent triangular actions on $\mathbb{C}^n$ can be badly behaved in general: the quotient may not be separated and may not be affine even if it is separated [15]. Nevertheless, this point of view does yield positive results in certain simple cases.

Theorem 2. [4, 14] *Let G be a linear algebraic group and let H be a closed subgroup. Assume that* $\dim R_u(H) \leq 1$ *or* $\dim R_u(G) \leq 3$. *Then G/H is affine if and only if H satisfies Matsushima's criterion.*

Matsushima's Criterion is not sufficient in general. The following example was discovered by Winklemann [15]: Let $M = \mathrm{SL}(6, \mathbb{C})$ and let

$$G = \left\{ \begin{bmatrix} 1 & 0 \\ z & A \end{bmatrix} \mid z \in \mathbb{C}^6, A \in M \right\} \cong M \times \mathbb{C}^6. \tag{1}$$

Let

$$H = R_u(H) = \left\{ \begin{bmatrix} 1 & & & & & & \\ 0 & 1 & & & & & \\ 0 & 0 & 1 & & & & \\ 0 & t & s & 1 & & & \\ 0 & 0 & t & 0 & 1 & & \\ t & t & 0 & 0 & 0 & 1 & \\ s + \frac{1}{2}t^2 & \frac{1}{2}t^2 & 0 & 0 & 0 & t & 1 \end{bmatrix} \mid s, t \in \mathbb{C} \right\}. \tag{2}$$

It is not hard to verify that H satisfies Matsushima's Criterion, and that the quotient $M\backslash G/H$ exists and is a smooth contractible quasi-affine variety. However, direct calculation with invariants shows that $X = G/H$ is not affine. In fact, $M\backslash G/H$ is a smooth four-dimensional affine quadric with a 2-codimensional subspace removed.

3 Geometric Description

We now take a closer look at the underlying geometry of an affine homogeneous space G/H. As before, we let M be a maximal reductive subgroup of G and let $L \subset M$ be a maximal reductive subgroup of H. Let $U = R_u(G)$ and $V = R_u(H)$ be the unipotent radicals of G and H, respectively. If G/H is affine, then we know by Lemma 1 that M acts freely on G/V and, equivalently, that V acts freely on $M\backslash G = U$. Since G/V is affine and M is a reductive group acting freely, Hilbert's theorem on invariants implies that the geometric quotient $M\backslash G/V$ is affine. In fact, the local description of $G/V \to Y$ in terms of "slices," shows that the quotient map is a locally trivial principal bundle with fiber M, see [7, 12]. Consider the following diagram

$$\begin{array}{ccc} G & \xrightarrow{V} & G/V \\ {\scriptstyle M}\downarrow & & \downarrow{\scriptstyle M} \\ U & \xrightarrow{V} & Y \end{array}$$

where the vertical maps are the quotients by M and the horizontal maps are the quotients by V. Using local sections of $G \to Y$, we see that the fibration $U \to Y$ is also a locally trivial principal V-bundle. Thus, Y is the base of a locally trivial fibration where both the total space $U \cong \mathbb{C}^n$ and the fiber $V \cong \mathbb{C}^m$ are affine spaces.

Proposition 1. *If G/H is affine, then G/V is M-equivariantly isomorphic to $M \times Y$ and U is V-equivariantly isomorphic to $Y \times V$.*

Proof. The principal V-bundle $U \to Y$ is topologically trivial because the structure group is contractible, and this immediately implies that the bundle is holomorphically trivial [5]. To see that it is also algebraically trivial, we proceed by induction on $\dim V$. If $\dim V = 1$, then the triangular action of V on U is equivalent to a translation, see [4, 14] which implies $U \cong Y \times V$. If $\dim V > 1$, then there is a normal subgroup $V_1 \subset V$ such that $\dim V/V_1 = 1$. The bundle $G/V_1 \to G/V$ with fiber $V/V_1 \cong \mathbb{C}$ is trivial because $H^1(G/V, \mathcal{O}) = 0$. In particular, G/V_1 is affine, and hence the quotient $Y_1 = M\backslash G/V_1$ exists and is affine. Since $H^1(Y, \mathcal{O}) = 0$, the principal V/V_1-bundle $Y_1 \to Y$ is also trivial. By induction, $U \cong Y_1 \times V_1$, and therefore, $U \cong Y \times V/V_1 \times V_1 \cong Y \times V$. Finally, composing a global section $Y \to U$ with the projection $G \to G/V$ gives a global section $Y \to G/V$, and this implies G/V is M-equivariantly isomorphic to $M \times Y$. □

3.1 Cancellation Problem

The Cancellation Problem is the following: if $\mathbb{C}^n \cong Y \times \mathbb{C}^m$, is $Y \cong \mathbb{C}^{n-m}$? Since $U \cong \mathbb{C}^n$ and $V \cong \mathbb{C}^m$, the isomorphism $U \cong Y \times V$ of Proposition 1 gives an example of the Cancellation Problem. This problem remains unsolved in general, but has a positive answer if $\dim Y \le 2$ [9].

Obviously, Y is a smooth contractible affine variety. If $\dim Y \ge 3$, then Y is in fact diffeomorphic to $\mathbb{C}^{n-m}$, [16]. If Y is not algebraically isomorphic to $\mathbb{C}^{n-m}$ then Y is called an *exotic* affine space. Exotic affine spaces are known to exist, although no examples are known in the context of the Cancellation Problem, [16]. The relative simplicity of the subgroup H in (2) leads one to believe that it may indeed be possible to create exotic affine spaces of the form $Y = M\backslash G/V = U/V$ which would provide a negative answer to the Cancellation Problem at the same time.

3.2 Homogeneous Bundle Structure

The isomorphisms of Proposition 1 provide a natural bundle structure on an affine homogeneous space G/H.

Theorem 3. *Let L be a maximal reductive subgroup of H and let M be a maximal reductive subgroup of G containing L. If $X = G/H$ is affine then X is isomorphic to a homogeneous bundle over M/L, $X \cong M \times_L Y \to M/L$, with fiber Y a smooth contractible affine variety.*

Proof. The reductive group L acts by conjugation on both $U = R_u(G)$ and $V = R_u(H)$ and these actions are isomorphic to a linear representations. By Proposition 1, the V-equivariant isomorphism $U \cong Y \times V$ yields a V-equivariant map $s : U \to V$ satisfying $s(uv) = s(u)v$ for all $u \in U$, $v \in V$.

If we average s over a maximal compact subgroup K of L,

$$\hat{s}(u) = \int_{k\in K} k^{-1}s(kuk^{-1})k\,dk, \quad u \in U$$

(where dk is some invariant measure on K), then $\hat{s}$ is still V-equivariant. Moreover, since K is Zariski-dense in L, $\hat{s}(lul^{-1}) = l\hat{s}(u)l^{-1}$ for all $l \in L$, $u \in U$. If we identify Y with the L-invariant subvariety $\hat{s}^{-1}(1) \subset U$, we obtain a natural action of L on Y and the isomorphism $U \cong Y \times V$ is L-equivariant. Moreover, the right L action on G/V preserves the decomposition of Proposition 1, $G/V \cong M \times Y$, so that G/H is isomorphic to the homogeneous bundle $M \times_L Y = M \times Y/\sim$ where $(m, y) \sim (ml^{-1}, l \cdot y)$, for all $m \in M$, $y \in Y$, $l \in L$. □

If the homogeneous bundle of Theorem 3 is a homogeneous vector bundle, then it is possible to "change" the groups G and H so that their maximal reductive subgroups and unipotent radicals are aligned.

Theorem 4. *Let $X = G/H$ be affine and let $X = M \times_L Y \to M/L$ be the homogeneous bundle of Theorem 3. If Y is isomorphic to a linear representation of L, then there exist linear algebraic groups $\hat{G} = M \cdot R_u(\hat{G})$ and $\hat{H} = L \cdot R_u(\hat{H})$ such that $G/H \cong \hat{G}/\hat{H}$ and $R_u(\hat{H}) \subset R_u(\hat{G})$.*

Proof. By Theorem 1, the sections $H^0(M/L, X) \cong Y\big|^M$ generate the bundle. Therefore, there exists a finite dimensional M-submodule $U \subset H^0(M/L, X)$ that spans the vector space fiber Y over the identity coset $z_0 \in M/L$. The semi-direct product $\hat{G} = M \cdot U$ then acts on $X = M \times_L Y$ by $(m, u) \cdot [m', y] = [mm', y + u(m')]$ for all $m, m' \in M$, $y \in Y$, and $u \in U$. (Recall that the section u is an L-equivariant map $u : M \to Y$, $u(ml^{-1}) = lu(m)$ for all $m \in M$, $l \in L$.) This action is clearly transitive, because the sections U span Y over the identity coset $z_0 \in M/L$. The isotropy subgroup of the point $[1, 0] \in M \times_L Y$ is easily computed to be the semi-direct product $\hat{H} = L \cdot V$ where $V = \{u \in U \mid u(L) = 0\}$. □

References

1. Barth, W., Otte, M.: Invariante holomorphe Functionen auf reduktiven Liegruppen. Math. Ann. **201** 97–112 (1973)
2. Białynicki-Birula, A., Hochschild, G. Mostow, G.D.: Extensions of representations of algebraic linear groups. Am. J. Math. **85** 131–144 (1963)
3. Borel, A.: On affine algebraic homogeneous spaces. Arch. Math. **45** 74–78 (1985)
4. Cline, E., Parshall B., Scott L.: Induced modules and affine quotients. Math. Ann. **230** 1–14 (1977)
5. Grauert, H.: Analytische Faserungen über holomorph-vollständingen Räumen. Math. Ann. **135** 263–273 (1958)
6. Hochschild, G. Mostow, G.: Representations and representative functions of Lie groups, III. Ann. Math. **70** 85–100 (1959)
7. Luna, D.: Slices étales. Bull. Soc. Math. France, Mémoire **33** 81–105 (1973)
8. Matsushima, Y.: Espaces homogènes de Stein des groupes de Lie complexes. Nagoya Math. J. **18** 153–164 (1961)
9. Miyanishi, M., Sugie, T.: Affine surfaces containing cylinderlike open sets. J. Math. Kyoto Univ. **20** 11–42 (1980)
10. Mumford, D., Fogarty, J.: *Geometric Invariant Theory*, 2nd edn. (Springer, Berlin, Heidelberg, New York 1982)
11. Richardson, R.: Affine coset spaces of affine algebraic groups. Bull. London. Math. Soc. **91** 38–41 (1977)
12. Snow, D.M.: Reductive group actions on Stein spaces. Math. Ann. **259** 79–97 (1982)
13. Snow, D.M.: Stein quotients of connected complex Lie groups. Manuscripta Math. **50** 185–214 (1985)
14. Snow, D.M.: Triangular actions on $\mathbf{C}^3$. Manuscripta Math. **60** 407–415 (1988)
15. Winkelmann, J.: On Stein homogeneous manifolds and free holomorphic $\mathbf{C}$ actions on $\mathbf{C}^n$. Math. Ann. **186** 593–612 (1990)
16. Zaidenberg, M.G.: On exotic algebraic structures on affine spaces. St. Petersburg Math. J. **11** 703–760 (2000)

Hermitian Characteristics Of Nilpotent Elements

Evgueni A. Tevelev

Department of Mathematics, University of Texas, Austin, Texas 78712, USA.
tevelev@math.utexas.edu

Abstract. We define and study several equivariant stratifications of the isotropy and coisotropy representations of a parabolic subgroup in a complex reductive group.

0 Introduction

Let G be a connected complex reductive algebraic group with a Levi subgroup $L \subset P$. Then the Lie algebra $\mathfrak{g}$ of G admits a *block decomposition* $\mathfrak{g} = \oplus_{\chi \in \widehat{Z}} \mathfrak{g}_\chi$, where $\widehat{Z}$ is the character group of the connected component Z of the center of L. We have $\mathfrak{g}_0 = \mathfrak{l}$, the Lie algebra of L. If $L = T$ is a maximal torus of G then a block decomposition is just a root decomposition. In this case the basic fact about semisimple Lie algebras is that any element $e \in \mathfrak{g}_\chi$, $\chi \neq 0$, can be embedded in a homogeneous $\mathfrak{sl}_2$-triple $\langle e, h, f \rangle$, where $f \in \mathfrak{g}_{-\chi}$, $h_\chi \in \mathfrak{t}_\mathbb{R}$, where $\mathfrak{t}_\mathbb{R}$ is the canonical real form of a Cartan subalgebra $\mathfrak{t}$. In this paper we try to generalize this fact in the following direction.

Let $P \subset G$ be a proper parabolic subgroup with a Levi decomposition $P = L \cdot N$, and P^- the opposite parabolic subgroup, so $P \cap P^- = L$. Let $\mathfrak{p}$, $\mathfrak{n}$ (the unipotent radical of $\mathfrak{p}$), $\mathfrak{p}^-$ denote the corresponding Lie algebras.

P is maximal if and only if $\widehat{Z}$-grading reduces to $\mathbb{Z}$-grading of the form $\oplus_{i=-k}^{k} \mathfrak{g}_i$, in which case $\mathfrak{p} = \oplus_{i=0}^{k} \mathfrak{g}_i$. In general, $\widehat{Z}$ admits a total ordering such that $\mathfrak{p} = \oplus_{\chi \geq 0} \mathfrak{g}_\chi$, $\mathfrak{n} = \oplus_{\chi > 0} \mathfrak{g}_\chi$. We shall also use the identification $\mathfrak{g}/\mathfrak{p}^- = \oplus_{\chi > 0} \mathfrak{g}_\chi$, which is an isomorphism of L-modules.

Any element $e \in \mathfrak{n}$ (resp. $e \in \mathfrak{g}/\mathfrak{p}^-$) has the weight decomposition $e = \sum_{\chi > 0} e_\chi$, where $e_\chi \in \mathfrak{g}_\chi$. Each e_χ is a homogeneous nilpotent element of $\mathfrak{g}$, and therefore can be embedded (not uniquely) in a homogeneous $\mathfrak{sl}_2$-triple $\langle e_\chi, h_\chi, f_\chi \rangle$, where $f_\chi \in \mathfrak{g}_{-\chi}$, $h_\chi \in \mathfrak{l}$. Here by an $\mathfrak{sl}_2$-triple $\langle e, f, h \rangle$ we mean a collection of possibly zero vectors such that $[e, f] = h$, $[h, e] = 2e$, $[h, f] = -2f$. The collection of elements $\{h_\chi\}$ is called *a multiple characteristic* of e (or of the collection $\{e_\chi\}$). A compact real form of the Lie algebra of a reductive subgroup in G is, by definition, the Lie algebra of a compact real form of this complex algebraic subgroup. We fix a compact real form $\mathfrak{k} \subset \mathfrak{l}$. The multiple characteristic $\{h_\chi\}$ is called *Hermitian* if any $h_\chi \in i\mathfrak{k}$.

In this paper we consider the following question: is it true that any $e \in \mathfrak{n}$ (resp. $e \in \mathfrak{g}/\mathfrak{p}^-$) admits a Hermitian multiple characteristic up to P-conjugacy (resp. up to P^--conjugacy)? Clearly, we may (and shall) suppose that G is simple. We can identify parabolic subgroups with coloured Dynkin diagrams. Black vertices correspond to simple roots such that the corresponding root subspaces belong to the Levi part of the parabolic subgroup. The following theorem is the main result of this paper.

Theorem 1. *Suppose that G is simple and not E_7 or E_8, or G is equal to E_7 or E_8, but P is not of type* 38–59 (*see the Table at the end of this paper*). *Then for any $e \in \mathfrak{n}$* (*resp. $e \in \mathfrak{g}/\mathfrak{p}^-$*) *there exists $p \in P$* (*resp. $p \in P^-$*) *such that* $\mathrm{Ad}(p)e$ *admits a Hermitian multiple characteristic.*

The proof involves many case-by-case considerations, and the open cases 38–59 are too complicated to be analyzed using methods presented here. The *adjoint case $e \in \mathfrak{g}/\mathfrak{p}^-$* and the *coadjoint case $e \in \mathfrak{n}$* are studied separately.

Theorem 1 has the following corollary valid over an arbitrary algebraically closed field k of characteristic 0. For any collection of elements $v_1, \ldots, v_n$ of an algebraic Lie algebra $\mathfrak{h}$ we denote by $\langle v_1, \ldots, v_n \rangle_{\mathrm{alg}}$ the minimal algebraic Lie subalgebra of $\mathfrak{h}$ containing $v_1, \ldots, v_n$.

Theorem 2. *For the pairs (G, P) satisfying Theorem 1 over $\mathbb{C}$, the following is true over k. For any $e \in \mathfrak{n}$* (*resp. $e \in \mathfrak{g}/\mathfrak{p}^-$*) *there exists $p \in P$* (*resp. $p \in P^-$*) *such that* $\mathrm{Ad}(p)e$ *admits a multiple characteristic $\{h_\chi\}$ with reductive $\langle h_\chi \rangle_{\mathrm{alg}}$.*

The following Theorem 3 (conjectured in [Te2]) was already proved in [Te1] for all parabolic subgroups that satisfy Theorem 2. Let $\mathfrak{R}_G$ denote the set of irreducible representations of G. For any $V \in \mathfrak{R}_G$, there exists a unique maximal proper P-submodule $M_V \subset V$. Therefore, we have a linear map $\Psi_V : \mathfrak{g}/\mathfrak{p} \to \mathrm{Hom}(M_V, V/M_V)$, $\Psi_V(x)v = x \cdot v \mod M_V$.

Theorem 3. *For the pairs (G, P) satisfying Theorem 2, there exists an algebraic P-invariant stratification $\mathfrak{g}/\mathfrak{p} = \sqcup_{i=1}^N X_i$ such that for any $V \in \mathfrak{R}_G$, the function $\mathrm{rk}\,\Psi_V$ is constant along each X_i. In other words, the linear span of functions $\mathrm{rk}\,\Psi_V$ in the algebra of constructible functions on $\mathfrak{g}/\mathfrak{p}$ is finite-dimensional.*

If the stratification is known explicitly it allows one to solve the following classical geometric problem: For any irreducible equivariant spanned vector bundle $\mathcal{L}$ on G/P, determine whether its generic global section has zeros. This class of geometric problems includes, for example, the exact estimates on the maximal possible dimension of a projective subspace in a generic projective hypersurface of given degree, or of an isotropic subspace of a generic skew-symmetric form of given degree. The corresponding algorithm and examples were given in [Te2].

The stratification from Theorem 3 provides an alternative to the orbit decomposition for the action of P^- on $\mathfrak{g}/\mathfrak{p}^-$. It is known that this action has an open orbit, see [LW]; however the number of orbits is usually infinite, see [PR], [BHR].

All necessary facts about complex and real Lie groups, Lie algebras, and algebraic groups used in this paper without specific references can be found in [OV]. In particular, we enumerate simple roots of simple Lie algebras as in [loc. cit.] If $\mathfrak{g}$ is a simple group of rank r then $\alpha_1, \ldots, \alpha_r$ denote its simple roots, $\varphi_1, \ldots, \varphi_r$ denote its fundamental weights. For any dominant weight λ we denote by $R(\lambda)$ the irreducible representation of highest weight λ.

This paper was written during my stay at the Erwin Schrödinger Institute in Vienna and at the University of Glasgow. I would like to thank my hosts, especially Sergei Merkulov and Roger Bielawski, for the warm hospitality. I am grateful to the referees for numerous corrections and improvements.

1 Hermitian Characteristics In Classical Groups

In this section we give some background information, prove Theorem 1 for classical groups, and prove Theorem 2. We recall the notion of the Moore–Penrose inverse of linear maps. Let V_1 and V_2 be vector spaces with Hermitian scalar products. For any linear map $F : V_1 \to V_2$ its Moore–Penrose inverse is a linear map $F^+ : V_2 \to V_1$ defined as follows. Consider $\operatorname{Ker} F \subset V_1$ and $\operatorname{Im} F \subset V_2$. Let $\operatorname{Ker}^\perp F \subset V_1$ and $\operatorname{Im}^\perp F \subset V_2$ be their orthogonal complements with respect to the Hermitian scalar products. Then F defines via restriction a bijective linear map $\tilde{F} : \operatorname{Ker}^\perp F \to \operatorname{Im} F$. Then $F^+ : V_2 \to V_1$ is a unique linear map such that $F^+|_{\operatorname{Im}^\perp F} = 0$ and $F^+|_{\operatorname{Im} F} = \tilde{F}^{-1}$. It is easy to see that F^+ is the unique solution of the following system of *Penrose equations*:

$$F^+FF^+ = F^+, \quad FF^+F = F, \quad FF^+ \text{ and } F^+F \text{ are Hermitian.} \qquad (*)$$

For example, if F is invertible then $F^+ = F^{-1}$.

Proof of Theorem 1 in case of $\mathrm{GL}(V)$ *and* $\mathrm{SL}(V)$. Suppose now that $\mathfrak{g} = \mathfrak{gl}(V)$ and take any decomposition $V = V_1 \oplus \ldots \oplus V_k$. Then we have a block decomposition $\mathfrak{gl}(V) = \oplus_{i,j=1}^k \operatorname{Hom}(V_i, V_j)$. The parabolic subalgebra $\mathfrak{p} = \oplus_{i \le j} \operatorname{Hom}(V_i, V_j)$ has the unipotent radical $\mathfrak{n} = \oplus_{i<j} \operatorname{Hom}(V_i, V_j)$ and any parabolic subalgebra in $\mathfrak{gl}(V)$ has this form. Consider the corresponding $\widehat{Z}$-grading. Then $\mathfrak{l} = \mathfrak{gl}(V)_0 = \oplus_{i=1}^k \operatorname{End}(V_i)$ and all other rectangular blocks $\operatorname{Hom}(V_i, V_j)$ get different grades.

We claim that for any collection $\{e_{ij}\}$, $e_{ij} \in \operatorname{Hom}(V_i, V_j)$, $i < j$, there exists a unique Hermitian multiple characteristic with respect to any compact real form of $\mathfrak{l}$. Indeed, we fix Hermitian scalar products in $V_1, \ldots, V_k$ and set $\mathfrak{k} = \oplus_{i=1}^k \mathfrak{u}(V_i)$, where $\mathfrak{u}(V_i)$ is the Lie algebra of skew-Hermitian operators.

Any compact real form of $\mathfrak{l}$ has this presentation. Now let us take Moore–Penrose inverses $f_{ji} = e_{ij}^+ \in \mathrm{Hom}(V_j, V_i)$ of elements e_{ij}. The equations $(*)$ are equivalent to the statement that $\langle e_{ij}, h_{ij} = [e_{ij}, f_{ji}], f_{ji}\rangle$ is an $\mathfrak{sl}_2$-triple containing e_{ij} with a Hermitian characteristic h_{ij}. Therefore, $\{h_{ij}\}_{i<j}$ is a unique Hermitian multiple characteristic of $\{e_{ij}\}$. The same argument applies for $\mathfrak{sl}(V)$. $\square$

Definitions.

- *An element $x \in \mathfrak{g}_\chi$ and its $\mathrm{Ad}(L)$-orbit $\mathcal{O}$ are called ample if for any $e_\chi \in \mathcal{O}$ there exists a homogeneous $\mathfrak{sl}_2$-triple $\langle e_\chi, h_\chi, f_\chi\rangle$ with $h_\chi \in i\mathfrak{k}$.*
- *The $\widehat{Z}$-grading of $\mathfrak{g}$ is ample in degree χ if all $\mathrm{Ad}(L)$-orbits in $\mathfrak{g}_\chi$ are ample.*
- *A parabolic subgroup P is called weakly ample if for every $e \in \mathfrak{n}$, there exists $l \in L$ such that $\mathrm{Ad}(l)e$ admits a Hermitian multiple characteristic.*

A non-zero weight $\chi \in \widehat{Z}$ is called *reduced* if $\mathfrak{g}_{2\chi} = 0$.

Basic Lemma. *If $\chi \in \widehat{Z}$ is reduced then the grading is ample in degree χ.*

This lemma was proved in [Te1]. It shows that almost all components of the grading are automatically ample.

Theorem 1 in the remaining classical cases follows from Theorem 4 below.

Theorem 4. *Any parabolic subgroup in $\mathrm{SO}_n(\mathbb{C})$ or $\mathrm{Sp}_n(\mathbb{C})$ is weakly ample.*

Proof of Theorem 4. Suppose that $V = \mathbb{C}^n$ is a complex vector space endowed with a non-degenerate bilinear form ω which is either symmetric or skew-symmetric. We denote by $G(\omega) \subset \mathrm{SL}(V)$ the corresponding special orthogonal or symplectic group of automorphisms of V preserving ω. Let $\mathfrak{g} \subset \mathfrak{sl}(V)$ be its Lie algebra of skew-symmetric operators with respect to ω. The subspace $U \subset V$ is called isotropic if ω vanishes on U. Let $0 = F_0 \subset F_1 \subset \ldots \subset F_k \subset V$ be a flag of isotropic subspaces. Then the stabilizer P of this flag in $G(\omega)$ is a parabolic subgroup and any parabolic subgroup has this form.

In order to fix the Levi subgroup of P, we choose subspaces $U_k^+ \subset F_k$ complementary to F_{k-1} and we choose an isotropic subspace G_k transversal to F_k such that the restriction of ω to $F_k \oplus G_k$ is non-degenerate. Then the subgroup L of $G(\omega)$ preserving both $U_1^+, \ldots, U_k^+$ and G_k is a Levi subgroup of P. F_k and G_k are naturally dual to each other with respect to the bilinear form: any $v \in G_k$ defines a linear functional $\omega(v, \cdot)$ on F_k. Let $G_k = U_1^- \oplus \ldots \oplus U_k^-$ be the decomposition dual to the decomposition $F_k = U_1^+ \oplus \ldots \oplus U_k^+$. Let $W = (F_k \oplus G_k)^\perp$. Then L automatically preserves $U_1^-, \ldots, U_k^-$ and W. The semisimple part of $\mathfrak{l}$ is isomorphic to $\mathfrak{sl}(U_1^-) \oplus \ldots \oplus \mathfrak{sl}(U_k^-) \oplus \mathfrak{g}(W)$, where $\mathfrak{g}(W)$ is the Lie algebra of skew-symmetric operators of W. Each $\mathfrak{sl}(U_i^-)$ acts naturally in U_i^-, dually on U_i^+, and trivially on W and on U_j^+, U_j^- for $j \neq i$. $\mathfrak{g}(W)$ acts naturally on W and trivially on $F_k \oplus G_k$. The center $\mathfrak{z}$ of $\mathfrak{l}$ acts on U_k^+ by scalar transformations $\lambda_k E$, on U_k^- by $-\lambda_k E$, and trivially on W. Here $\lambda_1, \ldots, \lambda_k$ is a basis of $\mathfrak{z}^*$.

The choice of a compact form in $\mathfrak{l}$ is equivalent to the choice of a compact form in each $\mathfrak{sl}(U_i^-)$ (i.e. the choice of a Hermitian scalar product in U_i^-) and a compact form $\mathfrak{k}(W)$ of $\mathfrak{g}(W)$. We shall vary Hermitian scalar products in U_i^- later, but the compact real form of $\mathfrak{g}(W)$ is to be fixed once and for all. We fix a basis $b_1, \ldots, b_m$ of W such that the matrix of ω in this basis is equal to I, where $I = \mathrm{Id}$ in the symmetric case and $I = \begin{pmatrix} 0 & \mathrm{Id} \\ -\mathrm{Id} & 0 \end{pmatrix}$ in the skew-symmetric case. We fix a standard Hermitian form in W by the formula $\{u, v\} = \omega(u, I^t \overline{v})$. Then the subalgebra $\mathfrak{k}(W)$ of skew-Hermitian operators in $\mathfrak{g}(W)$ is its compact real form.

Non-trivial $\widehat{Z}$-graded components of $\mathfrak{g}$ can be described as follows.

- Any linear operator $A_{ij} : U_i^+ \to U_j^+$, $i \neq j$ gives by duality the linear operator $A'_{ij} : U_j^- \to U_i^-$. We define the skew-symmetric operator $\tilde{A}_{ij}$ by setting $\tilde{A}_{ij}|_{U_i^+} = A_{ij}$, $\tilde{A}_{ij}|_{U_j^-} = -A'_{ij}$; the restriction of $\tilde{A}_{ij}$ to other components of the decomposition $V = \oplus_k U_k^+ \oplus \oplus_k U_k^- \oplus W$ is trivial. These operators form $\mathfrak{g}_{\lambda_j - \lambda_i}$, which belongs to $\mathfrak{n}$ if $i > j$.
- Any linear operator $A'_i : W \to U_i^+$ (resp. $A'_i : W \to U_i^-$) determines by duality the linear operator $A_i : U_i^- \to W$ (resp. $A_i : U_i^+ \to W$), where we identify W and W^*, $w \mapsto \omega(w, \cdot)$. We define the skew-symmetric operator $\tilde{A}_i$ by setting $\tilde{A}_i|_W = A'_i$, $\tilde{A}_i|_{U_i^\mp} = -A_i$; the restriction of $\tilde{A}_i$ to other components of the decomposition $V = \oplus_k U_k^+ \oplus \oplus_k U_k^- \oplus W$ is trivial. These operators form $\mathfrak{g}_{\lambda_i} \subset \mathfrak{n}$ (resp. $\mathfrak{g}_{-\lambda_i}$).
- Any linear operator $B_{ij} : U_i^- \to U_j^+$ (resp. $B_{ij} : U_i^+ \to U_j^-$), $i \neq j$ determines by duality the linear operator $B'_{ij} : U_j^- \to U_i^+$ (resp. $B'_{ij} : U_j^+ \to U_i^-$). We define the skew-symmetric operator $\tilde{B}_{ij}$ with respect to ω by setting $\tilde{B}_{ij}|_{U_i^\mp} = B_{ij}$, $\tilde{B}_{ij}|_{U_j^\pm} = -B'_{ij}$; the restriction of $\tilde{B}_{ij}$ to other components of the decomposition $V = \oplus_k U_k^+ \oplus \oplus_k U_k^- \oplus W$ is trivial. These operators form $\mathfrak{g}_{\lambda_i + \lambda_j} \subset \mathfrak{n}$ (resp. $\mathfrak{g}_{-\lambda_i - \lambda_j}$).
- Finally, any skew-symmetric linear operator $B_i : U_i^- \to U_i^+$ (resp. $B_i : U_i^+ \to U_i^-$) defines the skew-symmetric operator $\tilde{B}_i$ with respect to ω: $\tilde{B}_i|_{U_i^\mp} = B_i$; the restriction of $\tilde{B}_i$ to other components of the decomposition $V = \oplus_k U_k^+ \oplus \oplus_k U_k^- \oplus W$ is trivial. These operators form $\mathfrak{g}_{2\lambda_i} \subset \mathfrak{n}$ (resp. $\mathfrak{g}_{-2\lambda_i}$). This component is trivial if $\dim U_i^+ = 1$.

By the Basic Lemma, if $\mathfrak{g}_{2\chi} = 0$ then for any $e \in \mathfrak{g}_\chi$ and for any compact form $\mathfrak{k} \subset \mathfrak{l}$ there exists a homogeneous $\mathfrak{sl}_2$ triple $\langle e, h, f \rangle$ with a Hermitian characteristic $h \in i\mathfrak{k}$. Therefore, in our case it remains to prove that for any set of elements $e_i \in \mathfrak{g}_{\lambda_i}$, $i = 1, \ldots, k$ there exists a compact form $\mathfrak{k} \subset \mathfrak{l}$ and a Hermitian multiple characteristic $h_i \in i\mathfrak{k}$, $i = 1, \ldots, k$.

Any e_i is defined by a linear map $A_i : U_i^- \to W$, and $f_i \in \mathfrak{g}_{-\lambda_i}$ is defined by a linear map $B_i : W \to U_i^-$. Using the description of weighted components of $\mathfrak{g}$ given above, it is easy to see that e_i and f_i can be embedded in a homogeneous $\mathfrak{sl}_2$-triple if and only if A_i and B_i satisfy the following system of matrix equations

$$2A_i = 2A_iB_iA_i - (A_iB_i)^{\#}A_i, \quad 2B_i = 2B_iA_iB_i - B_i(A_iB_i)^{\#},$$

where for any $A \in \mathrm{Hom}(W,W)$ we denote by $A^{\#}$ its adjoint operator with respect to ω. Moreover, the characteristic of this $\mathfrak{sl}_2$-triple will be Hermitian if and only if

$$B_iA_i,\ A_iB_i - (A_iB_i)^{\#} \text{ are Hermitian operators.}$$

Therefore, it remains to prove the following lemma.

Lemma. *Suppose that $U = \mathbb{C}^n$, $W = \mathbb{C}^k$ are complex vector spaces. Let ω be a non-degenerate symmetric (resp. skew-symmetric) form on W with matrix $I = \mathrm{Id}$, resp. $I = \begin{pmatrix} 0 & \mathrm{Id} \\ -\mathrm{Id} & 0 \end{pmatrix}$. We fix a standard Hermitian form in W. Let $A \in \mathrm{Hom}(U,W)$. Then there exists a Hermitian form on U and an operator $B \in \mathrm{Hom}(W,U)$ such that*

$$2A = 2ABA - (AB)^{\#}A, \quad 2B = 2BAB - B(AB)^{\#}, \qquad (*)$$

$$BA,\ AB - (AB)^{\#} \text{ are Hermitian operators,} \qquad (**)$$

where for any $A \in \mathrm{Hom}(W,W)$ we denote by $A^{\#}$ its adjoint operator with respect to ω.

Proof. Let $W_0 = \mathrm{Ker}\,\omega|_{\mathrm{Im}\,A}$, let W_1 be the orthogonal complement to W_0 in $\mathrm{Im}\,A$ with respect to the Hermitian form. Let $U_2 = \mathrm{Ker}\,A$. Choose a subspace $\tilde{U} \subset U$ complementary to U_2. Then A defines a bijective linear map $\tilde{A} : \tilde{U} \to \mathrm{Im}\,A$. Let $U_0 = \tilde{A}^{-1}(W_0)$, $U_1 = \tilde{A}^{-1}(W_1)$. We fix a Hermitian form on U such that U_0, U_1, and U_2 are pairwise orthogonal and claim that the system of equations $(*, **)$ has a solution.

Let $W_2 = I\overline{W_0}$, where bar denotes the complex conjugation. Then $W_0 \cap W_2 = \{0\}$, the restriction of ω to $W_0 \oplus W_2$ is non-degenerate, and $W_0 \oplus W_2$ is orthogonal to W_1 both with respect to ω and with respect to the Hermitian form. Let W_3 be the orthogonal complement to $W_0 \oplus W_1 \oplus W_2$ with respect to the Hermitian form.

We define the operator B as follows:

$$B|_{W_0} = \tilde{A}^{-1},\ B|_{W_1} = 2\tilde{A}^{-1},\ B|_{W_2} = B|_{W_3} = 0.$$

Then we have

$$AB|_{W_0} = \mathrm{Id},\ AB|_{W_1} = 2\cdot\mathrm{Id},\ AB|_{W_2} = AB|_{W_3} = 0,$$

therefore

$$(AB)^{\#}|_{W_0} = 0,\ (AB)^{\#}|_{W_1} = 2\cdot\mathrm{Id},\ (AB)^{\#}|_{W_2} = \mathrm{Id},\ (AB)^{\#}|_{W_3} = 0.$$

Since W_0, W_1, W_2, and W_3 are pairwise orthogonal with respect to the Hermitian form, it follows that $AB - (AB)^{\#}$ is a Hermitian operator. It is easy to check that $(*)$ holds. Now, we have

$$BA|_{U_0} = \mathrm{Id},\ BA|_{U_1} = 2\cdot\mathrm{Id},\ BA|_{U_2} = 0.$$

Since U_0, U_1, and U_2 are pairwise orthogonal with respect to the Hermitian form, it follows that BA is also a Hermitian operator. □

Proof of Theorem 2. Until the end of this section "an algebraic subvariety" means "a union of locally closed subvarieties". To avoid repetition, we consider the coadjoint case only; the adjoint case is absolutely similar.

Let $e \in \mathfrak{n}$. We need to prove that there exists $p \in P$ such that $\mathrm{Ad}(p)e$ admits a multiple characteristic $\{h_\chi\}$ with reductive $\langle h_\chi\rangle_{\mathrm{alg}}$. We can substitute the base field k by an algebraic closure of a subfield generated by coefficients of e in a fixed basis compatible with the grading. This shows that, without loss of generality, we may assume that k is embedded into $\mathbb{C}$ By Theorem 1, and there exists $p' \in P(\mathbb{C})$ such that $\mathrm{Ad}(p')e$ admits a Hermitian multiple characteristic $\{h'_\chi\}$ (defined over $\mathbb{C}$). It easily follows (see [Te1]) that $\langle h'_\chi\rangle_{\mathrm{alg}}$ is reductive. Let $W \subset \mathfrak{n}$ be the subset of all points that admit a multiple characteristic $\{h_\chi\}$ with reductive $\langle h_\chi\rangle_{\mathrm{alg}}$. We are going to show that W is a subvariety defined over k. Then the argument above implies that $\mathrm{Ad}(P)e\cap W$ is non-empty over $\mathbb{C}$, hence non-empty over k, hence the Theorem.

Let n be the number of positive weights. Consider the variety of $\mathfrak{sl}_2$-triples

$$S \subset \bigoplus_{\chi<0}\mathfrak{g}_\chi \oplus \mathfrak{g}_0^{\oplus n} \oplus \bigoplus_{\chi>0}\mathfrak{g}_\chi,$$

$$S = \big\{f_\chi \in \mathfrak{g}_{-\chi},\ h_\chi \in \mathfrak{g}_0,\ e_\chi \in \mathfrak{g}_\chi \;\big|\; [e_\chi, f_\chi] = h_\chi,\ [h_\chi, e_\chi] = 2e_\chi,\ [h_\chi, f_\chi] = -2f_\chi\big\}.$$

Let π_1 denote the projection of S on $\mathfrak{g}_0^{\oplus n}$, π_2 denote the projection of S on $\mathfrak{n}$. These projections are defined over k. Let R denote the set of n-tuples of points $(x_1,\ldots,x_n) \in \mathfrak{g}_0^{\oplus n}$ such that $\langle x_1,\ldots,x_n\rangle_{\mathrm{alg}}$ is reductive. Then, clearly, $W = \pi_2(\pi_1^{-1}(R))$. Therefore, it suffices to show that R is a subvariety of $\mathfrak{g}_0^{\oplus n}$. By a theorem of Richardson [Ri], $\langle x_1,\ldots,x_n\rangle_{\mathrm{alg}}$ is reductive if and only if the orbit of $(x_1,\ldots,x_n) \in \mathfrak{g}_0^{\oplus n}$ is closed with respect to the diagonal action of $L = G_0$. It remains to notice that if a reductive group L acts on an affine variety X then the union of the closed orbits $Y \subset X$ is a subvariety (defined over the field of definition of the action). Indeed, one can argue using induction on $\dim X$ and the two following observations (their proofs can be found in [Vi]): If generic G-orbits in X are not closed then there exists a proper closed subvariety $X_0 \subset X$ that contains all closed orbits. If generic G-orbits in X are closed then there exists an open G-invariant subset $U \in X$ such that all orbits in U are closed in X. Then we may pass from X to $X \setminus U$.

2 Exceptional Groups

The proof of Theorem 1 for the classical groups given above relies on the fact that their block decompositions can be easily understood in terms of linear algebra. However, block decompositions of exceptional groups are much more complicated, diverse, and interesting. Here we collect some important results about these decompositions. The actual proof of Theorem 1 will be finished in Section 3.

2.0 Comparison Lemma

It follows from the Basic Lemma that if all positive weights $\chi \in \widehat{Z}$ are reduced except at most one, then the corresponding parabolic subgroup is weakly ample. Indeed, let $e \in \mathfrak{n}$ and let χ be the only non-reduced weight. Take any homogeneous characteristic h_χ of e_χ. Then there exists $l \in L$ such that $\mathrm{Ad}(l)h_\chi \in i\mathfrak{k}$ because all compact real forms of $\mathfrak{l}$ are conjugate, see also [Te1]. It follows that $\mathrm{Ad}(l)e$ admits a Hermitian multiple characteristic. Therefore, we need to prove Theorem 1 only for parabolic subgroups such that there exist two or more not reduced $\widehat{Z}$-weights. Moreover, some non-reduced weights may also correspond to ample components of the grading. The following lemma provides a lot of examples.

Comparison Lemma. *Let G' and G'' be reductive algebraic groups with Lie algebras $\mathfrak{g}'$ and $\mathfrak{g}''$, parabolic subgroups $P' \subset G'$ and $P'' \subset G''$, and Levi subgroups $L' \subset P'$ and $L'' \subset P''$. Let Z' and Z'' be the centers of L' and L''. Consider positive weights $\chi' \in \widehat{Z}'$ and $\chi'' \in \widehat{Z}''$. Suppose that $[L', L'] = [L'', L''] = H$ and the H-modules $\mathfrak{g}'_{\chi'}$ and $\mathfrak{g}''_{\chi''}$ coincide. Suppose that H is either simple or isomorphic to $\mathrm{SL}_n \times \mathrm{SL}_m$, in which case $\mathfrak{g}'_{\chi'} \simeq \mathfrak{g}''_{\chi''} \simeq \mathbb{C}^n \otimes \mathbb{C}^m$ as an H-module. Then the $\widehat{Z}'$-grading of $\mathfrak{g}'$ is ample in degree χ' iff the $\widehat{Z}''$-grading of $\mathfrak{g}''$ is ample in degree χ''.*

Proof. We may assume without loss of generality that P' and P'' are maximal parabolic subgroups. Let $\mathfrak{h}$ be the Lie algebra of H. Consider two local Lie algebras $\mathfrak{g}'_{-\chi'} \oplus \mathfrak{g}'_0 \oplus \mathfrak{g}'_{\chi'}$ and $\mathfrak{g}''_{-\chi''} \oplus \mathfrak{g}''_0 \oplus \mathfrak{g}''_{\chi''}$. We choose $c' \in \mathfrak{z}'$ and $c'' \in \mathfrak{z}''$ such that $\mathrm{ad}(c')|_{\mathfrak{g}'_{\pm\chi'}} = \pm\mathrm{Id}$ and $\mathrm{ad}(c'')|_{\mathfrak{g}''_{\pm\chi''}} = \pm\mathrm{Id}$. Then $\mathfrak{g}'_0 = \mathbb{C}c' \oplus \mathfrak{h}$ and $\mathfrak{g}''_0 = \mathbb{C}c'' \oplus \mathfrak{h}$. We fix compact real forms $\mathfrak{k}' \subset \mathfrak{g}'_0$ and $\mathfrak{k}'' \subset \mathfrak{g}''_0$. Then $\mathfrak{k}' = i\mathbb{R}c' \oplus \mathfrak{m}$ and $\mathfrak{k}'' = i\mathbb{R}c'' \oplus \mathfrak{m}$, where $\mathfrak{m}$ is a compact real form in $\mathfrak{h}$. Suppose that $e' \in \mathfrak{g}'_{\chi'}$ and the $\widehat{Z}'$-grading of $\mathfrak{g}'$ is ample in degree χ'. Then there exists $f' \in \mathfrak{g}'_{-\chi'}$ such that $\langle e', h' = [e', f'], f' \rangle$ is an $\mathfrak{sl}_2$-triple and $h' \in i\mathfrak{k}'$. Let $h' = h'_1 + h'_2$, where $h'_1 \in i\mathbb{R}c'$ and $h'_2 \in i\mathfrak{m}$. Let $\mathfrak{h} = \oplus \mathfrak{h}_p$ be the direct sum of simple Lie algebras. Then we have the corresponding decomposition $\mathfrak{m} = \oplus \mathfrak{m}_p$ and $h'_2 = \sum (h'_2)_p$. We can identify $\mathfrak{g}'_0$ and $\mathfrak{g}''_0$ as Lie algebras by assigning c'' to c'. Then we can identify $\mathfrak{g}'_{\chi'}$ with $\mathfrak{g}''_{\chi''}$ and $\mathfrak{g}'_{-\chi'}$ with $\mathfrak{g}''_{\chi''}$ as their modules. Let $e'' \in \mathfrak{g}''_{\chi''}$ correspond to e' and $f'' \in \mathfrak{g}''_{-\chi''}$ correspond to f'. Let $h'' = [e'', f''] = h''_1 + h''_2$, where $h''_1 \in \mathbb{C}c''$ and $h''_2 \in \mathfrak{h}$, $h''_2 = \sum (h''_2)_p$.

We claim that $h'' \in i\mathfrak{k}''$ and its real multiple is a characteristic of e''. It would follow that the $\widehat{Z}''$-grading of $\mathfrak{g}''$ is ample in degree χ''.

Let $\tilde{h}' = \tilde{h}'_1 + \tilde{h}'_2$ be the corresponding element of $\mathfrak{g}'_0$, where $\tilde{h}'_1 \in \mathbb{C}c'$ and $\tilde{h}'_2 \in \mathfrak{h}$, $\tilde{h}'_2 = \sum(\tilde{h}'_2)_p$. It is sufficient to prove that $\tilde{h}'_1 = \alpha h'_1$ and $(\tilde{h}'_2)_p = \beta_p(h'_2)_p$, where α and β are positive real numbers (here we use our restrictions to H). The commutator maps

$$\Phi' : \mathfrak{g}'_{-\chi'} \otimes \mathfrak{g}'_{\chi'} \to \mathfrak{g}'_0 \quad \text{and} \quad \Phi'' : \mathfrak{g}''_{-\chi''} \otimes \mathfrak{g}''_{\chi''} \to \mathfrak{g}''_0$$

can be decomposed as $\Phi' = \Omega' \circ \Psi'$, $\Phi'' = \Omega'' \circ \Psi''$, where Ψ' and Ψ'' are moment maps and $\Omega' : (\mathfrak{g}'_0)^* \to \mathfrak{g}'_0$, $\Omega'' : (\mathfrak{g}''_0)^* \to \mathfrak{g}''_0$ are pairings given by K' and K'', where K', K'' are restrictions of the Killing forms in $\mathfrak{g}'$ and $\mathfrak{g}''$ to $\mathfrak{g}'_0$ and $\mathfrak{g}''_0$. Under the identifications above we have $\Psi' = \Psi''$. It remains to notice that $\mathbb{C}c'$ is orthogonal to $\mathfrak{h}$ under K', $\mathbb{C}c''$ is orthogonal to $\mathfrak{h}$ under K'', the restrictions of K' and K'' to $(\mathfrak{h})_p$ coincide with the Killing form of $(\mathfrak{h})_p$ multiplied by a positive real number, and that $K'(c', c')$ and $K''(c'', c'')$ are both real and positive. □

For instance, take a maximal parabolic subgroup of F_4. Then the corresponding grading is $\mathfrak{g} = \oplus_{i=-N}^{N} \mathfrak{g}_i$, where N is equal to 2, 3, or 4. In the first and the second case all weights but one are reduced and everything follows from the Basic Lemma. If $N = 4$ then there are two non-reduced weights, namely 1 and 2. However, the representation of $[L, L]$ on $\mathfrak{g}_1$ in this case is locally isomorphic to the representation of $\mathrm{SL}_2 \times \mathrm{SL}_3$ on $\mathbb{C}^2 \otimes \mathbb{C}^3$ and the result follows from the Comparison Lemma.

Using these simple arguments, it is easy to verify that all parabolic subgroups not listed in the table at the end of the paper are weakly ample (in particular, all parabolic subgroups in G_2, F_4 and E_6 are weakly ample). Theorem 1 for the entries 1–37 from the table will be checked in the next sections. Entries 38–59 will be verified in the sequel of this paper.

2.1 Non-degenerate Deformations

A projective variety $X \subset \mathbb{P}(V)$ is called a 2-variety if it is cut out by quadrics, i.e. its homogeneous ideal $I \in \mathbb{C}[V]$ is generated by I_2. For example, if G is a semi-simple Lie group and V is an irreducible G-module then the projectivization of the cone of highest weight vectors in V is a 2-variety by a result of B. Kostant (see also [Li]). Suppose that $X \subset \mathbb{P}(V)$ is a 2-variety. A linear map $A : \mathbb{C}^2 \to V$ is called degenerate if $\dim \mathrm{Im}\, A = 2$ and $\mathbb{P}(\mathrm{Im}\, A) \cap X$ is a point.

Proposition 1. (A) *If $A : \mathbb{C}^2 \to V$ is degenerate, and $B \in \mathrm{Hom}(\mathbb{C}^2, \mathbb{C}^p)$ is non-trivial, then there exists $C \in \mathrm{Hom}(\mathbb{C}^p, V)$ such that $A + CB$ is non-degenerate.*

(B) *If $A, B : \mathbb{C}^2 \to V$ are degenerate, then there exists $E \in \mathrm{Hom}(\mathbb{C}^2, \mathbb{C}^2)$ such that $A + BE$ is non-degenerate.*

(C) *Suppose $A : \mathbb{C}^2 \to V$ is degenerate, and $v \in V$ does not belong to the cone over X. Then there exists $f \in (\mathbb{C}^2)^*$ such that $A + v \cdot f$ is non-degenerate.*

Proof. (A) Clearly, we can choose C such that $\dim \operatorname{Im}(A + CB) < 2$.

(B) If $(\operatorname{Im} A) \cap (\operatorname{Im} B) \neq 0$ then there exists $E \in \operatorname{Hom}(\mathbb{C}^2, \mathbb{C}^2)$ such that we have inequality $\dim \operatorname{Im}(A + BE) < 2$ and we are done. Otherwise, let $V_0 = (\operatorname{Im} A) + (\operatorname{Im} B)$, $X_0 = X \cap \mathbb{P}(V_0)$. Clearly, $\dim V_0 = 4$ and X_0 is a 2-variety in $\mathbb{P}(V_0)$. We take skew lines $l_0 = \mathbb{P}(\operatorname{Im} A)$ and $l_1 = \mathbb{P}(\operatorname{Im} B)$. Each of them intersects X_0 at a point. For any $E \in \operatorname{Hom}(\mathbb{C}^2, \mathbb{C}^2)$, $\dim \operatorname{Im}(A+BE) = 2$. A line $l \subset \mathbb{P}(V_0)$ is equal to $\mathbb{P}(\operatorname{Im}(A + BE))$ for some $E \in \operatorname{Hom}(\mathbb{C}^2, \mathbb{C}^2)$ if and only if $l \cap l_1 = \varnothing$.

Clearly, $\dim X_0 < 3$. If $\dim X_0 < 2$ then there exists a line $l \subset \mathbb{P}(V_0)$ such that $l \cap l_1 = l \cap X_0 = \varnothing$ and we are done. Otherwise, since $l_1 \not\subset X_0$, there exist hyperplanes $H_1, H_2 \subset \mathbb{P}(V)$ and points $p_1 \in H_1$, $p_2 \in H_2$ such that $l_1 = H_1 \cap H_2$, $p_1, p_2 \notin l_1$, $p_1, p_2 \in X_0$. Then the line l connecting p_1 and p_2 does not intersect l_1 and either intersects X_0 in two points or belongs to X_0.

(C) Indeed, if $v \in \operatorname{Im} A$, then we can find f such that $\dim \operatorname{Im}(A + v \cdot f) < 2$. Suppose that $v \notin \operatorname{Im} A$. Then any 2-dimensional subspace in $U = \langle v \rangle + \operatorname{Im} A$ not containing v can be realised as $\operatorname{Im}(A + v \cdot f)$ for some f. Let $Y = \mathbb{P}U \cap X$, $b \in \mathbb{P}U$ be the point corresponding to v. Then $b \notin Y$. We need to find a line l in $\mathbb{P}U$ such that $b \notin l$ and $l \cap Y$ is not a point. If $\dim Y = 0$, then we can find a line that does not intersect b and Y at all. If $\dim Y = 1$ and Y has a line as an irreducible component, then we can take this component as a line we are looking for. Finally, in the remaining cases a generic line in $\mathbb{P}U$ intersects Y in more then one (actually, two) points. □

2.2 Kac Theorem

Consider the $\widehat{Z}$-graded reductive Lie algebra $\mathfrak{g} = \oplus_{\chi \in \widehat{Z}} \mathfrak{g}_\chi$, $\chi \neq 0$. Let $\mathcal{O}$ denote the open L-orbit in $\mathfrak{g}_\chi$, $e \in \mathcal{O}$. Kac has proved in [Ka] that there exists an $\mathfrak{sl}_2$-triple $\langle e, h, f \rangle$ with $h \in \mathfrak{z}$ if and only if the complement of $\mathcal{O}$ has codimension 1. In particular, we have the following proposition:

Proposition 2. *If $\mathcal{O}$ is open and its complement has codimension 1, then $\mathcal{O}$ is ample.*

The list of representations of $[L, L]$ on $\mathfrak{g}_\chi$ with the above property can be found in [loc. cit.].

2.3 Witt's Theorem

The following proposition is well known as Witt's Theorem.

Proposition 3. *Consider a vector space W endowed with a non-degenerate quadratic form Q. Orbits of* $\operatorname{GL}(V) \times \operatorname{O}(W)$ *on* $\operatorname{Hom}(V, W)$ *are parametrized by pairs of integers (i, j) such that*

$$0 \leq i \leq \min(\dim V, \dim W), \quad 0 \leq j \leq \min(i, \dim W - i).$$

Here $A \in \mathrm{Hom}(V, W)$ *corresponds to the pair* $(\dim A, \dim \mathrm{Ker}\, Q|_{\dim A})$.

With respect to the action of $\mathrm{GL}(V) \times \mathrm{SO}(W)$, some of these orbits may split into two orbits. This action appears as the action of a Levi subgroup of a maximal parabolic subgroup in $G = \mathrm{SO}_n$ on $\mathfrak{g}_1$. It was shown in [Te1] that ample orbits correspond to pairs $(k, 0)$ and (k, k). More precisely, a linear map $A : \mathbb{C}^k \to \mathbb{C}^n$ (or the corresponding tensor $\tilde{A} \in (\mathbb{C}^k)^* \otimes \mathbb{C}^n$), where $\mathbb{C}^n$ is endowed with a non-degenerate scalar product, is called ample if the restriction of the scalar product in $\mathbb{C}^n$ to $\mathrm{Im}\, A$ is either non-degenerate or trivial.

Proposition 4. *Let* $V = \mathbb{C}^n$ *be endowed with a non-degenerate scalar product.*

(A) *Let* $k \leq 3$ *or* $k = n = 4$. *Suppose that the linear map* $A : \mathbb{C}^k \to V$ *is not ample, and the linear map* $B : \mathbb{C}^k \to \mathbb{C}^p$ *is not trivial. Then* $A + CB$ *is ample for some* $C \in \mathrm{Hom}(\mathbb{C}^p, V)$.

(B) *Let* $n \leq 4$. *Suppose that the linear map* $A : \mathbb{C}^k \to V$ *is not ample, and* $v \in V$ *is not trivial. Then there exists a linear function* $f : (\mathbb{C}^k)^*$ *such that* $A + v \cdot f$ *is ample.*

Proof. Simple calculation. □

The tensor $\tilde{A} \in \mathbb{C}^k \otimes \mathbb{C}^2 \otimes \mathbb{C}^2$ may be considered as a linear map $A_0 : (\mathbb{C}^k)^* \to (\mathbb{C}^2 \otimes \mathbb{C}^2)$. The space $\mathbb{C}^2 \otimes \mathbb{C}^2$ has a canonical quadratic form det. $\tilde{A}$ is called not ample if the restriction of det to $\mathrm{Im}\, A_0$ is degenerate but not trivial. $\tilde{A}$ may also be viewed as a linear map $A : (\mathbb{C}^2)^* \to (\mathbb{C}^2 \otimes \mathbb{C}^k)$. Then $\tilde{A}$ is not ample if and only if $\mathrm{Im}\, A$ is 2-dimensional and $(\mathrm{Im}\, A) \cap R$ is a line, where $R \subset \mathbb{C}^2 \otimes \mathbb{C}^k$ is the variety of rank 1 matrices.

Proposition 5. (A) *Suppose that the linear maps* $A : \mathbb{C}^2 \to (\mathbb{C}^2 \otimes \mathbb{C}^3)$ *and* $B : \mathbb{C}^2 \to (\mathbb{C}^2 \otimes \Lambda^2\mathbb{C}^3)$ *are not ample. Then there exists* $E \in \mathrm{End}(\mathbb{C}^2) \otimes \mathbb{C}^3$ *such that* $B + E \circ A$ *is ample; here we embed* $\mathbb{C}^3$ *in* $\mathrm{Hom}(\mathbb{C}^3, \Lambda^2\mathbb{C}^3)$ *in the obvious way.*

(B) *Suppose that the linear map* $A : \Lambda^2\mathbb{C}^3 \to (\mathbb{C}^2 \otimes \mathbb{C}^2)$ *is not ample, and* $B \in \mathrm{Hom}(\mathbb{C}^3, \mathbb{C}^2)$ *is not trivial. Then there exists a linear map* $C \in \mathrm{Hom}(\mathbb{C}^3, \mathbb{C}^2)$ *such that* $A + B \wedge C$ *is ample.*

(C) *Let* $v \in \mathbb{C}^3$, $v \neq 0$, $B \in (\Lambda^2\mathbb{C}^3) \otimes \mathbb{C}^2$. *Then there exists* $A \in \mathbb{C}^3 \otimes \mathbb{C}^2$ *such that* $B + v \wedge A \in (\Lambda^2\mathbb{C}^3) \otimes \mathbb{C}^2$ *has rank* 2.

Proof. (A) Direct calculation shows that there exists $E_0 \in \mathrm{End}(\mathbb{C}^2) \otimes \mathbb{C}^3$ such that $E_0 \circ A$ belongs to the open orbit in $\mathrm{Hom}(\mathbb{C}^2, \mathbb{C}^2 \otimes \Lambda^2\mathbb{C}^3)$ with respect to the group $G = \mathrm{GL}_2 \times \mathrm{GL}_2 \times \mathrm{GL}_3$. This complement of this orbit has codimension 1, therefore there exists $\lambda \in \mathbb{C}$ such that $B + \lambda E_0 \circ A$ belongs to this open orbit. Now we can take $E = \lambda E_0$ by Proposition 2.

(B) Suppose first that $\dim \mathrm{Im}\, B = 2$. We take a basis $\{e_1, e_2, e_3\}$ of $\mathbb{C}^3$ such that $B(e_3) = 0$. If we take C such that $C(e_3) = 0$, then $B \wedge C(e_1 \wedge e_3) =$

$B \wedge C(e_2 \wedge e_3) = 0$ and $B \wedge C(e_1 \wedge e_2)$ can be made arbitrary. Thus we are done by Proposition 4.A.

Suppose now that $\dim \operatorname{Im} B = 1$. Then the linear space of all possible maps of the form $B \wedge C$ consists of all linear maps $X : \Lambda^2\mathbb{C}^3 \to (\mathbb{C}^2 \otimes \mathbb{C}^2)$ such that $\operatorname{Ker} X \supset R$ and $\operatorname{Im} X \subset L$, where $R \subset \mathbb{C}^3$ is a fixed 1-dimensional subspace and $L \subset \mathbb{C}^2 \otimes \mathbb{C}^2$ is a fixed 2-dimensional subspace isotropic with respect to det. We need to consider two cases: $\dim \operatorname{Im} A = 2$ and $\dim \operatorname{Im} A = 3$.

Let $\dim \operatorname{Im} A = 2$. We fix a basis $\{e_1, e_2, e_3\}$ of $\Lambda^2\mathbb{C}^3$ such that $A(e_1) = 0$ and $v = A(e_2)$ spans the kernel of restriction of det to $\langle v, u\rangle$, where $u = A(e_3)$. Clearly, we may assume that R is spanned by e_1, e_2, or e_3. If $R = \mathbb{C}e_1$ and $L \cap \langle v, u\rangle = 0$ then we can choose X such that $\operatorname{Im}(A + X)$ is an arbitrary 2-dimensional subspace not intersecting L and the claim obviously follows. If $R = \mathbb{C}e_1$ and $L \cap \langle v, u\rangle \neq 0$ then we can choose X such that $\dim(A+X) < 2$ and we are done. If $R = \mathbb{C}e_2$ and $v \notin L$ then we can find $w \in L$ such that v and w are not orthogonal. Then we take the map X such that $X(e_1) = w$, $X(e_3) = 0$. $\operatorname{Im}(A + X)$ is 3-dimensional and non-degenerate with respect to det. If $R = \mathbb{C}e_2$ and $v \in L$ then we take X such that $X(e_1) = 0$ and $u + X(e_3)$ is isotropic with respect to det. This is possible, because the orthogonal complement to L is L itself, but $u \notin L$. Then $\operatorname{Im}(A+X)$ is 2-dimensional and isotropic with respect to det. If $R = \mathbb{C}e_3$ and $v \notin L$ then we can find $w \in L$ such that v and w are not orthogonal. Then we take the map X such that $X(e_1) = w$, $X(e_2) = 0$. $\operatorname{Im}(A + X)$ is 3-dimensional and non-degenerate with respect to det. If $R = \mathbb{C}e_3$ and $v \in L$ then we take X such that $X(e_2) = -v$ and $X(e_1)$ is not orthogonal to u with respect to det. This is possible, because the orthogonal complement to L is L itself, but $u \notin L$. Then $\operatorname{Im}(A + X)$ is 2-dimensional and non-degenerate with respect to det.

Now let $\dim \operatorname{Im} A = 3$. We fix a basis $\{e_1, e_2, e_3\}$ of $\Lambda^2\mathbb{C}^3$ such that $v = A(e_1)$ spans the kernel of the restriction of det to $\operatorname{Im} A$, $u = A(e_2)$ and $w = A(e_3)$ are isotropic. We may assume that R either belongs to $\langle e_2, e_3\rangle$ or coincides with $\mathbb{C}e_1$.

Let $R \subset \langle e_2, e_3\rangle$. If $v \notin L$, then we take $x \in L$ such that v is not orthogonal to x. We take an operator X such that $X(e_2) = X(e_3) = 0$, $X(e_1) = \lambda x$, $\lambda \in \mathbb{C}$. Then $\operatorname{Im}(A + X)$ is 3-dimensional and non-degenerate with respect to det for generic λ. If $v \in L$, then we take X such that $X(e_1) = -v$, $X(e_2) = X(e_3) = 0$. Then $\operatorname{Im}(A + X)$ is 2-dimensional and non-degenerate with respect to det.

Let $R = \mathbb{C}e_1$. If $v \notin L$, then we take $x \in L$ such that v is not orthogonal to x. We take an operator X such that $X(e_1) = 0$, $X(e_2) = X(e_3) = x$. Then $\operatorname{Im}(A + X)$ is 3-dimensional and non-degenerate with respect to det. If $v \in L$ then either $L = \langle v, u\rangle$ or $L = \langle v, w\rangle$. It is sufficient to consider only the first case. We take X such that $X(e_1) = 0$, $X(e_2) = -u$, $X(e_3) = 0$. Then $\operatorname{Im}(A + X) = \langle v, w\rangle$ is 2-dimensional and isotropic.

(C) We fix a basis $\{e_1, e_2, e_3\}$ of $\mathbb{C}^3$ such that $v = e_1$ and a basis $\{f_1, f_2\}$ of $\mathbb{C}^2$. We take $A = e_2 \otimes f_1 + e_3 \otimes f_2$. Then $v \wedge A = (e_1 \wedge e_2) \otimes f_1 + (e_1 \wedge e_3) \otimes f_2$ has rank 2. Therefore, $B + v \wedge \lambda A$ has rank 2 for some $\lambda \in \mathbb{C}$. □

A tensor $\tilde{A} \in \mathbb{C}^k \otimes \Lambda^2\mathbb{C}^4$ is called ample if the restriction of the quadratic form Pf to $\operatorname{Im} A$ is trivial or non-degenerate, where $A : (\mathbb{C}^k)^* \to \Lambda^2\mathbb{C}^4$ is the corresponding map.

Proposition 6. (A) *Suppose that the linear maps $A : \mathbb{C}^2 \to \Lambda^2\mathbb{C}^4$ and $B : \mathbb{C}^2 \otimes \mathbb{C}^2 \to \Lambda^3\mathbb{C}^4$ are not ample. Then there exists a map $C : \mathbb{C}^2 \to \mathbb{C}^4$ such that $B + A \wedge C$ is ample.*

(B) *Suppose that the linear maps $A : \Lambda^2\mathbb{C}^4 \to \mathbb{C}^2$ and $B : \mathbb{C}^4 \to (\mathbb{C}^2 \otimes \mathbb{C}^2)$ are not ample. Then there exists a map $C : \mathbb{C}^4 \to (\mathbb{C}^2)^* \otimes \mathbb{C}$ such that $A + B \wedge C$ is ample.*

(C) *Suppose that the linear map $A : \mathbb{C}^2 \to \Lambda^2\mathbb{C}^4$ is not ample, and $w \in \mathbb{C}^4$ is not trivial. Then there exists a map $B : \mathbb{C}^2 \to \mathbb{C}^4$ such that $A + w \wedge B$ is ample.*

(D) *Suppose that $A \in \mathbb{C}^4 \otimes \Lambda^2\mathbb{C}^4$ is not ample. If $B \in \Lambda^2\mathbb{C}^4 \otimes \Lambda^3\mathbb{C}^4$ is not ample, then there exists $C \in \mathbb{C}^4 \otimes \mathbb{C}^4$ such that $B + A \wedge C$ is ample.*

(E) *Suppose that $A \in \mathbb{C}^4 \otimes \Lambda^2\mathbb{C}^4$ is not ample. If $u \in \mathbb{C}^4$, $u \neq 0$, then there exists $C = \sum v_i \otimes w_i \in \mathbb{C}^4 \otimes \mathbb{C}^4$ such that $A + \sum v_i \otimes (w_i \wedge u)$ is ample.*

Proof. (A) It suffices to find C such that $A \wedge C$ belongs to the open orbit, because it is ample. We fix bases $\{e_1, e_2, e_3, e_4\}$ of $\mathbb{C}^4$, $\{f_1, f_2\}$ of the first $\mathbb{C}^2$, $\{g_1, g_2\}$ of the second $\mathbb{C}^2$. We may assume that $A(f_1) = e_1 \wedge e_2 + e_3 \wedge e_4$, $A(f_2) = e_1 \wedge e_3$. Consider C such that $C(g_1) = e_4$, $C(g_2) = e_2$. Then $A \wedge C$ is an isomorphism.

(B) If $\dim B = 3$, then there exists a map $C : \mathbb{C}^4 \to (\mathbb{C}^2)^* \otimes \mathbb{C}$ such that $A + B \wedge C$ belongs to the open orbit, hence it is ample. Let $\dim B = 2$. Simple calculations show that we need to prove the following claim.

Let $\{e_1, e_2, e_3, e_4\}$ (resp. $\{f_1, f_2\}$) be a basis of $\mathbb{C}^4$ (resp. of $\mathbb{C}^2$). Suppose that the map $A : \mathbb{C}^2 \to \Lambda^2\mathbb{C}^4$ is not ample. Then we claim that there exist vectors $v, u \in \mathbb{C}^4$ such that the map B is ample, where $B(f_1) = A(f_1) + e_1 \wedge u$, $B(f_2) = A(f_2) + e_2 \wedge u + e_1 \wedge v$. Indeed, $\dim \operatorname{Im} A = 2$ and we may assume that either $A(f_1)$ or $A(f_2)$ spans the kernel of the restriction of Pf to $\operatorname{Im} A$. Consider the first case. If $A(f_1) = e_1 \wedge w$, then we take $v = 0$, $u = -w$. If $A(f_1)$ is not of the form $e_1 \wedge w$, then there exists v such that $A(f_1) \wedge e_1 \wedge v \neq 0$. If we take $u = 0$, then B is ample. Consider the second case. Since $A(f_1) \wedge A(f_1) \neq 0$, we can find w such that $A(f_1) \wedge e_1 \wedge w \neq 0$. If $v = \lambda w$ and $u = 0$, then B is ample for generic $\lambda \in \mathbb{C}$.

(C) We fix a basis $\{e_1, e_2, e_3, e_4\}$ of $\mathbb{C}^4$ such that $w = e_1$ and a basis $\{f_1, f_2\}$ such that $A(f_1)$ spans the kernel of the restriction of Pf to $\operatorname{Im} A$. We need to prove that there exist $u, v \in \mathbb{C}^4$ such that the linear map $C : \mathbb{C}^2 \to \Lambda^2\mathbb{C}^4$ is ample, where $C(f_1) = A(f_1) + e_1 \wedge u$, $C(f_2) = A(f_2) + e_1 \wedge v$. If $A(f_1) = e_1 \wedge x$, then it suffices to take $u = -x$. Otherwise, we take $u = 0$ and we choose v such that $A(f_1) \wedge e_1 \wedge v \neq 0$. Then $\dim \operatorname{Im} C = 2$ and the restriction of Pf to $\operatorname{Im} C$ is non-degenerate.

(D) and (E). Let $\tilde{A} : (\mathbb{C}^4)^* \to \Lambda^2\mathbb{C}^4$ be the corresponding map. Since the restriction of Pf to $\operatorname{Im} \tilde{A}$ is degenerate but not trivial, in suitable bases A has one of the following forms:

$$A_1 = e_1 \otimes (f_1 \wedge f_2 + f_3 \wedge f_4) + e_2 \otimes (f_1 \wedge f_3),$$
$$A_2 = e_1 \otimes (f_1 \wedge f_2 + f_3 \wedge f_4) + e_2 \otimes (f_1 \wedge f_3) + e_3 \otimes (f_1 \wedge f_4),$$
$$A_3 = e_1 \otimes (f_1 \wedge f_2 + f_3 \wedge f_4) + e_2 \otimes (f_1 \wedge f_3) + e_3 \otimes (f_3 \wedge f_4),$$
$$A_4 = e_1 \otimes (f_1 \wedge f_2 + f_3 \wedge f_4) + e_2 \otimes (f_1 \wedge f_3) + e_3 \otimes (f_1 \wedge f_4) + e_4 \otimes (f_1 \wedge f_2),$$
$$A_5 = e_1 \otimes (f_1 \wedge f_2 + f_3 \wedge f_4) + e_2 \otimes (f_1 \wedge f_3) + e_3 \otimes (f_1 \wedge f_4) + e_4 \otimes (f_2 \wedge f_4).$$

(D) We take $C = e_3 \otimes f_2 + e_4 \otimes f_4$. Then

$$A \wedge C = (e_1 \wedge e_3) \otimes (f_2 \wedge f_3 \wedge f_4) - (e_2 \wedge e_3) \otimes (f_1 \wedge f_2 \wedge f_3)$$
$$+ (e_1 \wedge e_4) \otimes (f_1 \wedge f_2 \wedge f_4) + (e_2 \wedge e_4) \otimes (f_1 \wedge f_3 \wedge f_4).$$

The restriction of Pf to the 4-dimensional image of the corresponding map is non-degenerate, so $A \wedge C$ belongs to the open orbit, which is ample by the results of §3.2. Therefore, $B + A \wedge (\lambda C)$ is also ample for some $\lambda \in \mathbb{C}$.

(E) Let $A = A_1$. If $u \in \langle f_1, f_3 \rangle$, then we take $C = e_2 \otimes u'$ such that $u' \wedge u = -f_1 \wedge f_3$. If $u \notin \langle f_1, f_3 \rangle$, then there exists u' such that $f_1 \wedge f_3 \wedge u' \wedge u \neq 0$ and we take $C = e_1 \otimes u'$.

Let $A = A_2$ or A_4. If $u \notin \langle f_1, f_3 \rangle$, then there exists u' such that $f_1 \wedge f_3 \wedge u' \wedge u \neq 0$ and we take $C = e_3 \otimes u'$. If $u \notin \langle f_1, f_4 \rangle$, then there exists u' such that $f_1 \wedge f_4 \wedge u' \wedge u \neq 0$ and we take $C = e_2 \otimes u'$. Finally, if $u = \lambda f_1$, then we take $C = (e_2 \otimes f_3 + e_3 \otimes f_4)/\lambda$.

Let $A = A_3$. If $u \in \langle f_1, f_3 \rangle$, then we take $C = e_2 \otimes u'$ such that $u' \wedge u = -f_1 \wedge f_3$. If $u \notin \langle f_1, f_3 \rangle$, then there exists u' such that $f_1 \wedge f_3 \wedge u' \wedge u \neq 0$ and we take $C = e_3 \otimes u'$.

Let $A = A_5$. If $u \in \langle f_1, f_4 \rangle$, then we take $C = e_3 \otimes u'$ such that $u' \wedge u = -f_1 \wedge f_4$. If $u \notin \langle f_1, f_4 \rangle$, then there exists u' such that $f_1 \wedge f_4 \wedge u' \wedge u \neq 0$ and we take $C = e_1 \otimes u'$. □

2.4 Spinors

We recall the definition of half-spinor representations. Let $V = \mathbb{C}^{2m}$ be an even-dimensional vector space with a non-degenerate symmetric scalar product $(\cdot, \cdot)$. Let $\mathrm{Cl}(V)$ be the Clifford algebra of V. Recall that $\mathrm{Cl}(V)$ is in fact a superalgebra, $\mathrm{Cl}(V) = \mathrm{Cl}^0(V) \oplus \mathrm{Cl}^1(V)$, where $\mathrm{Cl}^0(V)$ (resp. $\mathrm{Cl}^1(V)$) is the linear span of elements of the form $v_1 \cdot \ldots \cdot v_r$, $v_i \in V$, where r is even (resp. r is odd). Then we have

$$\mathrm{Spin}(V) = \{a \in \mathrm{Cl}^0(V) \mid a = v_1 \cdot \ldots \cdot v_r,\ v_i \in V,\ Q(v_i, v_i) = 1\}.$$

$\mathrm{Spin}(V)$ is a connected simply connected algebraic group.

Given $a \in \mathrm{Cl}(V)$, $a = v_1 \cdot \ldots \cdot v_r$, $v_i \in V$, let $\overline{a} = (-1)^r v_r \cdot \ldots \cdot v_1$. This is a well-defined involution of $\mathrm{Cl}(V)$. Using it, we may define an action R of $\mathrm{Spin}(V)$ on V by the formula $R(a)v = a \cdot v \cdot \overline{a}$. Then R is a double covering $\mathrm{Spin}(V) \to \mathrm{SO}(V)$.

Let $U \subset V$ be a maximal isotropic subspace, hence $\dim U = m$. Take also any maximal isotropic subspace U' such that $U \oplus U' = V$. For any $v \in U$ we define an operator $\rho(v) \in \mathrm{End}(\Lambda^* U)$ by the formula

$$\rho(v) \cdot u_1 \wedge \ldots \wedge u_r = v \wedge u_1 \wedge \ldots u_r.$$

For any $v \in U'$ we define an operator $\rho(v) \in \mathrm{End}(\Lambda^* U)$ by the formula

$$\rho(v) \cdot u_1 \wedge \ldots \wedge u_r = \sum_{i=1}^{r} (-1)^{i-1} (v, u_i) \wedge u_1 \wedge \ldots u_{i-1} \wedge u_{i+1} \wedge \ldots \wedge u_r.$$

These operators are well-defined and therefore by linearity we have a linear map $\rho : V \to \mathrm{End}(\Lambda^* U)$. It is easy to check that for any $v \in V$ we have $\rho(v)^2 = (v, v)\mathrm{Id}$. Therefore we have a homomorphism of associative algebras $\mathrm{Cl}(V) \to \mathrm{End}(\Lambda^* U)$, i.e., $\Lambda^* U$ is a $\mathrm{Cl}(V)$-module, easily seen to be irreducible. Therefore, $\Lambda^* U$ is also a $\mathrm{Spin}(V)$-module (spinor representation), now reducible. However, the even part $\Lambda^{\mathrm{ev}} U$ is an irreducible $\mathrm{Spin}(V)$-module, called the half-spinor module S^+. Another irreducible half-spinor module S^- is defined as $\Lambda^{\mathrm{od}} U$.

The map ρ defines a morphism of $\mathrm{Spin}(V)$ modules $V \otimes S^{\pm} \to S^{\mp}$. If m is odd, then S^+ is dual to S^-. If m is even, then the spinor representation $\Lambda^* U$ admits an interesting non-degenerate bilinear form $(\cdot, \cdot)$ defined as follows. Let $\det \in \Lambda^m U$ be a fixed non-trivial element. Then (u, v) is equal to the coefficient of det of the element $(-1)^{\left[\frac{\deg u}{2}\right]} u \wedge v$. Obviously, S^+ is orthogonal to S^- and the restriction of (u, v) to $S^{\pm}$ is orthogonal if $m = 4k$ and symplectic if $m = 4k + 2$. In particular, S^+ and S^- are self-dual if m is even.

If $m = 4$, then V, S^+ and S^- are twisted forms of each other with respect to outer isomorphisms of Spin_8 (triality principle). Let $V = \mathbb{C}^8$ be a vector space equipped with a non-degenerate scalar product. We denote by S^+, S^- the corresponding half-spinor modules. If R denotes V, S^+, or S^-, then a tensor $A \in \mathbb{C}^k \otimes R$ is called ample if the restriction of the scalar product of R to $\mathrm{Im}\, \tilde{A}$ is trivial or non-degenerate, where $\tilde{A} : (\mathbb{C}^k)^* \to R$ is the corresponding map.

Proposition 7. (A) *Let $k = 2, 3$. If $A = \sum a_i \otimes s_i \in \mathbb{C}^k \otimes S^+$ and $B \in \mathbb{C}^k \otimes S^-$ are not ample, then there exists $v \in V$ such that $C = B + \sum a_i \otimes \rho(v) s_i$ is ample.*

(B) *If $A = \sum a_i \otimes s_i \in \mathbb{C}^3 \otimes S^+$ and $B \in \Lambda^2 \mathbb{C}^3 \otimes S^-$ are not ample, then there exists $x = \sum v_i \otimes s_i \in \mathbb{C}^3 \otimes V$ such that $C = B + \sum_{i,j} (a_i \wedge v_j) \otimes \rho(x_j) s_i$ is ample.*

(C) *If $A \in \mathbb{C}^3 \otimes S^-$ is not ample and $s \in S^+$ is not trivial, then there exists $B = \sum a_i \otimes w_i \in \mathbb{C}^3 \otimes V$ such that $C = A + \sum a_i \otimes \rho(w_i) s$ is ample.*

Proof. (A) Let $k = 2$. We choose a basis f_1, f_2 of $\mathbb{C}^2$ and e_1, e_2, e_3, e_4 of U (maximal isotropic subspace of V) such that

$$A = f_1 \otimes 1 + f_2 \otimes (e_1 \wedge e_2 + e_3 \wedge e_4), \quad B = f_1 \otimes x_1 + f_2 \otimes x_2.$$

If $x_1 \in U \subset \Lambda^{\mathrm{od}} U$ then we take $v = -x_1$. Otherwise, there exists $v' \in U'$ such that $\rho(v')(e_1 \wedge e_2 + e_3 \wedge e_4)$ is not orthogonal to x_1. Then we take $v = \lambda v'$: for generic λ the restriction of $(\cdot,\cdot)$ to $\operatorname{Im}\tilde{C}$ is non-degenerate; hence C is ample.

Let $k = 3$. We choose a basis $\{f_1, f_2, f_3\}$ of $\mathbb{C}^3$ and $\{e_1, e_2, e_3, e_4\}$ of U (maximal isotropic subspace of V) such that A has one of the following forms:

$$A_1 = f_1 \otimes (e_1 \wedge e_2 + e_3 \wedge e_4) + f_2 \otimes 1,$$
$$A_2 = f_1 \otimes 1 + f_2 \otimes (e_1 \wedge e_2) + f_3 \otimes (e_3 \wedge e_4),$$
$$A_3 = f_1 \otimes (1 + e_1 \wedge e_2 \wedge e_3 \wedge e_4) + f_2 \otimes e_1 \wedge e_2 + f_3 \otimes e_1 \wedge e_3.$$

Let

$$B = f_1 \otimes x_1 + f_2 \otimes x_2 + f_3 \otimes x_3.$$

Let $\mathbf{A} = \mathbf{A_1}$. We need to show that there exist $u, v \in U$ such that the following element if ample:

$$C = f_1 \otimes \big(x_1 + u + v \wedge (e_1 \wedge e_2 + e_3 \wedge e_4)\big) + f_2 \otimes (x_2 + v) + f_3 \otimes x_3.$$

If $x_3 \notin U$, then there exists $v' \in U$ such that $(v, x_3) \neq 0$ and $u' \in U$ such that $(e_1 \wedge e_2 + e_3 \wedge e_4) \wedge u' \wedge v' \neq 0$. We take $v = \lambda v'$, $u = \lambda u'$. Then for generic λ the image of $\tilde{C}$ is 3-dimensional and non-degenerate; hence C is ample.

Suppose now that $x_3 \in U$. If $x_3 = 0$, then we finish the proof as in the case $k = 2$.

Let $x_3 \neq 0$. If $x_2 \notin U$, then there exist $v' \in U$ such that $(x_3, v' \wedge (e_1 \wedge e_2 + e_3 \wedge e_4)) \neq 0$, $x_2 \wedge v' \neq 0$. We take $u = 0$, $v = \lambda v'$. Then for generic λ the image of $\tilde{C}$ is 3-dimensional and non-degenerate; hence C is ample.

Suppose now that $x_2 \in U$. Then we choose u and v such that $u + v \wedge (e_1 \wedge e_2 + e_3 \wedge e_4) = -x_1$. Then the image of $\tilde{C}$ is isotropic.

Let $\mathbf{A} = \mathbf{A_2}$. We need to show that there exist $v \in U$, $u \in \langle e_1, e_2 \rangle$, $w \in \langle e_3, e_4 \rangle$ such that the following element is ample:

$$C = f_1 \otimes (x_1 + v) + f_2 \otimes (x_2 + v \wedge (e_1 \wedge e_2) + u) + f_3 \otimes (x_3 + v \wedge (e_3 \wedge e_4) + w).$$

If $x_1 \notin U$, then there exists $v' \in U$ such that $(v, x_1) \neq 0$ and $u \in \langle e_1, e_2 \rangle$ (or $w \in \langle e_3, e_4 \rangle$) such that $u \wedge v \wedge e_3 \wedge e_4 \neq 0$ (or $w \wedge v \wedge e_1 \wedge e_2 \neq 0$) We take $v = \lambda v'$, $u = \lambda u'$, $w = 0$ (or $w = \lambda w'$, $u = 0$). Then for generic λ the image of $\tilde{C}$ is 3-dimensional and non-degenerate; hence C is ample.

Suppose now that $x_1 \in U$. Then we may suppose that $x_1 = 0$ after taking $v = -x_1$. Now we are going to find u and w.

If x_3 is not perpendicular to $\langle e_1, e_2 \rangle$ (resp. x_2 is not perpendicular to $\langle e_3, e_4 \rangle$), then we take u' such that $(u', x_3) \neq 0$, $w' = 0$ (resp. take w' such that $(w', x_2) \neq 0$, $u' = 0$) and set $u = \lambda u'$, $w = \lambda w'$. Then $\operatorname{Im}\tilde{C}$ is 2-dimensional and non-degenerate, hence ample, for generic λ.

Let $x_3 \perp \langle e_1, e_2 \rangle$, $x_2 \perp \langle e_3, e_4 \rangle$.

If $(x_3, x_3) = 0$, then $(x_2, x_3) = 0$, otherwise B is ample. Then $(x_2, x_2) \neq 0$, otherwise B is ample. Hence x_2 is not perpendicular to $\langle e_1, e_2 \rangle$ and we take $w = 0$ and v such that $(x_2, x_2) + 2(x_2, u) = 0$.

Suppose, therefore, that $(x_3, x_3) \neq 0$, and, similarly, that $(x_2, x_2) \neq 0$. It follows that x_2 is not perpendicular to $\langle e_1, e_2 \rangle$, x_3 is not perpendicular to $\langle e_3, e_4 \rangle$. We take u', w' such that $(x_2, u) \neq 0$, $(x_3, w') \neq 0$. We set $u = \lambda u'$, $w = \lambda w'$. Then $\operatorname{Im} \tilde{C}$ is 2-dimensional and non-degenerate for generic λ, hence C is ample for these values of λ.

Let $\mathbf{A} = \mathbf{A_3}$. There exists $x \in V$ such that $\rho(x)(1+e_1 \wedge e_2 \wedge e_3 \wedge e_4) = -x_1$. Therefore, without loss of generality we may assume that $x_1 = 0$. If B is ample, then there is nothing to prove. Otherwise, the restriction of a scalar form R to $\langle x_2, x_3 \rangle$ has one-dimensional kernel. After a suitable change of basis we may assume that this kernel is spanned by x_2. There exists $x' \in V$ such that $\rho(x')(1 + e_1 \wedge e_2 \wedge e_3 \wedge e_4)$ is not orthogonal to x_2. We take $x = \lambda x'$. Then for generic λ the image of $\tilde{C}$ is three-dimensional and non-degenerate with respect to R.

(B) We choose a basis $\{f_1, f_2, f_3\}$ of $\mathbb{C}^2$ and $\{e_1, e_2, e_3, e_4\}$ of U (maximal isotropic subspace of V) such that A has one of the following forms:

$$A_1 = f_1 \otimes (e_1 \wedge e_2 + e_3 \wedge e_4) + f_2 \otimes 1,$$
$$A_2 = f_1 \otimes 1 + f_2 \otimes (e_1 \wedge e_2) + f_3 \otimes (e_3 \wedge e_4),$$
$$A_3 = f_1 \otimes (1 + e_1 \wedge e_2 \wedge e_3 \wedge e_4) + f_2 \otimes e_1 \wedge e_2 + f_3 \otimes e_1 \wedge e_3.$$

Let

$$B = (f_2 \wedge f_3) \otimes x_1 + (f_1 \wedge f_3) \otimes x_2 + (f_1 \wedge f_2) \otimes x_3.$$

We take an isotropic subspace $U' \subset V$ complementary to U and choose a basis $\{e_1^*, e_2^*, e_3^*, e_4^*\}$ of U' such that $(e_i, e_j^*) = \delta_{ij}$.

Let $\mathbf{A} = \mathbf{A_3}$. We take

$$x = \lambda(f_3 \otimes e_1 + f_1 \otimes e_4 + f_2 \otimes e_1^* + f_3 \otimes e_3^*).$$

Then the image of $\tilde{D}$ is 3-dimensional and non-degenerate, where $D = \sum_{i,j} (a_i \wedge v_j) \otimes \rho(x_j) s_i$. Therefore, for generic λ, the image of $\tilde{C}$ is 3-dimensional and non-degenerate, hence C is ample.

Let $\mathbf{A} = \mathbf{A_2}$. The same proof as above, but for

$$x = \lambda(f_3 \otimes e_1^* + f_2 \otimes e_3^* + f_1 \otimes (e_1 + e_2 + e_3 + e_1^*)).$$

Let $\mathbf{A} = \mathbf{A_1}$. The same proof as above, but for

$$x = \lambda(f_3 \otimes e_1 + f_3 \otimes e_1^* + f_2 \otimes e_2).$$

(C) If $(s, s) \neq 0$, then $\rho(V)s = S^-$, otherwise, $\rho(V)s = U_0$, where $U_0 \subset S^-$ is a maximal isotropic subspace. It is clear now that we can find B such that $\operatorname{Im} \tilde{C}$ is isotropic. □

3 E_7 and E_8 : The Zoo

We associate to each parabolic subgroup its coloured Dynkin diagram. To each graded component $\mathfrak{g}_\chi$, $\chi > 0$ we associate the coloured Dynkin diagram of the corresponding parabolic subgroup with vertices indexed as follows. The lattice $\widehat{Z}$ is isomorphic to a sublattice in the root lattice spanned by simple roots marked white on the coloured Dynkin diagram. We write the corresponding coefficient of χ near each white vertex. The representation of $[\mathfrak{l}, \mathfrak{l}]$ on $\mathfrak{g}_\chi$ is irreducible and the Dynkin diagram of $[\mathfrak{l}, \mathfrak{l}]$ is equal to the black subdiagram of the coloured Dynkin diagram. We write the numerical labels of χ near each black vertex (omitting zeroes). We shall pick several positive weights and call them *twisting weights.* All reduced positive weights of the form $\chi + \sum \pm\mu_i$, where $\chi > 0$ is not reduced and all $\mu_i > 0$ are twisting, are called *rubbish weights.* We shall draw a diagram with the set of vertices given by not reduced and rubbish weights and arrows indexed by twisting weights: an arrow μ has a tail χ_1 and a head χ_2 if and only if $[\mathfrak{g}_{\chi_1}, \mathfrak{g}_\mu] = \mathfrak{g}_{\chi_2}$. For any $e \in \mathfrak{n}$ (resp. $e \in \mathfrak{g}/\mathfrak{p}^-$) we shall try to find an element $u \in U$ (resp. $u \in U^-$) such that all non-reduced graded components of the element $\mathrm{Ad}(u)e$ belong to ample orbits, except for at most one. Here $U \subset P$ is the subgroup with a Lie algebra generated by all twisting weights components, and $U^- \subset P^-$ is the subgroup with a Lie algebra generated by all graded components $\mathfrak{g}_\chi$, where $-\chi$ is twisting. The standard strategy will be to decrease the number of non-reduced components by applying elements of the form $\exp p$, where p belongs to the graded component of some twisting weight. These 'elementary transformations' will be made in a special order, because they may change some other components as well. We shall use big Latin letters for non-reduced weights, small Latin letters for rubbish weights, and numbers for twisting weights. To save space, we shall use abbreviations of the form

$$L \underset{X.Y}{\overset{n}{\Longrightarrow}} M.$$

This means that if $x_L \subset \mathfrak{g}_L$ is not ample (or not trivial if L is a reduced weight), then we can apply the element $\exp(p_n)$, where $p_n \in \mathfrak{g}_n$, to make $x_M \in \mathfrak{g}_M$ ample, and the reason for this is given in Proposition X.Y. Parabolic subgroups from the table will be subdivided into groups; in each group the proof is similar.

Case 1A

This case includes parabolic subgroups with numbers 1 and 2. We shall consider the one with number 1.

Not reduced weights: Twisting weights: Rubbish weights: Diagram:

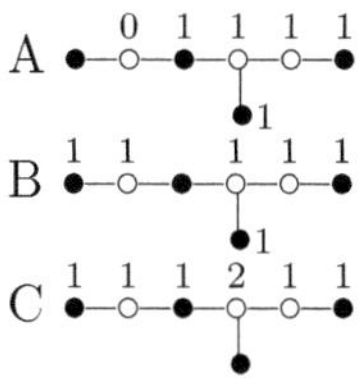

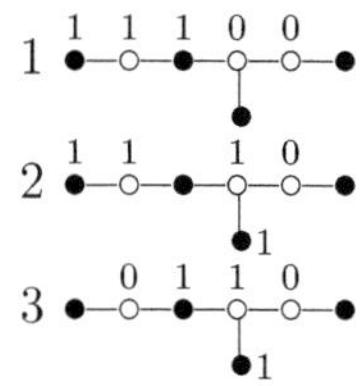

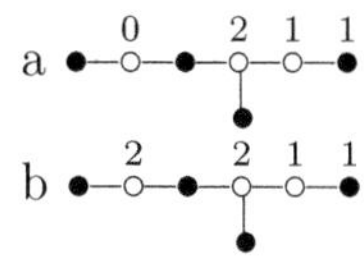

$$\begin{array}{ccccc} A & \xrightarrow{1} & B & & \\ \downarrow 3 & \searrow 2 & \downarrow 3 & \searrow 2 & \\ a & \xrightarrow{1} & C & \xrightarrow{1} & b \end{array}$$

Consider the coadjoint case, $x \in \mathfrak{n}$. Then $A \underset{1.B}{\overset{1}{\Longrightarrow}} B$, $A \underset{1.B}{\overset{2}{\Longrightarrow}} C$. If x_A is ample, then $B \underset{1.B}{\overset{3}{\Longrightarrow}} C$.

Adjoint case, $x \in \mathfrak{g}/\mathfrak{p}^-$. $b \underset{1.A}{\overset{-2}{\Longrightarrow}} B$, $b \underset{1.A}{\overset{-1}{\Longrightarrow}} C$. Let $x_b = 0$. Then $C \underset{1.B}{\overset{-3}{\Longrightarrow}} B$, $C \underset{1.B}{\overset{-2}{\Longrightarrow}} A$. If x_C is ample, then $B \underset{1.B}{\overset{-1}{\Longrightarrow}} A$.

Case 1B

This case includes parabolic subgroups with numbers 3, 4, 5, 6, 7, 8, 9. The proof is similar to the previous case (using Proposition 1.A and 1.B) but the combinatorics is more involved. To save space, we omit this proof and refer the reader to the even more complicated Case 5.

Case 2A

Parabolic subgroup with number 19.

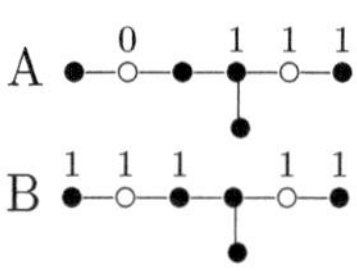

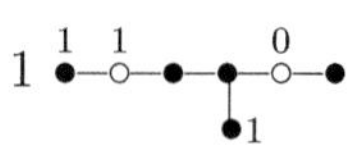

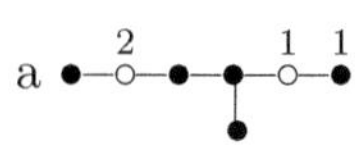

$$A \xrightarrow{1} B \xrightarrow{1} a$$

Adjoint and coadjoint cases follow from Propositions 6.A, 6.B and 1.A.

Case 2B

This case includes parabolic subgroups with numbers 20 and 21. We shall consider the one with number 20.

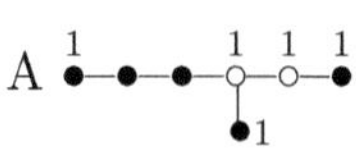

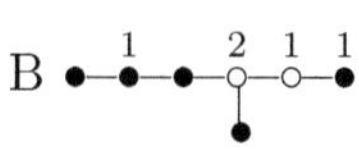

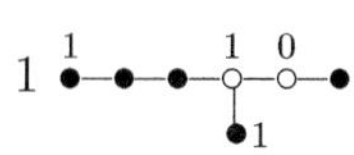

$$A \xrightarrow{1} B$$

Both the adjoint and coadjoint cases follow from Proposition 6.A and Proposition 6.B.

3.1 Case 2C

This case includes parabolic subgroups with numbers 22, 23, 24, 25, 26, 27. We shall consider the one with number 26.

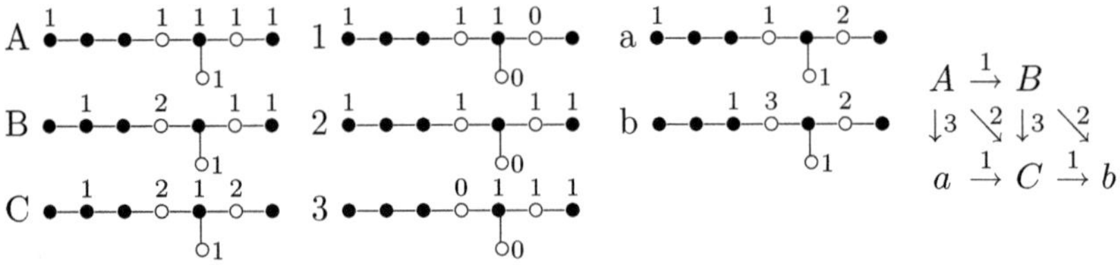

Coadjoint case, $x \in \mathfrak{n}$. Then $A \xRightarrow[6.B]{1} B$, $A \xRightarrow[6.B]{2} C$. If x_A is ample, then $B \xRightarrow[1.B]{3} C$.

Adjoint case, $x \in \mathfrak{g}/\mathfrak{p}^-$. Then $b \xRightarrow[6.C]{-2} B$ and $b \xRightarrow[6.C]{-1} C$. Let $x_b = 0$. Then $C \xRightarrow[1.B]{-3} B$ and $C \xRightarrow[6.A]{-2} A$. Let x_C be ample. Then $B \xRightarrow[6.A]{-1} A$.

Case 2D

This case includes parabolic subgroups with numbers 27, 28, 29. We shall consider the one with number 29.

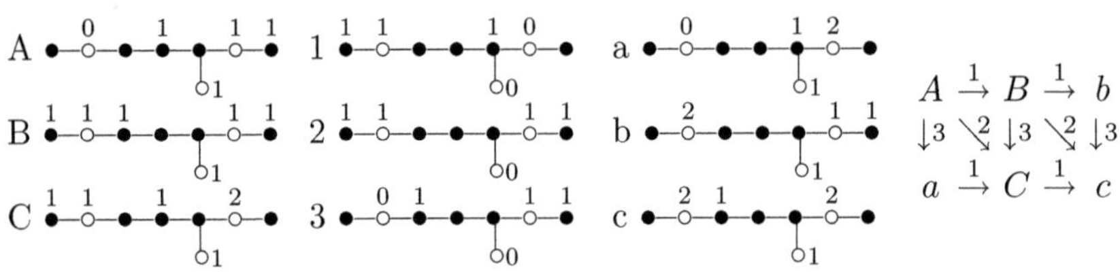

Coadjoint case, $x \in \mathfrak{n}$. Then $A \xRightarrow[6.A]{1} B$, $A \xRightarrow[1.B]{2} C$. If x_A is ample, then $B \xRightarrow[6.B]{3} C$.

Adjoint case, $x \in \mathfrak{g}/\mathfrak{p}^-$. Then $c \xRightarrow[6.C]{-1} C$ and $c \xRightarrow[4.A]{-2} B$. Let $x_c = 0$. Then $C \xRightarrow[6.A]{-3} B$ and $C \xRightarrow[1.B]{-2} A$. Let x_C be ample. If $x_b = 0$, then $B \xRightarrow[6.B]{-1} A$. If $x_b \neq 0$ then $b \xRightarrow[1.A]{-1} B$.

Case 2E

This case includes parabolic subgroups with numbers 30, 31. We shall consider the one with number 31.

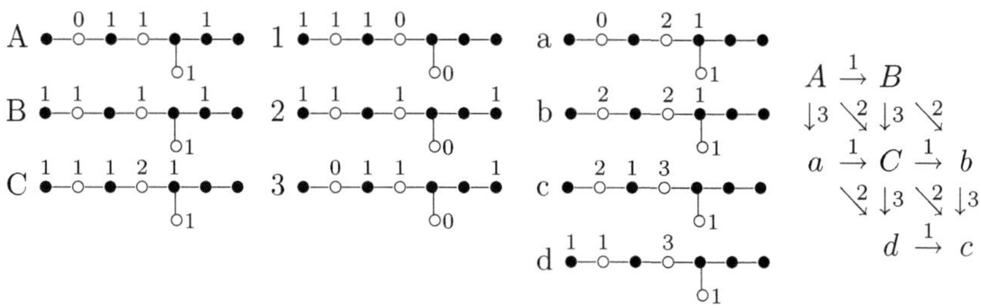

Coadjoint case, $x \in \mathfrak{n}$. Then $A \overset{1}{\underset{1.B}{\Longrightarrow}} B$ and $A \overset{2}{\underset{6.A}{\Longrightarrow}} C$. If x_A is ample, then $B \overset{3}{\underset{6.A}{\Longrightarrow}} C$.

Adjoint case, $x \in \mathfrak{g}/\mathfrak{p}^-$. If $x_c \neq 0$ or $x_b \neq 0$, then we first apply $\exp(p_{-3})$ to make $x_b \neq 0$ if necessary. Then $b \overset{-2}{\underset{6.C}{\Longrightarrow}} B$ and $b \overset{-1}{\underset{4.A}{\Longrightarrow}} C$. If $x_c, x_b = 0$ then $d \overset{-3}{\underset{1.A}{\Longrightarrow}} C$ and $B \overset{-1}{\underset{1.B}{\Longrightarrow}} A$. If $x_c, x_b, x_d = 0$ then $C \overset{-3}{\underset{6.B}{\Longrightarrow}} B$ and $C \overset{-2}{\underset{6.B}{\Longrightarrow}} A$. If x_C is ample, then $B \overset{-1}{\underset{1.B}{\Longrightarrow}} A$.

Case 3

This case includes parabolic subgroups with numbers 32, 33. We shall consider the one with number 33.

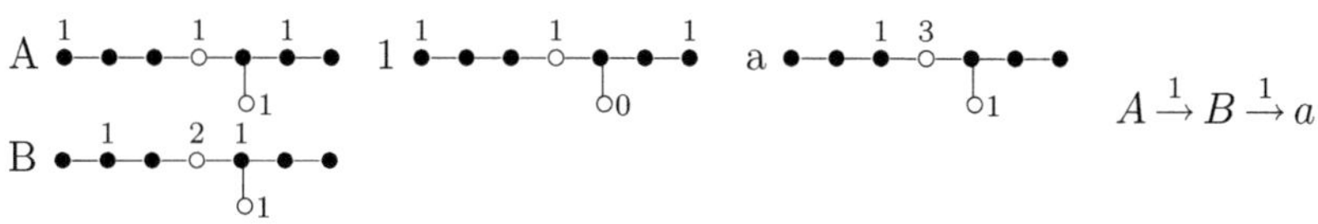

The adjoint and coadjoint cases follow from Proposition 6.D and Proposition 6.E.

Case 4A

In this case we study the parabolic subgroup number 34.

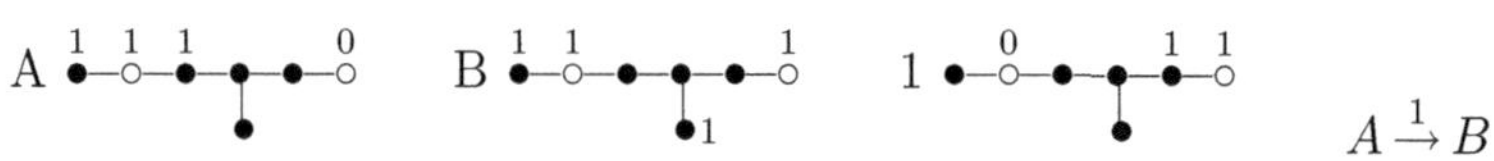

Both the adjoint and coadjoint cases follow from Proposition 7.A.

Case 4B

This case includes parabolic subgroups with numbers 35, 36. We shall consider the one with number 36.

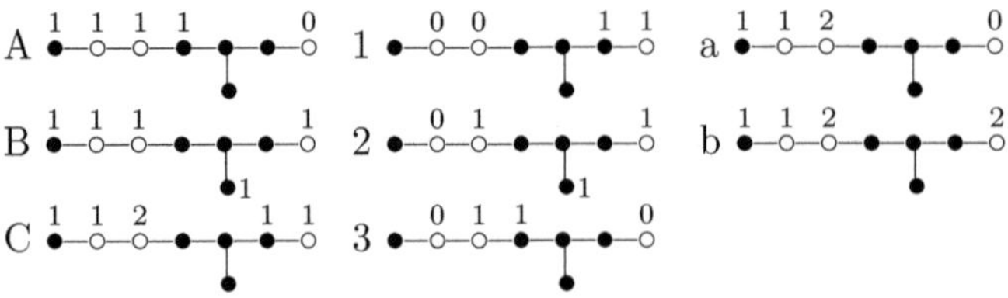

$$\begin{array}{ccccc} A & \xrightarrow{1} & B & & \\ \downarrow 3 & \searrow^{2} & \downarrow 3 & \searrow^{2} & \\ a & \xrightarrow{1} & C & \xrightarrow{1} & b \end{array}$$

Easy digram-chasing shows that everything follows from Propositions 7.A and 1.A.

Case 4C

This case includes the parabolic subgroup with number 37.

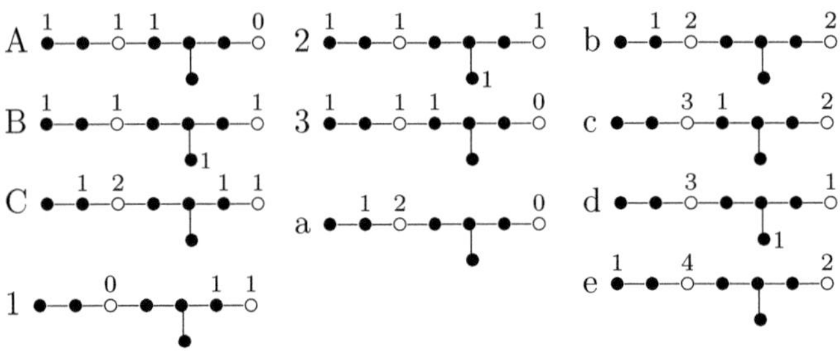

$$\begin{array}{ccccc} A & \xrightarrow{1} & B & & \\ \downarrow 3 & \searrow^{2} & \downarrow 3 & \searrow^{2} & \\ a & \xrightarrow{1} & C & \xrightarrow{1} & b \\ & \searrow^{2} & \downarrow 3 & \searrow^{2} & \downarrow 3 \\ & & d & \xrightarrow{1} & c \\ & & & \searrow^{2} & \downarrow 3 \\ & & & & e \end{array}$$

The coadjoint case follows from Proposition 7.A and Proposition 7.B. The adjoint case easily follows from Proposition 7.A, Proposition 7.B, Proposition 4.A, and Proposition 7.C.

Case 5A

This case contains parabolic subgroup with number 10.

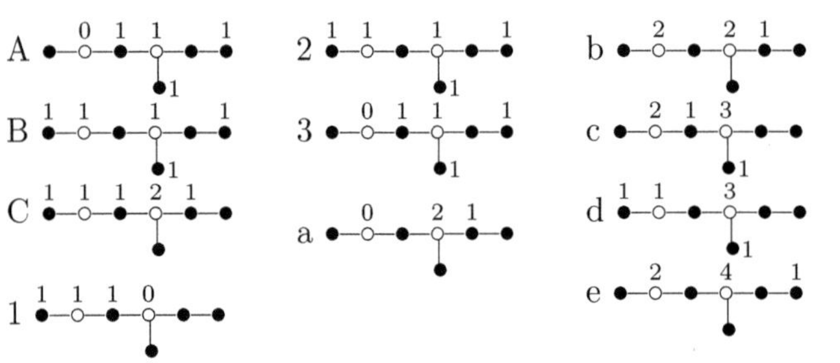

$$\begin{array}{ccccc} A & \xrightarrow{1} & B & & \\ \downarrow 3 & \searrow^{2} & \downarrow 3 & \searrow^{2} & \\ a & \xrightarrow{1} & C & \xrightarrow{1} & b \\ & \searrow^{2} & \downarrow 3 & \searrow^{2} & \downarrow 3 \\ & & d & \xrightarrow{1} & c \\ & & & \searrow^{2} & \downarrow 3 \\ & & & & e \end{array}$$

Consider the coadjoint case, let $x \in \mathfrak{n}$. Then $A \underset{1.B}{\overset{1}{\Longrightarrow}} B$, $A \underset{5.A}{\overset{2}{\Longrightarrow}} C$. If x_A is ample, then $B \underset{5.A}{\overset{3}{\Longrightarrow}} C$.

Adjoint case, $x \in \mathfrak{g}/\mathfrak{p}^-$. Suppose first that either x_b, x_c, or x_e is not trivial. Applying $\exp(p_{-3})$ if necessary, we may assume that $x_b \neq 0$. Then $b \underset{4.A}{\overset{-2}{\Longrightarrow}} B$,

$b \underset{4.A}{\overset{-1}{\Longrightarrow}} C$. Let $x_b = x_c = x_e = 0$. Then $d \underset{1.A}{\overset{-3}{\Longrightarrow}} C$, $B \underset{1.B}{\overset{-1}{\Longrightarrow}} A$. Let $x_d = 0$. Then $C \underset{5.A}{\overset{-3}{\Longrightarrow}} B$, $C \underset{1.B}{\overset{-2}{\Longrightarrow}} A$. Let x_C be ample, then $B \underset{1.B}{\overset{-1}{\Longrightarrow}} A$.

Case 5B

This case includes parabolic subgroups with numbers 11, 12, 13. We shall consider the one with number 11.

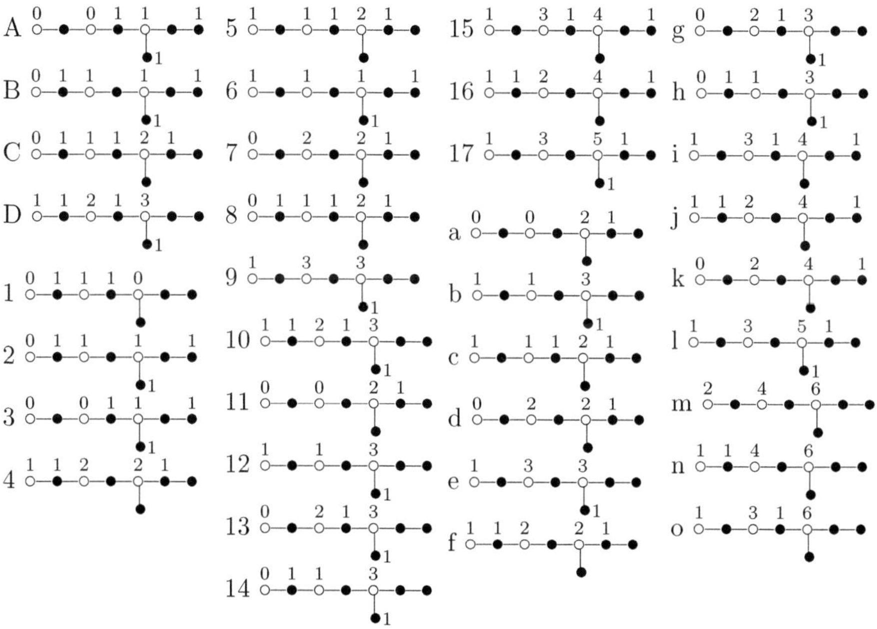

$A \xrightarrow{1} B \quad a \xrightarrow{1} C \xrightarrow{1} d \quad b \xrightarrow{1} D \xrightarrow{1} e \quad c \xrightarrow{1} f \quad h \xrightarrow{1} g \quad j \xrightarrow{1} i \quad o \xrightarrow{1} n$
$A \xrightarrow{2} C \xrightarrow{2} g \quad B \xrightarrow{2} d \quad c \xrightarrow{2} D \xrightarrow{2} i \quad a \xrightarrow{2} h \xrightarrow{2} k \quad b \xrightarrow{2} j \xrightarrow{2} l \xrightarrow{2} n \quad f \xrightarrow{2} e$
$A \xrightarrow{3} a \quad B \xrightarrow{3} C \xrightarrow{3} h \quad f \xrightarrow{3} D \xrightarrow{3} j \quad c \xrightarrow{3} b \quad d \xrightarrow{3} g \xrightarrow{3} k \quad e \xrightarrow{3} i \xrightarrow{3} l \xrightarrow{3} o$
$A \xrightarrow{4} D \quad B \xrightarrow{4} e \quad C \xrightarrow{4} i \quad a \xrightarrow{4} j \xrightarrow{4} m \quad h \xrightarrow{4} l \quad k \xrightarrow{4} n$
$A \xrightarrow{5} b \quad B \xrightarrow{5} D \quad C \xrightarrow{5} j \quad d \xrightarrow{5} i \xrightarrow{5} m \quad g \xrightarrow{5} l \quad k \xrightarrow{5} o$
$A \xrightarrow{6} c \quad B \xrightarrow{6} f \quad C \xrightarrow{6} D \quad a \xrightarrow{6} b \quad d \xrightarrow{6} e \quad g \xrightarrow{6} i \quad h \xrightarrow{6} j \quad k \xrightarrow{6} l \xrightarrow{6} m$
$A \xrightarrow{7} g \quad a \xrightarrow{7} k \quad b \xrightarrow{7} l \quad c \xrightarrow{7} i \quad j \xrightarrow{7} n$
$A \xrightarrow{8} h \quad B \xrightarrow{8} g \quad C \xrightarrow{8} k \quad D \xrightarrow{8} l \quad c \xrightarrow{8} j \xrightarrow{8} o \quad f \xrightarrow{8} i \xrightarrow{8} n$
$A \xrightarrow{9} i \quad a \xrightarrow{9} l \quad b \xrightarrow{9} m \quad h \xrightarrow{9} n$
$A \xrightarrow{10} j \quad B \xrightarrow{10} i \quad C \xrightarrow{10} l \quad D \xrightarrow{10} m \quad g \xrightarrow{10} n \quad h \xrightarrow{10} o$
$B \xrightarrow{11} h \quad d \xrightarrow{11} k \quad e \xrightarrow{11} l \quad f \xrightarrow{11} j \quad i \xrightarrow{11} o \qquad B \xrightarrow{12} j \quad d \xrightarrow{12} l \quad e \xrightarrow{12} m \quad g \xrightarrow{12} o$
$A \xrightarrow{13} k \quad D \xrightarrow{13} n \quad b \xrightarrow{13} o \quad c \xrightarrow{13} l \qquad B \xrightarrow{14} k \quad D \xrightarrow{14} o \quad e \xrightarrow{14} n \quad f \xrightarrow{14} l$
$A \xrightarrow{15} l \quad C \xrightarrow{15} n \quad a \xrightarrow{15} o$
$c \xrightarrow{15} m \qquad B \xrightarrow{16} l \quad C \xrightarrow{16} o \quad d \xrightarrow{16} n \quad f \xrightarrow{16} m \qquad A \xrightarrow{17} o \quad B \xrightarrow{17} n$

Consider the coadjoint case, $x \in \mathfrak{n}$. Then $A \overset{1}{\underset{1.B}{\Longrightarrow}} B$, $A \overset{2}{\underset{5.A}{\Longrightarrow}} C$, $A \overset{4}{\underset{4.B}{\Longrightarrow}} D$. Suppose that x_A is ample, then $B \overset{3}{\underset{5.A}{\Longrightarrow}} C$, $B \overset{5}{\underset{4.B}{\Longrightarrow}} D$. Let x_B be ample, then $C \overset{6}{\underset{4.B}{\Longrightarrow}} D$.

Adjoint case, $x \in \mathfrak{g}/\mathfrak{p}^-$. Then $n \overset{-13}{\underset{1.A}{\Longrightarrow}} D$, $n \overset{-15}{\underset{1.A}{\Longrightarrow}} C$, $n \overset{-17}{\underset{1.A}{\Longrightarrow}} B$. Let $x_n = 0$, then $o \overset{-14}{\underset{1.A}{\Longrightarrow}} D$, $o \overset{-16}{\underset{1.A}{\Longrightarrow}} C$, $o \overset{-17}{\underset{1.A}{\Longrightarrow}} A$. Let $x_o = 0$, $x_l \neq 0$ or $x_m \neq 0$ (in the latter case we apply $\exp(p_{-6})$ if necessary to make $x_l \neq 0$). Then $l \overset{-10}{\underset{4.A}{\Longrightarrow}} C$, $l \overset{-15}{\underset{5.C}{\Longrightarrow}} A$, $l \overset{-16}{\underset{5.C}{\Longrightarrow}} B$. Let $x_l = x_m = 0$, then $i \overset{-2}{\underset{1.A}{\Longrightarrow}} D$, $i \overset{-4}{\underset{5.C}{\Longrightarrow}} C$, $i \overset{-10}{\underset{4.A}{\Longrightarrow}} B$. Let $x_i = 0$, then $j \overset{-3}{\underset{1.A}{\Longrightarrow}} D$, $j \overset{-5}{\underset{5.C}{\Longrightarrow}} C$, $j \overset{-10}{\underset{4.A}{\Longrightarrow}} A$. Let $x_j = 0$, then $k \overset{-8}{\underset{4.A}{\Longrightarrow}} C$, $k \overset{-13}{\underset{4.A}{\Longrightarrow}} A$, $k \overset{-14}{\underset{4.A}{\Longrightarrow}} B$. Let $x_k = 0$, then $g \overset{-2}{\underset{1.A}{\Longrightarrow}} C$, $g \overset{-8}{\underset{1.A}{\Longrightarrow}} B$, $g \overset{-7}{\underset{4.B}{\Longrightarrow}} A$. Let $x_g = 0$, then $h \overset{-3}{\underset{1.A}{\Longrightarrow}} C$, $h \overset{-8}{\underset{1.A}{\Longrightarrow}} A$, $h \overset{-11}{\underset{4.B}{\Longrightarrow}} B$. Let $x_h = 0$, $x_e \neq 0$. Applying $\exp(p_{-6})$ if necessary, we also make $x_d = 0$. Then $e \overset{-1}{\underset{4.A}{\Longrightarrow}} D$, $e \overset{-4}{\underset{1.A}{\Longrightarrow}} B$, $C \overset{-2}{\underset{5.A}{\Longrightarrow}} A$, Let $x_e = 0$, then $D \overset{-6}{\underset{4.B}{\Longrightarrow}} C$, $D \overset{-5}{\underset{4.B}{\Longrightarrow}} B$, $D \overset{-4}{\underset{4.B}{\Longrightarrow}} A$. Let x_D be ample, then $d \overset{-1}{\underset{4.A}{\Longrightarrow}} C$, $d \overset{-2}{\underset{4.A}{\Longrightarrow}} B$. Let $x_d = 0$, then $C \overset{-2}{\underset{5.A}{\Longrightarrow}} A$, $C \overset{-3}{\underset{5.A}{\Longrightarrow}} B$. Let x_C be ample. Then $B \overset{-1}{\underset{1.B}{\Longrightarrow}} A$.

3.2 Case 5C

This case includes parabolic subgroups with numbers 14, 15. We shall consider the one with number 14.

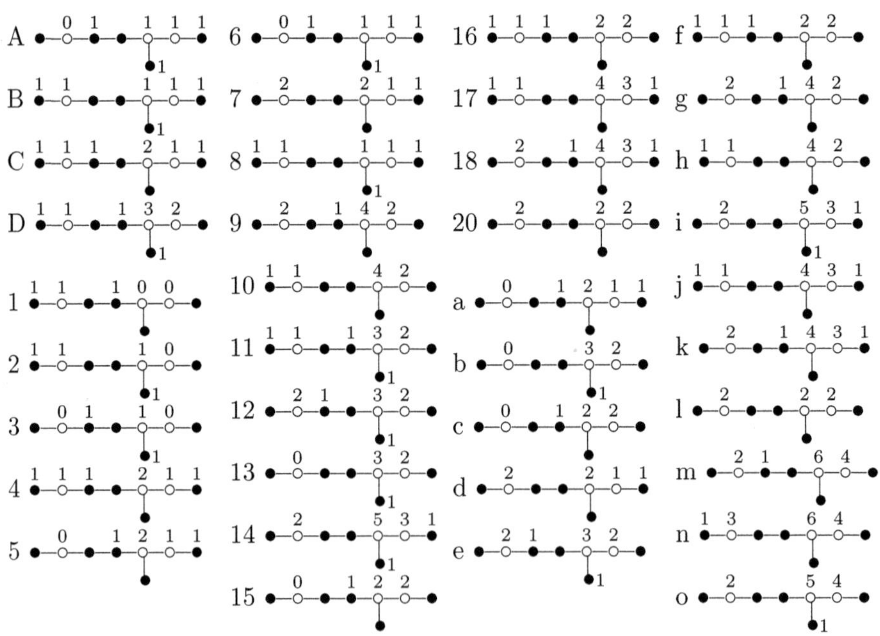

$A\xrightarrow{1}B \quad a\xrightarrow{1}C\xrightarrow{1}d \quad b\xrightarrow{1}D\xrightarrow{1}e \quad c\xrightarrow{1}f \quad f\xrightarrow{1}l \quad h\xrightarrow{1}g \quad j\xrightarrow{1}k \quad m\xrightarrow{1}n$
$A\xrightarrow{2}C \quad B\xrightarrow{2}d \quad b\xrightarrow{2}h \quad c\xrightarrow{2}D\xrightarrow{2}g \quad f\xrightarrow{2}e \quad j\xrightarrow{2}i \quad o\xrightarrow{2}n$
$A\xrightarrow{3}a \quad B\xrightarrow{3}C \quad c\xrightarrow{3}b \quad f\xrightarrow{3}D\xrightarrow{3}h \quad k\xrightarrow{3}i \quad l\xrightarrow{3}e\xrightarrow{3}g \quad o\xrightarrow{3}m$
$A\xrightarrow{4}D\xrightarrow{4}i \quad B\xrightarrow{4}e \quad C\xrightarrow{4}g \quad a\xrightarrow{4}h \quad c\xrightarrow{4}j\xrightarrow{4}m \quad f\xrightarrow{4}k\xrightarrow{4}n$
$A\xrightarrow{5}b \quad B\xrightarrow{5}D \quad C\xrightarrow{5}h \quad d\xrightarrow{5}g \quad e\xrightarrow{5}i \quad f\xrightarrow{5}j \quad l\xrightarrow{5}k\xrightarrow{5}m$
$A\xrightarrow{6}c \quad B\xrightarrow{6}f \quad C\xrightarrow{6}D\xrightarrow{6}j \quad a\xrightarrow{6}b \quad d\xrightarrow{6}e\xrightarrow{6}k\xrightarrow{6}o \quad g\xrightarrow{6}i \quad i\xrightarrow{6}m$
$A\xrightarrow{7}e \quad a\xrightarrow{7}g \quad b\xrightarrow{7}i \quad c\xrightarrow{7}k \quad j\xrightarrow{7}n$
$A\xrightarrow{8}f \quad B\xrightarrow{8}l \quad C\xrightarrow{8}e \quad a\xrightarrow{8}D\xrightarrow{8}k \quad b\xrightarrow{8}j\xrightarrow{8}o \quad h\xrightarrow{8}i\xrightarrow{8}n$
$A\xrightarrow{9}i \quad c\xrightarrow{9}m \quad f\xrightarrow{9}n \qquad B\xrightarrow{10}i \quad f\xrightarrow{10}m \quad l\xrightarrow{10}n$
$A\xrightarrow{11}j \quad B\xrightarrow{11}k \quad C\xrightarrow{11}i \quad D\xrightarrow{11}m \quad e\xrightarrow{11}n \quad f\xrightarrow{11}o$
$A\xrightarrow{12}k \quad D\xrightarrow{12}n \quad a\xrightarrow{12}i \quad b\xrightarrow{12}m \quad c\xrightarrow{12}o \qquad B\xrightarrow{13}j \quad d\xrightarrow{13}i \quad e\xrightarrow{13}m \quad l\xrightarrow{13}o$
$A\xrightarrow{14}m \quad B\xrightarrow{14}n \qquad C\xrightarrow{15}j \quad d\xrightarrow{15}k \quad e\xrightarrow{15}o \quad g\xrightarrow{15}m$
$C\xrightarrow{16}k \quad D\xrightarrow{16}o \quad a\xrightarrow{16}j \quad g\xrightarrow{16}n \quad h\xrightarrow{16}m \qquad B\xrightarrow{17}o \quad C\xrightarrow{17}m \quad d\xrightarrow{17}n$
$A\xrightarrow{18}o \quad C\xrightarrow{18}n \quad a\xrightarrow{18}m \qquad a\xrightarrow{20}k \quad b\xrightarrow{20}o \quad h\xrightarrow{20}n$

Consider the coadjoint case, let $x \in \mathfrak{n}$. Then $A \overset{1}{\underset{4.B}{\Longrightarrow}} B$, $A \overset{2}{\underset{1.B}{\Longrightarrow}} C$, $A \overset{4}{\underset{5.A}{\Longrightarrow}} D$. Let x_A be ample, then $B \overset{3}{\underset{4.B}{\Longrightarrow}} C$, $B \overset{5}{\underset{4.B}{\Longrightarrow}} D$. Let x_B be ample, then $C \overset{6}{\underset{5.A}{\Longrightarrow}} D$.

Adjoint case, $x \in \mathfrak{g}/\mathfrak{p}^-$. Then $n \overset{-12}{\underset{1.A}{\Longrightarrow}} D$, $n \overset{-14}{\underset{1.A}{\Longrightarrow}} B$, $n \overset{-18}{\underset{1.A}{\Longrightarrow}} C$. Let $x_n = 0$, then $m \overset{-11}{\underset{4.A}{\Longrightarrow}} D$, $m \overset{-14}{\underset{4.A}{\Longrightarrow}} A$, $m \overset{-17}{\underset{4.A}{\Longrightarrow}} C$. Let $x_m = 0$, then $o \overset{-16}{\underset{1.A}{\Longrightarrow}} D$, $o \overset{-17}{\underset{1.A}{\Longrightarrow}} B$, $o \overset{-18}{\underset{1.A}{\Longrightarrow}} A$. Let $x_o = 0$, then $i \overset{-4}{\underset{1.A}{\Longrightarrow}} D$, $i \overset{-11}{\underset{1.A}{\Longrightarrow}} C$, $i \overset{-9}{\underset{4.B}{\Longrightarrow}} A$. Let $x_i = 0$, then $k \overset{-8}{\underset{4.A}{\Longrightarrow}} D$, $k \overset{-11}{\underset{1.A}{\Longrightarrow}} B$, $k \overset{-12}{\underset{5.C}{\Longrightarrow}} A$. Let $x_k = 0$, then $j \overset{-6}{\underset{1.A}{\Longrightarrow}} D$, $j \overset{-11}{\underset{1.A}{\Longrightarrow}} A$, $j \overset{-15}{\underset{4.B}{\Longrightarrow}} C$. Let $x_j = 0$. If $x_g \neq 0$, then by Proposition 5.E, we can apply $\exp(p_{-3})$ to force x_e to have rank 2. Afterwards, $e \overset{-1}{\underset{5.C}{\Longrightarrow}} D$, $e \overset{-4}{\underset{1.A}{\Longrightarrow}} B$, $e \overset{-7}{\underset{1.C}{\Longrightarrow}} A$. Let $x_g = 0$, then $e \overset{-1}{\underset{5.C}{\Longrightarrow}} D$, $e \overset{-8}{\underset{4.A}{\Longrightarrow}} C$, $e \overset{-4}{\underset{1.A}{\Longrightarrow}} B$. Let $x_e = 0$, then $h \overset{-3}{\underset{1.A}{\Longrightarrow}} D$, $h \overset{-5}{\underset{1.A}{\Longrightarrow}} C$; if $x_d \neq 0$ then $d \overset{-2}{\underset{1.A}{\Longrightarrow}} B$, if $x_d = 0$ then $B \overset{-1}{\underset{4.B}{\Longrightarrow}} A$. Let $x_h = 0$, then $D \overset{-6}{\underset{5.A}{\Longrightarrow}} C$, $D \overset{-5}{\underset{4.B}{\Longrightarrow}} B$, $D \overset{-4}{\underset{5.A}{\Longrightarrow}} A$. Let x_D be ample, then $d \overset{-1}{\underset{1.A}{\Longrightarrow}} C$, $d \overset{-2}{\underset{1.A}{\Longrightarrow}} B$. Let $x_d = 0$, then $C \overset{-3}{\underset{4.B}{\Longrightarrow}} B$, $C \overset{-2}{\underset{1.B}{\Longrightarrow}} A$. Let x_C be ample, then $B \overset{-1}{\underset{4.B}{\Longrightarrow}} A$.

Case 5D

In this case we study parabolic subgroup with number 16.

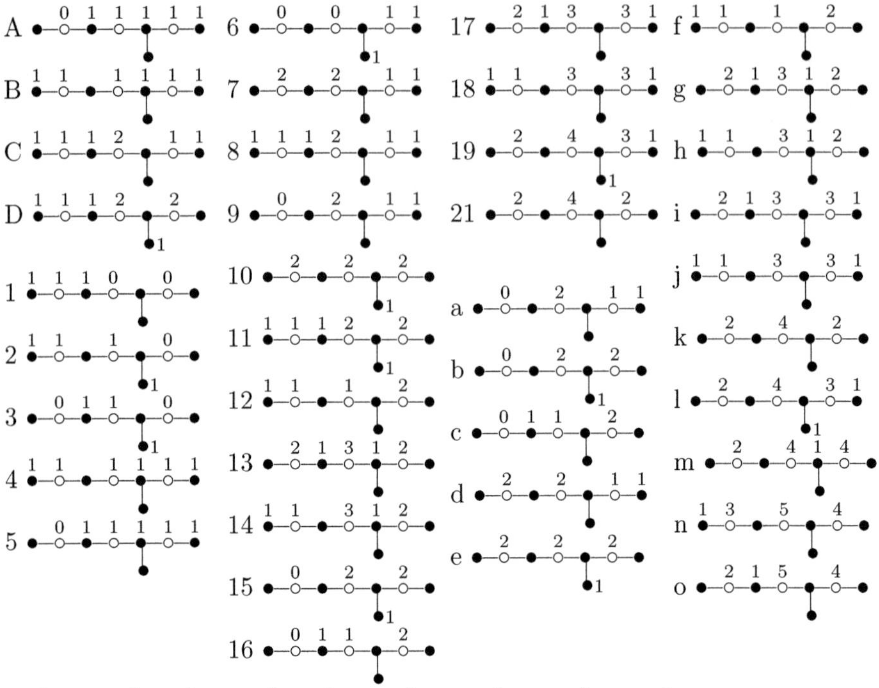

$A \xrightarrow{1} B \quad a \xrightarrow{1} C \xrightarrow{1} d \quad b \xrightarrow{1} D \xrightarrow{1} e \quad c \xrightarrow{1} f \quad h \xrightarrow{1} g \quad j \xrightarrow{1} i \quad o \xrightarrow{1} n$

$A \xrightarrow{2} C \quad B \xrightarrow{2} d \quad b \xrightarrow{2} h \xrightarrow{2} k \quad c \xrightarrow{2} D \xrightarrow{2} g \quad f \xrightarrow{2} e \quad j \xrightarrow{2} l \quad m \xrightarrow{2} n$

$A \xrightarrow{3} a \quad B \xrightarrow{3} C \quad c \xrightarrow{3} b \quad e \xrightarrow{3} g \xrightarrow{3} k \quad f \xrightarrow{3} D \xrightarrow{3} h \quad i \xrightarrow{3} l \quad m \xrightarrow{3} o$

$A \xrightarrow{4} D \xrightarrow{4} i \quad B \xrightarrow{4} e \quad C \xrightarrow{4} g \quad a \xrightarrow{4} h \xrightarrow{4} l \xrightarrow{4} n \quad b \xrightarrow{4} j \xrightarrow{4} m$

$A \xrightarrow{5} b \quad B \xrightarrow{5} D \xrightarrow{5} j \quad C \xrightarrow{5} h \quad d \xrightarrow{5} g \xrightarrow{5} l \xrightarrow{5} o \quad e \xrightarrow{5} i \xrightarrow{5} m$

$A \xrightarrow{6} c \quad B \xrightarrow{6} f \quad C \xrightarrow{6} D \quad a \xrightarrow{6} b \quad d \xrightarrow{6} e \quad g \xrightarrow{6} i \quad h \xrightarrow{6} j \quad k \xrightarrow{6} l \xrightarrow{6} m$

$A \xrightarrow{7} g \quad a \xrightarrow{7} k \quad b \xrightarrow{7} l \quad c \xrightarrow{7} i \quad j \xrightarrow{7} n$

$A \xrightarrow{8} h \quad B \xrightarrow{8} g \quad C \xrightarrow{8} k \quad D \xrightarrow{8} l \quad c \xrightarrow{8} j \xrightarrow{8} o \quad f \xrightarrow{8} i \xrightarrow{8} n$

$B \xrightarrow{9} h \quad d \xrightarrow{9} k \quad e \xrightarrow{9} l \quad f \xrightarrow{9} j \quad i \xrightarrow{9} o$

$A \xrightarrow{10} i \quad a \xrightarrow{10} l \quad b \xrightarrow{10} m \quad h \xrightarrow{10} n$

$A \xrightarrow{11} j \quad B \xrightarrow{11} i \quad C \xrightarrow{11} l \quad D \xrightarrow{11} m \quad g \xrightarrow{11} n \quad h \xrightarrow{11} o$

$C \xrightarrow{12} i \quad a \xrightarrow{12} j \quad h \xrightarrow{12} m \quad k \xrightarrow{12} n \qquad A \xrightarrow{13} l \quad D \xrightarrow{13} n \quad b \xrightarrow{13} o \quad c \xrightarrow{13} m$

$B \xrightarrow{14} l \quad D \xrightarrow{14} o \quad e \xrightarrow{14} n \quad f \xrightarrow{14} m \qquad B \xrightarrow{15} j \quad d \xrightarrow{15} l \quad e \xrightarrow{15} m \quad g \xrightarrow{15} o$

$C \xrightarrow{16} j \quad d \xrightarrow{16} i \quad g \xrightarrow{16} m \quad k \xrightarrow{16} o \qquad A \xrightarrow{17} m \quad C \xrightarrow{17} n \quad a \xrightarrow{17} o$

$B \xrightarrow{18} m \quad C \xrightarrow{18} o \quad d \xrightarrow{18} n \qquad A \xrightarrow{19} o \quad B \xrightarrow{19} n \qquad c \xrightarrow{21} o \quad f \xrightarrow{21} n$

Coadjoint case, $x \in \mathfrak{n}$. Then $A \overset{1}{\underset{1.B}{\Longrightarrow}} B$, $A \overset{2}{\underset{4.B}{\Longrightarrow}} C$, $A \overset{4}{\underset{5.A}{\Longrightarrow}} D$. Let x_A be ample. Then $B \overset{3}{\underset{4.B}{\Longrightarrow}} C$, $B \overset{5}{\underset{5.A}{\Longrightarrow}} D$. Let x_B be ample. Then $C \overset{6}{\underset{4.B}{\Longrightarrow}} D$.

Adjoint case, $x \in \mathfrak{g}/\mathfrak{p}^-$. Then $n \overset{-13}{\underset{1.A}{\Longrightarrow}} D$, $n \overset{-17}{\underset{1.A}{\Longrightarrow}} C$, $n \overset{-19}{\underset{1.A}{\Longrightarrow}} B$. Let $x_n = 0$. Then $o \overset{-14}{\underset{1.A}{\Longrightarrow}} D$, $o \overset{-18}{\underset{1.A}{\Longrightarrow}} C$, $o \overset{-19}{\underset{1.A}{\Longrightarrow}} A$. Let $x_o = 0$. Then $m \overset{-11}{\underset{4.A}{\Longrightarrow}} D$, $m \overset{-17}{\underset{4.A}{\Longrightarrow}} A$, $m \overset{-18}{\underset{4.A}{\Longrightarrow}} B$. Let $x_m = 0$. Then $l \overset{-8}{\underset{4.A}{\Longrightarrow}} D$, $l \overset{-11}{\underset{1.A}{\Longrightarrow}} C$, $l \overset{-13}{\underset{5.C}{\Longrightarrow}} A$. Let $x_l = 0$. Then $i \overset{-4}{\underset{1.A}{\Longrightarrow}} D$, $i \overset{-11}{\underset{1.A}{\Longrightarrow}} B$, $i \overset{-10}{\underset{4.B}{\Longrightarrow}} A$. Let $x_i = 0$. Then $j \overset{-5}{\underset{1.A}{\Longrightarrow}} D$, $j \overset{-11}{\underset{1.A}{\Longrightarrow}} A$, $j \overset{-15}{\underset{4.B}{\Longrightarrow}} B$. Let $x_j = 0$, $x_k \neq 0$. We can apply $\exp(p_{-3})$ to force x_g to have rank 2, then $g \overset{-2}{\underset{5.C}{\Longrightarrow}} D$, $g \overset{-4}{\underset{1.A}{\Longrightarrow}} C$, $g \overset{-7}{\underset{1.C}{\Longrightarrow}} A$. Let $x_k = 0$, then $g \overset{-2}{\underset{5.C}{\Longrightarrow}} D$, $g \overset{-4}{\underset{1.A}{\Longrightarrow}} C$, $g \overset{-8}{\underset{4.A}{\Longrightarrow}} B$. Let $x_g = 0$, then $h \overset{-3}{\underset{5.C}{\Longrightarrow}} D$, $h \overset{-5}{\underset{1.A}{\Longrightarrow}} C$, $h \overset{-8}{\underset{4.A}{\Longrightarrow}} A$. Let $x_h = 0$, $x_e \neq 0$. Applying $\exp(p_{-6})$ if necessary, we make $x_d = 0$, then $e \overset{-1}{\underset{4.A}{\Longrightarrow}} D$, $e \overset{-4}{\underset{4.A}{\Longrightarrow}} B$, $C \overset{-2}{\underset{4.B}{\Longrightarrow}} A$. Let $x_e = 0$. Then $D \overset{-6}{\underset{4.B}{\Longrightarrow}} C$, $D \overset{-5}{\underset{5.A}{\Longrightarrow}} B$, $D \overset{-4}{\underset{5.A}{\Longrightarrow}} A$. Suppose x_D is ample. Then $d \overset{-1}{\underset{1.A}{\Longrightarrow}} C$, $d \overset{-2}{\underset{1.A}{\Longrightarrow}} B$. Let $x_d = 0$. Then $C \overset{-3}{\underset{4.B}{\Longrightarrow}} B$, $C \overset{-2}{\underset{4.B}{\Longrightarrow}} A$. Let x_C be ample, then $B \overset{-1}{\underset{1.B}{\Longrightarrow}} A$.

Case 5E

This case consists of parabolic subgroups with numbers 17 and 18. We shall consider the one with number 17.

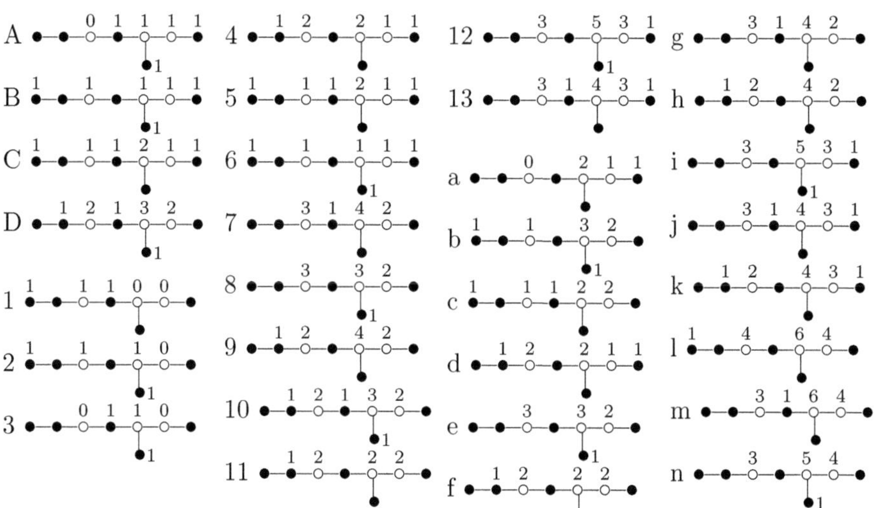

$A \xrightarrow{1} B \quad a \xrightarrow{1} C \xrightarrow{1} d \quad b \xrightarrow{1} D \xrightarrow{1} e \quad c \xrightarrow{1} f \quad h \xrightarrow{1} g \quad k \xrightarrow{1} j \quad m \xrightarrow{1} l$

$A \xrightarrow{2} C \quad B \xrightarrow{2} d \quad b \xrightarrow{2} h \quad c \xrightarrow{2} D \xrightarrow{2} g \quad f \xrightarrow{2} e \quad k \xrightarrow{2} i \quad n \xrightarrow{2} l$

$A \xrightarrow{3} a \quad B \xrightarrow{3} C \quad c \xrightarrow{3} b \quad e \xrightarrow{3} g \quad f \xrightarrow{3} D \xrightarrow{3} h \quad j \xrightarrow{3} i \quad n \xrightarrow{3} m$

$A \xrightarrow{4} D \quad B \xrightarrow{4} e \quad C \xrightarrow{4} g \quad a \xrightarrow{4} h \quad b \xrightarrow{4} i \quad c \xrightarrow{4} j \quad k \xrightarrow{4} l$

$A \xrightarrow{5} b \quad B \xrightarrow{5} D \xrightarrow{5} i \quad C \xrightarrow{5} h \quad c \xrightarrow{5} k \xrightarrow{5} m \quad d \xrightarrow{5} g \quad f \xrightarrow{5} j \xrightarrow{5} l$

$A\xrightarrow{6}c$ $B\xrightarrow{6}f$ $C\xrightarrow{6}D\xrightarrow{6}j$ $a\xrightarrow{6}b\xrightarrow{6}k\xrightarrow{6}n$ $d\xrightarrow{6}e$ $h\xrightarrow{6}i\xrightarrow{6}l$
$A\xrightarrow{7}i$ $c\xrightarrow{7}l$ $A\xrightarrow{8}j$ $a\xrightarrow{8}i$ $b\xrightarrow{8}l$
$B\xrightarrow{9}i$ $c\xrightarrow{9}m$ $f\xrightarrow{9}l$ $A\xrightarrow{10}k$ $B\xrightarrow{10}j$ $C\xrightarrow{10}i$ $D\xrightarrow{10}l$ $b\xrightarrow{10}m$ $c\xrightarrow{10}n$
$C\xrightarrow{11}j$ $a\xrightarrow{11}k$ $b\xrightarrow{11}n$ $h\xrightarrow{11}l$ $A\xrightarrow{12}m$ $B\xrightarrow{12}l$
$A\xrightarrow{13}n$ $C\xrightarrow{13}l$ $a\xrightarrow{13}m$

Consider the coadjoint case, let $x\in\mathfrak{n}$. Then $A\overset{1}{\underset{4.B}{\Longrightarrow}}B$, $A\overset{2}{\underset{4.B}{\Longrightarrow}}C$, $A\overset{4}{\underset{4.B}{\Longrightarrow}}D$. Let x_A be ample. Then $B\overset{3}{\underset{1.B}{\Longrightarrow}}C$, $B\overset{5}{\underset{5.A}{\Longrightarrow}}D$. Suppose that x_B is also ample. Then $C\overset{6}{\underset{5.A}{\Longrightarrow}}D$.

Adjoint case, let $x\in\mathfrak{g}/\mathfrak{p}^-$ Then $l\overset{-10}{\underset{4.A}{\Longrightarrow}}D$, $l\overset{-12}{\underset{4.A}{\Longrightarrow}}B$, $l\overset{-13}{\underset{4.A}{\Longrightarrow}}C$. Let $x_l=0$. Then $i\overset{-5}{\underset{1.A}{\Longrightarrow}}D$, $i\overset{-10}{\underset{1.A}{\Longrightarrow}}C$, $i\overset{-9}{\underset{4.B}{\Longrightarrow}}B$. Let $x_i=0$. Then $j\overset{-6}{\underset{1.A}{\Longrightarrow}}D$, $j\overset{-10}{\underset{1.A}{\Longrightarrow}}B$, $j\overset{-11}{\underset{4.B}{\Longrightarrow}}C$.. Let $x_j=0$, $x_g\neq0$. Then we first apply $\exp(p_{-3})$ and $\exp(p_{-5})$ to make $x_e=x_d=0$, then $g\overset{-2}{\underset{1.A}{\Longrightarrow}}D$, $g\overset{-4}{\underset{1.A}{\Longrightarrow}}C$, $B\overset{-1}{\underset{4.B}{\Longrightarrow}}A$. Let $x_g=0$, $x_e\neq0$. First, we apply $\exp(p_{-6})$ to make $x_d=0$, then $e\overset{-1}{\underset{1.A}{\Longrightarrow}}D$, $e\overset{-4}{\underset{1.A}{\Longrightarrow}}B$, $C\overset{-2}{\underset{4.B}{\Longrightarrow}}A$. Let $x_e=0$. Then $d\overset{-1}{\underset{5.C}{\Longrightarrow}}C$, $d\overset{-2}{\underset{5.C}{\Longrightarrow}}B$, $D\overset{-4}{\underset{4.B}{\Longrightarrow}}A$. Let $x_d=0$. Then $h\overset{-3}{\underset{4.A}{\Longrightarrow}}D$, $h\overset{-5}{\underset{4.A}{\Longrightarrow}}C$, $B\overset{-1}{\underset{4.B}{\Longrightarrow}}A$. Let $x_h=0$. Then $D\overset{-6}{\underset{5.A}{\Longrightarrow}}C$, $D\overset{-5}{\underset{5.A}{\Longrightarrow}}B$, $D\overset{-4}{\underset{4.B}{\Longrightarrow}}A$. Let x_D be ample. Then $C\overset{-3}{\underset{1.B}{\Longrightarrow}}B$, $C\overset{-2}{\underset{4.B}{\Longrightarrow}}A$. Let x_C be ample, then $B\overset{-1}{\underset{4.B}{\Longrightarrow}}A$.

4 Table

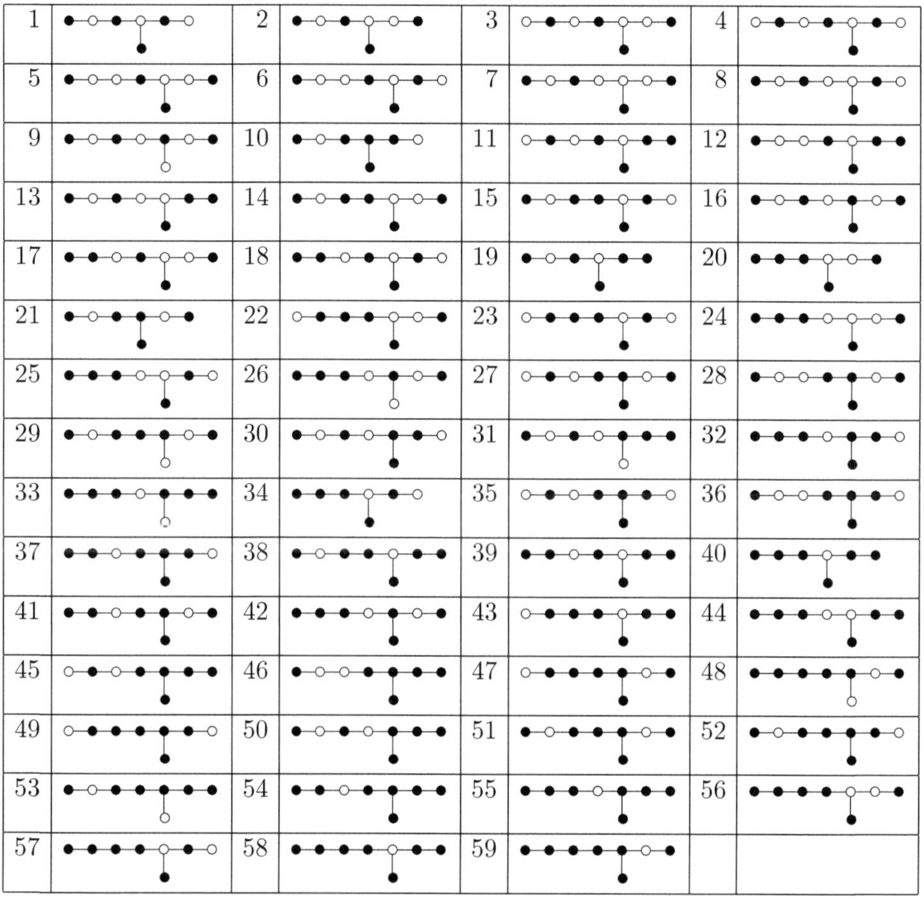

References

[BHR] Brüstle, T., Hille, L., Röhrle, G.: Finiteness for parabolic group actions in classical groups. Arch. Math. **76**(2) 81–87 (2001)

[Ka] Kac, V.G.: Some remarks on nilpotent orbits. J. Algebra **64** 190–213 (1980)

[Li] Lichtenstein, W.: A system of quadrics describing the orbit of the highest weight vector. Proc. A.M.S. **84** 605–608 (1982)

[LW] Lipsman, R., Wolf, J.: Canonical semi-invariants and the Plancherel formula for parabolic groups. Trans. Amer. Math. Soc. **269**(1) 111–131 (1982)

[PR] Popov, V., Röhrle, G.: On the number of orbits of a parabolic subgroup on its unipotent radical. Algebraic Groups and Lie Groups. Lehrer G.I. (Ed.): Austr. Math. Soc. Lecture Series. **9** 1997

[Ri] Richardson, R.W.: Conjugacy classes of n-tuples in Lie algebras and algebraic groups. Duke Math. J. **57** 1–35 (1988)

[Te1] Tevelev, E.: Moore–Penrose inverse, parabolic subgroups, and Jordan pairs. J. Lie Theory **12**(2) 461–481 (2002)

[Te2] Tevelev, E.: Isotropic subspaces of multilinear forms. Matematicheskie Zametki **69**(6) 925–933 (2001) (in Russian)

[OV] Onishchik, A.L., Vinberg, E.B.: *Lie Groups and Algebraic Groups* (Springer, Berlin, Heidelberg, New York 1990)

[Vi] Vinberg, E.: On stability of actions of reductive algebraic groups. In: Yuen, F., Mikhalev, A.A., Zelmanov, E. (Eds.): *Lie Algebras, Rings and Related Topics* (Springer, Hong Kong 2000) pp. 188–202

Determinants of Projective Varieties and their Degrees

Fyodor L. Zak*

Central Economics Mathematical Institute of the Russian Academy of Sciences, 47 Nakhimovskiĭav., Moscow 119418, Russia
Independent University of Moscow, 11 B. Vlas'evskiĭ, Moscow 121002, Russia.
zak@mccme.ru

To Christian Peskine, one of the few who still value elegance in mathematics — and in life

Introduction

Consider the vector space of square matrices of order r and the corresponding projective space $\mathbb{P} = \mathbb{P}^{r^2-1}$. The points of $\mathbb{P}$ are in a one-to-one correspondence with the square matrices modulo multiplication by a nonzero constant. Consider the Segre subvariety $X = \mathbb{P}^{r-1} \times \mathbb{P}^{r-1} \subset \mathbb{P}$ corresponding to the matrices of rank one and a filtration

$$X \subset X^2 \subset \cdots \subset X^{r-1} \subset X^r = \mathbb{P},$$

where, for $1 \le i \le r$, X^i denotes the i-th join of X with itself. We recall that by definition X^i (also called the $(i-1)$-st secant variety of X) is the closure of the subvariety of $\mathbb{P}$ swept out by the linear spans of general collections of i points of X. Thus in our case X^i corresponds to the cone of matrices whose rank does not exceed i. In other words, if M is a matrix and $z_M \in \mathbb{P}$ is the corresponding point, then $\operatorname{rk} M = \operatorname{rk}_X z_M$, where for $z \in \mathbb{P}$

$$\operatorname{rk}_X z = \min\{i \mid z \in X^i\}.$$

It is clear that $X^{r-1} \subset \mathbb{P}$ is the hypersurface of degree r defined by vanishing of determinant, which gives a method to define determinant (up to multiplication by a nonzero constant) in purely geometric terms.

Alternatively, one can consider the dual variety $X^* \subset \mathbb{P}^*$ whose points correspond to hyperplanes tangent to X. It is not hard to see that in the case

* Research partially supported by RFBR grant 01-01-00803

when $X = \mathbb{P}^{r-1} \times \mathbb{P}^{r-1}$ the variety X^* is a hypersurface of degree r in the dual space $\mathbb{P}^*$ which is also defined by vanishing of determinant. This gives another way to define determinant in geometric terms.

It is tempting to study similar notions and interrelations between them for arbitrary projective varieties. The idea is to associate, in a natural way, to any projective variety X a hypersurface (or at least a variety of small codimension) X^{ass} from which one should be able to reconstruct X (or at least some of its essential features). Then geometric and numerical invariants of X^{ass} would yield important information on X itself. On the other hand, varieties of low codimension (and particularly hypersurfaces) are easier to study from algebraic and analytic points of view. A classical way to realize this idea is to consider generic projections, and, indeed, invariants of multiple loci of such projections provide a useful tool for the study of projective varieties. However, generic projections are not canonical, and they preserve the dimension of variety while changing the dimension of the ambient linear space. In contrast to that, the above two constructions of hypersurfaces associated to the Segre variety $X = \mathbb{P}^{r-1} \times \mathbb{P}^{r-1}$ are well defined and preserve the dimension of the ambient space (we recall that $X^{\mathrm{ass}} = X^{r-1}$ for the first construction and $X^{\mathrm{ass}} = X^*$ for the second one). These constructions generalize to arbitrary projective varieties, and in the present paper we study these generalizations of determinant called respectively *join determinant* and *discriminant.* In particular, we give lower bounds for the degree of associated varieties in these two cases.

As we already observed, in the above examples the degree of associated hypersurface is equal to the order of matrix, and it turns out that, if we define the *order* of an arbitrary nondegenerate variety $X \subset \mathbb{P}^N$ by the formula $\operatorname{ord} X = \operatorname{rk}_X z$, where $z \in \mathbb{P}^N$ is a general point (cf. Definition 1.3), then the degree of associated variety (or determinant) in the above two senses is at least $\operatorname{ord} X$, so that $\operatorname{ord} X$ is the lowest possible value of degree of determinant. Furthermore, for the varieties on the boundary associated variety is a hypersurface, and, in the case of discriminant, we give a complete classification of varieties for which $\deg X^* = \operatorname{ord} X$. It seems that the lower the degree of "determinant" the more the points of the ambient space resemble matrices; in particular, the corresponding varieties tend to be homogeneous, which can be viewed as a generalization of multiplicativity of matrices.

Even though the notion of order of projective variety is quite natural, it is not widely used. We also give another lower bound for the degree of discriminant in terms of dimension and codimension. This bound is also sharp and, rather surprisingly, the varieties on the boundary seem to be the same as for the bound in terms of order. This second approach is based on a study of Hessian matrices of homogeneous polynomials, and, although this topic apparently pertains to pure algebra, it has numerous classical and modern links with fields ranging from differential equations and differential geometry to approximation theory and mathematical physics.

The paper is organized as follows. Section 1 is devoted to join determinants. We obtain a lower bound for their degree in terms of order and consider numerous examples. In section 2 we study dual varieties (discriminants) and obtain a lower bound for their degree (called *codegree*) in terms of order. We also consider various examples and classify varieties on the boundary. To put the problem in a proper perspective, in section 3 (which is of an expository nature) we collect various known results on varieties of small degree and codegree. In section 4 we study Jacobian linear systems and Hessian matrices. As an application, we obtain a lower bound for codegree in terms of dimension and codimension and consider varieties for which this bound is sharp.

To avoid unnecessary complications, throughout the note we deal with algebraic varieties over the field $\mathbb{C}$ of complex numbers.

Acknowledgements. I discussed the contents of the present paper, particularly that of Section 4, with many mathematicians. I am especially grateful to C. Ciliberto, J.-M. Hwang, J. Landsberg, L. Manivel, Ch. Peskine, F. Russo, E. Tevelev and M. Zaidenberg for useful comments. This paper is a considerably reworked and expanded version of my talk at the conference in Vienna, and I am grateful to V. L. Popov who patiently but firmly managed the process of organizing the conference as well as publishing its proceedings and who insightfully observed the connection between my work and the theory of transformation groups (this connection with homogeneous and prehomogeneous varieties is not quite evident in the present paper, but will hopefully be clarified later on). Large portions of the present paper were written during my stay at IRMA in Strasbourg, and I am grateful to CNRS and particularly to O. Debarre for inviting me there.

1 Join Determinants

Let $X \subset \mathbb{P}^N$, $\dim X = n$ be a nondegenerate projective variety (i.e., X is not contained in a hyperplane).

For an integer $k \geq 1$ we put

$$\widetilde{X^k} = \overline{\{(x_1, \ldots, x_k, u) \in \underbrace{X \times \cdots \times X}_{k} \times \mathbb{P}^N \mid \dim \langle x_1, \ldots, x_k \rangle = k-1, u \in \langle x_1, \ldots, x_k \rangle\}},$$

where, for a subset $A \subset \mathbb{P}^N$, we denote by $\langle A \rangle$ the linear span of A in $\mathbb{P}^N$ and bar denotes projective closure. We denote by φ^k the projection of $\widetilde{X^k}$ to $\mathbb{P}^N$.

Definition 1.1. The variety $X^k = \varphi^k(\widetilde{X^k})$ is called the *k-th join of X with itself.*

It is clear that if $\mathbb{P}^N = \mathbb{P}(V)$, where V is a vector space of dimension $N+1$, and $C_X \subset V$ is the cone corresponding to X, then X^k is the variety corresponding to the cone $\underbrace{C_X + \cdots + C_X}_{k}$. Furthermore, X^k is the join of X with X^{k-1} and $X^{k+1} = X^k$ if and only if $X^k = \mathbb{P}^N$.

It is often necessary to compute tangent spaces to joins.

Proposition 1.2. (Terracini lemma)

(a) *Let* $x_1, \ldots, x_k \in X$, *and let* $u \in \langle x_1, \ldots, x_k \rangle \subset X^k$. *Then*

$$T_{X^k,u} \supset \langle T_{X,x_1}, \ldots, T_{X,x_k} \rangle,$$

where $T_{X,x}$ (resp. $T_{X^k,u}$) *is the tangent space to the variety* X *at the point* x (resp. to the variety X^k at the point u).

(b) *If* $x_1, \ldots, x_k$ *is a general collection of points of* X, *and* u *is a general point of the linear subspace* $\langle x_1, \ldots, x_k \rangle$, *then*

$$T_{X^k,u} = \langle T_{X,x_1}, \ldots, T_{X,x_k} \rangle.$$

Proof. This is a special case of [Z2, Chapter V, Proposition 1.4].

Definition 1.3. For a point $z \in \mathbb{P}^N$ we put $\operatorname{rk}_X z = \min\{k \mid z \in X^k\}$. The number $\operatorname{rk}_X z$ is called the *rank of* z *with respect to the variety* X.

The number $\operatorname{ord} X = \min\{k \mid X^k = \mathbb{P}^N\} = \max\limits_{z \in \mathbb{P}^N}\{\operatorname{rk}_X z\}$ is called the *order* of the variety X.

The difference $\operatorname{cork}_X z = \operatorname{ord} X - \operatorname{rk} z$ is called the *corank of* z *with respect to the variety* X.

Thus $1 \le \operatorname{rk}_X z \le \operatorname{ord} X$, $\operatorname{rk}_X x = 1$ if and only if $x \in X$, $\operatorname{rk}_X z = \operatorname{ord} X$ for a general point $z \in \mathbb{P}^N$ ($z \notin X^{\operatorname{ord} X - 1}$), and we get a strictly ascending filtration

$$X \subset X^2 \cdots \subset X^{\operatorname{ord} X - 1} = X^J \subset X^{\operatorname{ord} X} = \mathbb{P}^N. \tag{1.3.1}$$

Definition 1.4. The filtration (1.3.1) is called the *rank filtration*, and the variety $X^J = X^{\operatorname{ord} X - 1}$ the *join determinant* of X.

The number $\operatorname{codim}_{\mathbb{P}^N} X^J - 1$ is called the *join defect* of X and is denoted by $\operatorname{jodef} X$; it is clear that $0 \le \operatorname{jodef} X \le \dim X = n$.

The degree $\deg X^J$ of the join determinant is called the *jodegree* of X and is denoted by $\operatorname{jodeg} X$.

Examples 1.5. 1) If $X = \mathbb{P}^{r-1} \times \mathbb{P}^{r-1} \subset \mathbb{P}$, then one gets the standard notions of order and rank discussed above. In particular, $\operatorname{ord} X = r$, X^J is the locus of degenerate matrices, $\operatorname{jodeg} X = \operatorname{ord} X$, and X^J is defined by vanishing of determinant.

2) If $X = \mathbb{P}^{a-1} \times \mathbb{P}^{b-1} \subset \mathbb{P}^{ab-1}$, $a \le b$ and $z_M \in \mathbb{P}^{ab-1}$ is the point corresponding to an $(a \times b)$-matrix M, then $\operatorname{rk} z_M = \operatorname{rk} M$. Furthermore, $\operatorname{ord} X = a$ and $\operatorname{codim}_{\mathbb{P}^{ab-1}} X^J = b - a + 1$. By a formula due to Giambelli, $\operatorname{jodeg} X = \frac{b!}{(a-1)!(b-a+1)!}$ (cf. [Ful, 14.4.14]), which is much larger than $a = \operatorname{ord} X$ if $b \ne a$.

3) Let $X = \mathbb{P}^1 \times \mathbb{P}^1 \times \mathbb{P}^1 \subset \mathbb{P}^7$, so that the ambient $\mathbb{P}^7$ can be interpreted as the space of cubic 2-matrices up to multiplication by a nonzero constant. Then $X^J = X$, $\operatorname{ord} X = 2$ and $\operatorname{jodeg} X = \deg X = 6$.
4) Let $C = v_m(\mathbb{P}^1) \subset \mathbb{P}^m$ be a rational normal curve. The points of the ambient space $\mathbb{P}^m$ can be interpreted as binary forms of degree m modulo multiplication by a nonzero constant. Our definition of rank again coincides with the usual definition for binary forms. Furthermore, $\operatorname{ord} C = \left[\frac{m+2}{2}\right]$, where brackets denote integral part,

$$\operatorname{codim}_{\mathbb{P}^m} C^J = \begin{cases} 1, & m \equiv 0 \pmod 2, \\ 2, & m \equiv 1 \pmod 2 \end{cases}$$

and

$$\operatorname{jodeg} C = \begin{cases} \operatorname{ord} C = \left[\frac{m+2}{2}\right], & m \equiv 0 \pmod 2, \\ \frac{(m+1)(m+3)}{8}, & m \equiv 1 \pmod 2 \end{cases}$$

(this can be easily computed basing on Sylvester's theory of binary forms; cf. [Syl]). Thus the jodegree of C is equal to the order of C for m even and is much larger than the order for m odd.
5) Let $X = v_2(\mathbb{P}^{r-1}) \subset \mathbb{P}^{\frac{r(r+1)}{2}-1}$ be the Veronese variety. The points of the ambient linear space correspond to symmetric matrices of order r, and if M is a matrix and $z_M \in \mathbb{P}^{\frac{r(r+1)}{2}-1}$ is the corresponding point, then $\operatorname{rk} z_M = \operatorname{rk} M$. Here the variety X is a (special) linear section of the Segre variety from 1), and, as in the case 1), $\operatorname{ord} X = r$ and X^J is the hypersurface of degree r defined by vanishing of determinant.
6) Let $X = G(r-1, 1) \subset \mathbb{P}^{\binom{r}{2}-1}$ be the Grassmann variety of lines in $\mathbb{P}^{r-1}$. Then $n = \dim X = 2(r-2)$ and the points of $\mathbb{P}^{\binom{r}{2}-1}$ correspond to skew-symmetric matrices of order r. If M is such a matrix and $z_M \in \mathbb{P}^{\binom{r}{2}-1}$ is the corresponding point, then $\operatorname{rk} z_M$ is equal to $\frac{1}{2} \operatorname{rk} M$ (i.e., to the pfaffian rank of M). Furthermore, $\operatorname{ord} X = \left[\frac{r}{2}\right] = \left[\frac{n}{4}\right] + 1$,

$$\operatorname{codim}_{\mathbb{P}^{\binom{r}{2}-1}} X^J = \begin{cases} 1, & r \equiv 0 \pmod 2, \\ 3, & r \equiv 1 \pmod 2 \end{cases}$$

and

$$\operatorname{jodeg} X = \begin{cases} \operatorname{ord} X = \frac{r}{2}, & r \equiv 0 \pmod 2, \\ \frac{1}{4}\binom{r+1}{3}, & r \equiv 1 \pmod 2 \end{cases}$$

(cf. [HT]). Thus the jodegree of X is equal to $\operatorname{ord} X$ for r even and is much larger than $\operatorname{ord} X$ for r odd. In this case X is a (special) linear section of the Segre variety from 1), and X^J corresponds to the skew-symmetric matrices whose rank is less than maximal. In particular, for r even the hypersurface X^J is defined by vanishing of Pfaffian.

Theorem 1.6. *Let $X \subset \mathbb{P}^N$ be a nondegenerate variety. Then* $\operatorname{jodeg} X \geq \operatorname{ord} X + \operatorname{jodef} X \geq \operatorname{ord} X$. *In particular, the jodegree is not less than the order, and equality is possible only if X^J is a hypersurface.*

Sketch of proof. Let $\{x_1, \ldots, x_{\operatorname{ord} X}\}$ be a general collection of points of X. By the definition of $\operatorname{ord} X$, the $(\operatorname{ord} X - 1)$-dimensional linear space $\langle x_1, \ldots, x_{\operatorname{ord} X}\rangle$ does not lie in X^J, but contains $\operatorname{ord} X$ linear subspaces of the form $\langle x_{i_1}, \ldots, x_{i_{\operatorname{ord} X-1}}\rangle$, $\dim \langle x_{i_1}, \ldots, x_{i_{\operatorname{ord} X-1}}\rangle = \operatorname{ord} X - 2$, $i_1 < \cdots < i_{\operatorname{ord} X-1}$. Thus a general line $l \subset \langle x_1, \ldots, x_{\operatorname{ord} X}\rangle$ is not contained in X^J, but meets it in at least $\operatorname{ord} X$ points. Furthermore, adding points, it is easy to construct a linear subspace $L \subset \mathbb{P}^N$, $\dim L = \operatorname{codim}_{\mathbb{P}^N} X^J = \operatorname{jodef} X + 1$ which is not contained in X^J, but meets it in at least $\operatorname{ord} X + \operatorname{jodef} X$ points. Thus $\deg X^J \geq \operatorname{ord} X + \operatorname{jodef} X$, and our claim follows. □

Remarks 1.7. (i) The same argument shows that

$$\deg X^k \geq k + \operatorname{codim}_{\mathbb{P}^N} X^k \tag{1.7.1}$$

usual bound for degree; cf. Theorem 3.4, (i)).

(ii) For $1 \leq k < \operatorname{ord} X$, the variety X^k is not contained in a hypersurface of degree k or less. In fact, suppose that $X^k \subset W$, where $W \subset \mathbb{P}^N$ is a hypersurface. Arguing as in the proof of Theorem 1.6, we see that, for a general collection $\{x_1, \ldots, x_{k+1}\}$ of points of X, the intersection $\langle x_1, \ldots, x_{k+1}\rangle \cap X^k$ contains $k+1$ hyperplanes of the form $\langle x_{i_1}, \ldots, x_{i_k}\rangle$, $i_1 < \cdots < i_k$. Then either $\deg W = \deg W \cap \langle x_1, \ldots, x_{k+1}\rangle \geq k+1$ or $W \supset \langle x_1, \ldots, x_{k+1}\rangle$. In the first case we are done, and in the second case $W \supset X^{k+1}$, so that we arrive at a contradiction by induction.

(iii) Arguing as in the proof of Theorem 1.6, one can show that, for an arbitrary point $z \in X^k$, one has

$$\operatorname{mult}_z X^k \geq k - \operatorname{rk} z + 1, \tag{1.7.2}$$

where $\operatorname{mult}_z X^k$ denotes the multiplicity of X^k at the point z (compare with the bound for multiplicity of points of dual varieties given in Section 2 (cf. (2.11.1)). We observe that, for $z \in X$, (iii) implies (i). For $k = \operatorname{ord} X - 1$ the inequality (1.7.2) assumes the form

$$\operatorname{mult}_z X^J \geq \operatorname{cork}_X z, \qquad z \in X^J. \tag{1.7.3}$$

This bound (which should be compared with the bound for multiplicity in Proposition 2.9) can be used to give a different proof of Theorem 1.6 (compare with the use of Proposition 2.8 in the proof of Theorem 2.7 below).

(iv) Since any variety of degree d and codimension e is contained in a hypersurface of degree $d - e + 1$ (and is in fact a set theoretic intersection of such hypersurfaces; to see this it suffices to project the variety from the linear span of a general collection of $e - 1$ points on it and take the cone over the image), (ii) also yields Theorem 1.6 and, more generally, (i).

(v) Theorem 1.6 and the remarks thereafter were known to the author for many years. Proofs of some special cases can also be found in [Ge, Lecture 7] and [C-J].

Theorem 1.6 gives a nice bound for jodegree in terms of order, but the notion of order might seem a bit unusual. So, it is desirable to give a bound in more usual terms, such as dimension and codimension. One has the following lower bound for order.

Proposition 1.8. *Let $X^n \subset \mathbb{P}^N$ be a nondegenerate variety. Then* $\operatorname{ord} X \geq \dfrac{N - \operatorname{jodef} X}{n+1} + 1 \geq \dfrac{N+1}{n+1}$.

Proof. In fact, for each natural k, $\dim X^k \leq \dim \widetilde{X^k} = kn+k-1$. In particular,

$$N - \operatorname{jodef} X - 1 = \dim X^J \leq (\operatorname{ord} X - 1)(n+1) - 1,$$

which yields the proposition. □

In the case when $C \subset \mathbb{P}^N$ is a curve, one always has $\operatorname{ord} C = \left[\frac{N+2}{2}\right]$, where brackets denote integral part, $\dim C^i = 2i - 1$, $1 \leq i \leq \operatorname{ord} C - 1$ and

$$\operatorname{jodef} C = \begin{cases} 0, & N \equiv 0 \pmod 2, \\ 1, & N \equiv 1 \pmod 2) \end{cases}$$

(cf., e.g., [Z2, chapter V, Example 1.6]). Thus the bound for order given in Proposition 1.8 is sharp in this case.

Theorem 1.9. *Let $C \subset \mathbb{P}^N$ be a nondegenerate curve. Then* $\operatorname{ord} C = \left[\frac{N+2}{2}\right]$ *and* $\operatorname{jodeg} C \geq \operatorname{ord} C$. *Furthermore,* $\operatorname{jodeg} C = \operatorname{ord} C$ *if and only if N is even and C is a rational normal curve* (cf. Example 1.5, 4).

Sketch of proof. The first claim follows from Theorem 1.6 in view of the computation of $\operatorname{codim}_{\mathbb{P}^m} C^J$ made above.

From Proposition 1.2 and the trisecant lemma (cf. [HR, 2.5], [Ha, Chapter IV, §3] or [Mum, §7 B]) it is easy to deduce that the map $\varphi^i : \widetilde{C^i} \to C^i$ is birational for $i < \operatorname{ord} C$. Suppose that $\operatorname{jodeg} C = \operatorname{ord} C$. Then, as we already observed, $N = 2k$, where k is a natural number, and $\operatorname{ord} C = \left[\frac{N+2}{2}\right] = k+1$. From Theorem 1.6 and Remark 1.7 it follows that $\operatorname{mult}_x C^J \geq k$ for each point $x \in C$. Thus, projecting C from the point x to $\mathbb{P}^{2k-1}$, we get a curve $C' \subset \mathbb{P}^{2k-1}$ such that $\deg C' = \deg C - 1$, $\operatorname{ord} C' = k$, and the projection $C^k \dashrightarrow {C'}^k = \mathbb{P}^{2k-1}$ is birational. Hence the map ${\varphi'}^k : \widetilde{{C'}^k} \to {C'}^k = \mathbb{P}^{2k-1}$ is also birational. Thus to prove our claim it suffices to show that any curve $C' \subset \mathbb{P}^{2k-1}$ with this property is a normal rational curve.

To this end, we take a general point $z \in {C'}^{k-1}$ and consider the projection $\pi_z : C' \to \mathbb{P}^1$ with center at the linear subspace $T_{{C'}^{k-1},z}$. From Proposition 1.2 it easily follows that $\deg \pi_z = \deg C' - 2(k-1)$, and so it suffices to verify that the map π_z is an isomorphism. If this were not so, then the map π_z would be ramified and there would exist a point $x \in C$ such that $T_{C,x} \cap T_{C'^{k-1},z} \neq \varnothing$. By the Terracini lemma, the line $\langle x, z \rangle$ lies in the branch locus of the map $\varphi'^k : \widetilde{C'^k} \to \mathbb{P}^{2k-1}$. Varying z in C'^{k-1}, it is easy to see that the branch locus is a hypersurface in $\mathbb{P}^{2k-1}$. The proof is completed by recalling that φ'^k is birational, and so in our case the branch locus has codimension at least two (cf. also [C-J]). □

For higher dimensions there is little chance to obtain a classification of varieties for which the inequality in Theorem 1.6 turns into equality (some of these varieties were listed in Examples 1.5). One of the reasons is that variety X is not uniquely determined by its join determinant X^J. Here follows a typical example of this phenomenon.

Example 1.10. Let $v_2(\mathbb{P}^2) \subset \mathbb{P}^5 \subset \mathbb{P}^6$ be the Veronese surface (cf. Example 1.5, 5) for $r = 3$), let $y \in \mathbb{P}^6$ be a point, let $Y \subset \mathbb{P}^6$ be the cone over $v_2(\mathbb{P}^2)$ with vertex y, and let X be the intersection of Y with a general hypersurface of degree $d \geq 2$. Then $X \subset Y \subset \mathbb{P}^6$ is a nondegenerate surface, $\dim X^2 = 5 = \dim Y^2$, and therefore $X^J = Y^J$ and $\operatorname{jodeg} X = \operatorname{ord} X = \operatorname{ord} Y = \operatorname{jodeg} Y = 3$.

Thus a lot of different surfaces have the same order and the same join determinant.

Remark 1.11. Of course, one can construct similar examples starting from other varieties whose higher self-joins have dimension smaller than expected (cf., e.g., Examples 1.5). Thus, in the case of Segre variety $Y = \mathbb{P}^{r-1} \times \mathbb{P}^{r-1}$ or Grassmann variety $Y = G(r-1, 1)$ with r even (Examples 1.5, 1), 6)) any general subvariety $X \subset Y$ of sufficiently small codimension has the same order and join determinant as Y. This leads to the notion of *constrained varieties* (cf. [Å]) which deserves to be studied in detail. It should be possible to classify "maximal" (nonconstrained) varieties with jodegree equal to order; then arbitrary varieties with this property should be (constrained) subvarieties of the maximal ones.

It should be noted however that even the lists of nonconstrained varieties of a fixed jodegree are rather big. For example, in view of the above observation, the list of nonconstrained varieties of jodegree three is much longer than the corresponding lists of varieties of degree three (cf. Theorem 3.3, (iii)) or codegree three (cf. Theorem 3.5, (iii)). Still, imposing additional natural conditions, such as inextensibility, can help to make this list shorter (cf. also Remark 2.13, (ii)).

Remark 1.12. A nice series of examples of varieties of small jodegree illustrating Remark 1.11 can be constructed as follows. Let $Y \subset \mathbb{P}^N$ be a nondegenerate variety, and let $X = v_2(Y)$ be the quadratic Veronese reembedding of Y. Then $\operatorname{ord} X \geq N+1$ (cf. [Z3, Theorem 2.7]), and in [Z3] we construct numerous examples of varieties for which $\operatorname{ord} X = N+1$ and $X^N = Z^N \cap \langle X \rangle$, where $Z = v_2(\mathbb{P}^N)$ and $\langle X \rangle$ is the linear span of X. Since Z^N is a hypersurface of degree $N+1$, for such a variety one has $\operatorname{jodeg} X = \operatorname{ord} X$, so that X is on the boundary of Theorem 1.6.

Example 1.13. A well known example highlighting Remark 1.12 is due to Clebsch (cf. [Cl] and [Z3, Remark 6.2]). Let $X = v_4(\mathbb{P}^2) \subset \mathbb{P}^{14}$. Then $\operatorname{jodeg} X = \operatorname{ord} X = 6$, and the join determinant of X is the locus of quartics representable as a sum of at most five bisquares.

2 Discriminants

Let $X \subset \mathbb{P}^N$ be a projective variety. Consider the subvariety

$$\mathcal{P}_X = \overline{\{(x, \alpha) \mid x \in \operatorname{Sm} X, L_\alpha \supset T_{X,x}\}} \subset X \times \mathbb{P}^{N*},$$

where $\operatorname{Sm} X$ denotes the subset of nonsingular points of X, $L_\alpha \subset \mathbb{P}^N$ is the hyperplane corresponding to a point $\alpha \in \mathbb{P}^{N*}$, $T_{X,x}$ is the (embedded) tangent space to X at x, and bar denotes the Zariski closure. Let $p : \mathcal{P}_X \to X$ and $\pi : \mathcal{P}_X \to \mathbb{P}^{N*}$ be the projections onto the two factors.

Definition 2.1. $\mathcal{P}_X$ is called the *conormal variety* of X, and $X^* = \pi(\mathcal{P}_X) \subset \mathbb{P}^{N*}$ is called the *dual variety* or the *discriminant* of X.

In most cases one can assume that X is nondegenerate because if $X \subset L$, where $L \subset \mathbb{P}^N$ is a linear subspace and X_L^* is the dual of X viewed as a subvariety of L, then X^* is the cone over X_L^* with vertex at the linear subspace ${}^\perp L \subset \mathbb{P}^{N*}$ of all the hyperplanes in $\mathbb{P}^N$ passing through L. Conversely, if $X \subset \mathbb{P}^N$ is a cone with vertex L over a variety $Y \subset \mathbb{P}^M$, $M = N - \dim L - 1$, then $X^* = Y^* \subset \mathbb{P}^{M*} = {}^\perp L \subset \mathbb{P}^{N*}$.

Unlike join determinant, discriminant allows to reconstruct the original variety, viz. one has the following

Theorem 2.2. $X^{**} = X$, *i.e., projective varieties are reflexive with respect to the notion of duality introduced in Definition* 2.1.

We refer to [Tev] for this and other results on dual varieties that are used in this paper.

Typically, for a nonsingular variety X, the dual variety is a hypersurface (in view of Theorem 2.2, one cannot expect this to be true for arbitrary varieties).

Definition 2.3. The number $\operatorname{codim}_{\mathbb{P}^{N*}} X^* - 1$ is called the (dual) *defect* of X and is denoted by $\operatorname{def} X$.

The degree of the dual variety X^* is called the *codegree* or (if $\operatorname{def} X = 0$) *class* of X and is denoted by $\operatorname{codeg} X$ (thus $\operatorname{codeg} X = \deg X^*$).

Proposition 2.4. *Let $X \subset \mathbb{P}^N$ be a projective variety, let $L \subset \mathbb{P}^N$ be a general linear subspace, $L \cap X = \varnothing$, let $\pi_L : \mathbb{P}^N \dashrightarrow \mathbb{P}^M$, $M = N - \dim L - 1$ be the projection with center at L, and let $Y = \pi_L(X) \subset \mathbb{P}^M$. Then $Y^* = X^* \cap {}^{\perp}L$, where ${}^{\perp}L \subset \mathbb{P}^{N*}$ parameterizes the hyperplanes in $\mathbb{P}^N$ passing through L and thus is naturally isomorphic to $\mathbb{P}^{M*}$.*

In the case when $\operatorname{def} X > 0$ one can apply Proposition 2.4 (combined with Theorem 2.2) to X^* to show that the dual of a general linear section of X is a general projection of X^*. This yields a reduction to the case of varieties of defect zero (of course, the degree of X^* is stable under general projections).

The Terracini lemma 1.2 yields useful information on the structure of X^*, viz. one gets a descending filtration

$$X^* \supset (X^2)^* \supset \cdots \supset (X^{\operatorname{ord} X - 1})^* \tag{2.4.1}$$

corresponding to the ascending filtration!(1.3.1) (cf. Proposition 4.2 below for a more precise statement).

Definition 2.5. The filtration (2.4.1) is called the *corank filtration.*

Consider the varieties described in Examples 1.5 from the point of view of structure of their duals.

Examples 2.6. 1) If $X = \mathbb{P}^{r-1} \times \mathbb{P}^{r-1} \subset \mathbb{P}$, then X^* is the hypersurface defined by vanishing of determinant. Hence $\operatorname{codeg} X = r = \operatorname{jodeg} X = \operatorname{ord} X$. In this case (2.4.1) is just the filtration by the corank of matrix.

2) If $X = \mathbb{P}^{a-1} \times \mathbb{P}^{b-1} \subset \mathbb{P}^{ab-1}$, $a \le b$, then X^* corresponds to matrices of rank smaller than a in the dual space, $\operatorname{def} X = b - a$, X^* is projectively isomorphic to X^J and, as we saw in Example 1.5, 2), $\operatorname{ord} X = a$ and $\operatorname{codeg} X = \frac{b!}{(a-1)!(b-a+1)!}$ (cf. [Ful, 14.4.14]), which is larger than $a = \operatorname{ord} X$ if $b \ne a$. Here again (2.4.1) is the filtration by corank.

3) Let $X = \mathbb{P}^1 \times \mathbb{P}^1 \times \mathbb{P}^1 \subset \mathbb{P}^7$ (cubic $(2 \times 2 \times 2)$-matrices). Then it is easy to see that $\operatorname{def} X = 0$ and X^* is a hypersurface of degree four (we recall that $\operatorname{jodeg} X = 6$; cf. Example 1.5, 3)). The discriminant (or hyperdeterminant) in this case was first computed by Cayley [Ca]. The variety X is a special case of symmetric Legendrean varieties all of which have codegree four (cf. [Mu], [LM] and Remark 3.6 below). More generally, if $X = \mathbb{P}^{n_1} \times \cdots \times \mathbb{P}^{n_k} \subset \mathbb{P}^{(n_1+1)\cdots(n_k+1)-1}$, $n_1 \le n_2 \le \cdots \le n_k$, then it is easy to see that $\operatorname{def} X = \max\{n_k - \sum_{i=1}^{k-1} n_i, 0\}$, and one can use tools from combinatorics to compute the codegree (cf. [GKZ, Chapter 14B] for the case of $\operatorname{def} X = 0$), which, in our opinion, does not make much sense; anyhow, the codegree is very large.

4) Let $C = v_m(\mathbb{P}^1) \subset \mathbb{P}^m$ be a rational normal curve. Then C^* is the hypersurface in $\mathbb{P}^{m*}$ swept out by the osculating $(m-2)$-spaces to the curve in $\mathbb{P}^{m*}$ parameterizing the osculating hyperplanes to C and defined by vanishing of discriminant of binary form. Thus, for $m > 2$, $\operatorname{codeg} C = 2m-2$ is

larger than $\operatorname{ord} C = \left[\frac{m+2}{2}\right]$, but (2.4.1) is again the filtration by the corank of binary form. It should be noted that in this case $\operatorname{codeg} C > \operatorname{jodeg} C$ for $m > 2$ even while $\operatorname{codeg} C < \operatorname{jodeg} C$ for $m > 3$ odd (cf. Example 1.5, 4)).

5) If $X = v_2(\mathbb{P}^{r-1}) \subset \mathbb{P}^{\frac{r(r+1)}{2}-1}$ is a Veronese variety, then X^* is the hypersurface of degenerate matrices in the dual space of symmetric matrices of order r defined by vanishing of determinant; in particular, (2.4.1) is the filtration by corank and $\operatorname{codeg} X = \operatorname{ord} X$. Furthermore, the dual variety X^* is projectively isomorphic to X^J and, in particular, $\operatorname{codeg} X = \operatorname{jodeg} X$.

6) Let $X = G(r-1, 1) \subset \mathbb{P}^{\binom{r}{2}-1}$ be the Grassmann variety of lines in $\mathbb{P}^{r-1}$. Then

$$\operatorname{def} X = \begin{cases} 0, & r \equiv 0 \pmod 2, \\ 2, & r \equiv 1 \pmod 2, \end{cases}$$

and (2.4.1) is the filtration by pfaffian corank. Furthermore,

$$\operatorname{codeg} X = \begin{cases} \operatorname{ord} X = \frac{r}{2}, & r \equiv 0 \pmod 2, \\ \frac{1}{4}\binom{r+1}{3}, & r \equiv 1 \pmod 2 \end{cases}$$

(cf. [HT]). The dual variety corresponds to the skew-symmetric matrices whose rank is less than maximal. In particular, if r is even, then the hypersurface X^* is defined by vanishing of Pfaffian. Thus the codegree is equal to $\operatorname{ord} X$ if r is even and is much larger than $\operatorname{ord} X$ if r is odd. We observe that also in this case X^* is projectively isomorphic to X^J and, in particular, $\operatorname{codeg} X = \operatorname{jodeg} X$.

The following result is an analogue of Theorem 1.6 for codegree.

Theorem 2.7. *Let $X^n \subset \mathbb{P}^N$ be a nondegenerate variety. Then $d^* = \operatorname{codeg} X \geq \operatorname{ord} X$. Furthermore, if X is not a cone, then $d^* \geq \operatorname{ord} X + \operatorname{def} X \geq \operatorname{ord} X$. In particular, the codegree is not less than the order, and equality is possible only if X^* is a hypersurface in its linear span $\langle X^* \rangle$.*

Sketch of proof. The idea is to produce points of high multiplicity in the dual variety X^*. For a point $\alpha \in X^*$ we denote by $\operatorname{mult}_\alpha X^*$ the multiplicity of X^* at α. If Λ is a general $(\operatorname{def} X + 1)$-dimensional linear subspace of $\mathbb{P}^{N*}$ passing through the point α, then Λ meets X^* at α and $d^* - \operatorname{mult}_\alpha X^*$ other points. Furthermore, if U is a small neighborhood of α in X^* and Λ' is a general $(\operatorname{def} X + 1)$-dimensional subspace of $\mathbb{P}^{N*}$ sufficiently close to Λ, then Λ' meets U in $\operatorname{mult}_\alpha X^*$ nonsingular points.

The multiplicity defines a stratification of the dual variety. To wit, put

$$X_k^* = \{\alpha \in X^* \mid \operatorname{mult}_\alpha X^* \geq k\},$$

and let

$$k_m = \max\{k \mid X_k^* \neq \varnothing\}.$$

Then

$$X^* = X_1^* \supset X_2^* \supset \cdots \supset X_{k_m}^*.$$

To give a lower bound for d^* it suffices to bound the number k_m from below. In fact, it is easy to prove the following

Proposition 2.8. *If X is not a cone, then $d^* \geq k_m + \operatorname{def} X + 1$.*

Proof. If $\operatorname{def} X = 0$, then there is nothing to prove. Suppose that $\operatorname{def} X > 0$, and let $\alpha \in X_{k_m}^*$ be a general point. Let $\varpi : \mathbb{P}^{N*} \dashrightarrow \mathbb{P}^{N-1*}$ be the projection with center at α, and let $X^{*\prime} = \varpi(X^*)$. Since X is nondegenerate, X^* is not a cone, and so $\dim X^{*\prime} = \dim X^*$ and $\deg X^{*\prime} = \deg X^* - \operatorname{mult}_\alpha X^* = d^* - k_m$. Moreover, $\operatorname{codim}_{\mathbb{P}^{N-1*}} X^{*\prime} = \operatorname{codim}_{\mathbb{P}^{N*}} X^* - 1 = \operatorname{def} X$. Since, by our assumption, X^* is nondegenerate, $X^{*\prime}$ has the same property, and so $\deg X^{*\prime} \geq \operatorname{def} X + 1$ (cf. Theorem 3.4, (i) below). Thus $d^* - k_m \geq \operatorname{def} X + 1$ and we are done. □

To give a bound for k_m, we consider the corank filtration (2.4.1) introduced in Definition 2.5.

Proposition 2.9. *For each natural number k, $k < \operatorname{ord} X$ one has $(X^k)^* \subset X_k^*$.*

Sketch of proof. To prove the claim one can argue by induction. Let $(x_1, \ldots, x_k) \in \underbrace{X \times \cdots \times X}_{k}$ be a general collection of k points of X, let $\mathcal{P}_{x_i} = \{\alpha \mid L_\alpha \supset T_{X,x_i}\} \subset X^*$, $i = 1, \ldots, k$ (we recall that L_α denotes the hyperplane corresponding to α), and let $\alpha_k \in \mathcal{P}_{x_1} \cap \cdots \cap \mathcal{P}_{x_k}$. By the Terracini lemma, it suffices to show that $\operatorname{mult}_{\alpha_k} X^* \geq k$. Suppose that, for a general point $\alpha_{k-1} \in \mathcal{P}_{x_1} \cap \cdots \cap \mathcal{P}_{x_{k-1}}$, in a neighborhood of α_k we already know that $\operatorname{mult}_{\alpha_{k-1}} X^* \geq k-1$. Then, joining α_{k-1} with a general point $\alpha \in \mathcal{P}_{x_k}$ close to α_k, adding $\operatorname{def} X$ general points of $\mathbb{P}^{N*}$, taking the linear span of these $\operatorname{def} X + 2$ points (which, since X is nondegenerate, meets X^* in finitely many points), and passing to a limit, one can show that $\operatorname{mult}_{\alpha_k} X^* \geq \operatorname{mult}_{\alpha_{k-1}} X^* + \operatorname{mult}_\alpha X^* \geq k$. □

Corollary 2.10. $k_m \geq \operatorname{ord} X - 1$.

Theorem 2.7 immediately follows from Proposition 2.8 and Corollary 2.10. □

Remarks 2.11. (i) Arguing as in the proof of Proposition 2.9, one can show that, for $1 \leq i \leq k < \operatorname{ord} X$ and an arbitrary point $\alpha \in (X^k)^*$, one has

$$\operatorname{mult}_\alpha (X^i)^* \geq k - i + 1. \tag{2.11.1}$$

This is an analogue of (1.7.2).

(ii) In view of Proposition 1.8, from Theorem 2.7 it follows that

$$d^* \geq \frac{N - \operatorname{jodef} X}{n+1} + 1 \geq \frac{N+1}{n+1}. \tag{2.11.2}$$

Furthermore, if $N' = N - \operatorname{def} X$, $\mathbb{P}^{N'} \subset \mathbb{P}^N$ is a general linear subspace and $X' = \mathbb{P}^{N'} \cap X$, so that $n' = \dim X' = n - \operatorname{def} X$, X'^* is a general projection of X^* to $\mathbb{P}^{N'*}$ (cf. Proposition 2.4) and $\operatorname{codeg} X' = \operatorname{codeg} X = d^*$, then the same argument shows that

$$\operatorname{ord} X' \geq \frac{N'+1}{n'+1}. \tag{2.11.3}$$

Thus from Theorem 2.7 it follows that

$$d^* = \operatorname{codeg} X \geq \frac{N'+1}{n'+1} \geq \frac{N+1}{n+1}. \tag{2.11.4}$$

Suppose further that

$$(X_l'^*)^2 \not\subset X'^* \tag{2.11.5}$$

for some l such that

$$X_l'^* \neq \varnothing. \tag{2.11.6}$$

Then one can choose points $\alpha, \beta \in X_l'^*$ so that the line $\langle \alpha, \beta \rangle$ is not contained in X'^*, from which it follows that

$$d^* \geq 2l. \tag{2.11.7}$$

By Proposition 1.8, $\frac{N'-n'}{n'+1} \leq \operatorname{ord} X' - 1$, and thus there exists an $l \geq \frac{N'-n'}{n'+1}$ satisfying (2.11.6) If (2.11.5) holds for such an l, then (2.11.7) yields

$$d^* \geq 2l \geq 2 \cdot \frac{N'-n'}{n'+1} \geq 2 \cdot \frac{N-n}{n+1}, \tag{2.11.8}$$

which is almost twice better than (2.11.3).

Inequalities (2.11.4) and (2.11.8) give lower bounds for the codegree in terms of dimension and codimension. We will obtain a still better (universal and sharp) bound of this type in Section 4 (cf. Theorem 4.7).

(iii) Our proof of Theorem 2.7 is based on producing points of high multiplicity in the dual variety. More precisely, in Proposition 2.9 we showed that the points in X^* corresponding to k-tangent hyperplanes have multiplicity at least k. There is another natural way to produce highly singular points in X^*. To wit, if the hyperplane section of X corresponding to a point $\alpha \in X^*$ has only isolated singular points $x_1, \dots, x_l \in \operatorname{Sm} X$, then it is easy to show that $\alpha \in X^*_\mu \setminus X^*_{\mu+1}$, where $\mu = \sum_{i=1}^l \mu_i$ and μ_i is the Milnor number of x_i. In fact, since the singularities are isolated, in this case X^* is a hypersurface, and if Λ is a general line passing through α and Λ' is a line sufficiently close to Λ, then, in a small neighborhood of α, Λ' meets X^* in $\operatorname{mult}_\alpha X^*$ nonsingular points corresponding to hyperplane sections having a unique nondegenerate quadratic singularity. Now our claim follows from the definition of Milnor number as the number of ordinary quadratic points for a small deformation of function (cf. [Dim] for a less direct proof).

This method allows to show that d^* is quite high (in particular, higher then all the bounds that we discuss here) if all (or at least sufficiently many, in a certain sense) hyperplane sections of X have only isolated singularities (which, for example, is the case when X is a nonsingular complete intersection). However, we cannot assume this, and indeed in our examples "on the boundary" (cf. Examples 2.6, 1), 5), 6) (for r even) and Theorem 3.5 (iii), II.4) all singularities of hyperplane sections are either nondegenerate quadratic or nonisolated (cf. Condition $(*)$ in [Z1]). So, we will not further discuss this method here.

Conjectured Theorem 2.12. *Let $X^n \subset \mathbb{P}^N$ be a nondegenerate variety. Then the following conditions are equivalent:*

(i) $d^* = \operatorname{codeg} X = \operatorname{ord} X$;

(ii) X *is* (a cone over) *one of the following varieties:*

I. X *is a quadric,* $N = n+1$, $d^* = 2$;

II. X *is a Scorza variety* (cf. [Z2, Chapter VI]). *More precisely, in this case there are the following possibilities*:

II.1. $X = v_2(\mathbb{P}^n)$ *is a Veronese variety,* $N = \frac{n(n+3)}{2}$, $d^* = n+1$;

II.2. $X = \mathbb{P}^a \times \mathbb{P}^a$, $a \geq 2$ *is a Segre variety,* $n = 2a$, $N = a(a+2) = \frac{n(n+4)}{4}$, $d^* = a+1$;

II.3. $X = G(2m+1, 1)$ *is the Grassmann variety of lines in* $\mathbb{P}^{2m+1}$, $m \geq 2$, $n = 4m$, $N = m(2m+3) = \frac{n(n+6)}{8}$, $d^* = m+1$;

II.4. $X = E$ *is the variety corresponding to the orbit of highest weight vector in the lowest dimensional nontrivial representation of the group of type* E_6, $n = 16$, $N = 26$, $d^* = 3$.

Idea of proof. The implication (ii)⇒(i) is immediate (cf., e.g., [Z2, Chapter VI]). Thus we only need to show that (i)⇒(ii).

The case $\operatorname{ord} X = 2$ being obvious, one can argue by induction on $\operatorname{ord} X$. Given a variety $X \subset \mathbb{P}^N$ with $d^* = \operatorname{ord} X$, we consider its projection $X' \subset \mathbb{P}^{N-n-1}$ from the (embedded) tangent space $T_{X,x}$ at a general point $x \in X$. Then $\operatorname{ord} X' = \operatorname{ord} X - 1$, and it can be shown that $\operatorname{codeg} X' = \operatorname{codeg} X - 1$. Thus X' also satisfies the conditions of the theorem, and, reducing the order, we finally arrive at the case $\operatorname{ord} X = 3$, which can be dealt with directly (for smooth varieties, cf. Theorem 3.5, (iii)). Then a close analysis allows to reconstruct X. □

Remarks 2.13. (i) Of course, the above proof is not complete, particularly in the nonsmooth case. Moreover, some details have not been verified yet. A complete proof will hopefully be given elsewhere.

(ii) Even without giving a complete classification of varieties satisfying (i), one can show that for such a variety $k_m = \operatorname{ord} X - 1$, $X^*_{k_m} = (X^{\operatorname{ord} X - 1})^*$ and $(X^*_{k_m})^{k_m} = X^*$. Thus $\operatorname{ord} X^*_{k_m} = \operatorname{ord} X$, and so $X^*_{k_m}$ is on the boundary of Theorem 1.6 for jodegree. In other words, in the extremal case the join determinant of $X^*_{k_m}$ coincides with the discriminant of X, and so classification of varieties of minimal codegree is "contained" in that of varieties of minimal jodegree. A posteriori, $X^*_{k_m}$ is projectively isomorphic to X, and so X also has minimal jodegree. On the other hand, there exist varieties for which jodegree is minimal while codegree is far from the boundary; cf., e.g., Examples 1.5, 4) and 2.6, 4).

(iii) It is worthwhile to observe that all the varieties in Theorem 2.12 are homogeneous. More precisely, varieties II.1–II.4 are the so called *Scorza varieties* (cf. [Z2, Chapter VI]). Thus in the case II the points of the ambient linear space correspond to (Hermitean) matrices over composition algebras (cf. [Z2, Chapter VI, Remark 5.10 and Theorem 5.11] and also [Ch]), and the points of the variety correspond to matrices of rank one. Furthermore, homogeniety is induced by multiplication of matrices. Thus the true meaning of Theorem 2.12 is that, in a certain sense, the varieties of Hermitean matrices of rank one over composition algebras are characterized by the property that their discriminant has minimal possible degree.

(iv) In Theorem 2.12 we give classification of varieties of minimal codegree, i.e., varieties X for which $\operatorname{codeg} X = \operatorname{ord} X$. This classification "corresponds" to classification of varieties of minimal degree (cf. Theorem 3.4 below). The next step in classification of varieties of small degree is to describe (linearly normal) varieties whose degree is close to minimal. The difference $\Delta_X = \deg X - \operatorname{codim} X - 1$ is called the Δ-*genus* of X. Thus, variety has minimal degree if and only if its Δ-genus vanishes. Classification of varieties of Δ-genus one (Del Pezzo varieties) is well known; much is also known about varieties whose Δ-genus is small (cf., e.g., [Fu]). Similarly, one can define ∇-*genus* of X by putting $\nabla_X = \operatorname{codeg} X - \operatorname{ord} X$ (or $\nabla_X = \operatorname{codeg} X - \operatorname{ord} X - \operatorname{def} X$). An interesting problem is then to give classification of projective varieties of small ∇-genus.

3 Varieties of Small Degree and Codegree

Theorems 2.7 and 2.12 give a nice idea of what it means for a variety to have small codegree. However, the answer is given in terms of the notion of order of variety which is not very common. A similar question about varieties of small degree has been studied for a century and a half. To put the problem in a proper perspective, in this section (which is of an expository nature) we collect various known results on varieties of small degree and codegree.

We start with giving sharp bounds for the codegree of nonsingular curves and surfaces in terms of standard invariants, such as codimension and degree.

In the case of curves one gets the following counterpart of Theorem 1.9.

Proposition 3.1. (i) *Let $C \subset \mathbb{P}^N$ be a nondegenerate nonsingular curve of degree d and codegree d^*. Then $d^* \geq 2d - 2$ with equality holding if and only if C is rational.*

(ii) *Let* $C \subset \mathbb{P}^N$ *be an arbitrary nondegenerate curve. Then* $d^* = \operatorname{codeg} C \geq 2 \operatorname{codim} C = 2N - 2$ *with equality holding if and only if* C *is a normal rational curve.*

Proof. First of all, it is clear that $\operatorname{def} C = 0$ for an arbitrary curve C. For any nonsingular curve C of genus g and degree d one has

$$\operatorname{codeg} C = 2(g + d - 1) \tag{3.1.1}$$

(this is essentially the Riemann–Hurwitz formula; its generalization to singular curves is immediate, but does not yield the inequality in (i)).

Assertion (i) immediately follows from (3.1.1) ((i) is trivially false without the assumption of smoothness; to see this, it suffices to consider the case $N = 2$ and use Theorem 2.2).

In the case of nonsingular curves, assertion (ii) follows from (i) since $d \geq N$ for an arbitrary nondegenerate curve C with equality holding if and only if C is a normal rational curve (this is a very special case of Theorem 3.4 below).

If $C \subset \mathbb{P}^N$ is a nondegenerate, but possibly singular curve and (x, y) is a general pair of points of C, then it is easy to see that the points α and β on the hypersurface C^* corresponding to the osculating hyperplanes at x and y respectively have multiplicity at least $N - 1$ (cf. also Remark 2.11, (iii)). From the Bertini theorem it follows that the line $\langle \alpha, \beta \rangle$ joining the points α and β in $\mathbb{P}^{N*}$ is not contained in C^* (in other words, the self-join of the dual (osculating) curve of C is not contained in the dual hypersurface). This proves the inequality $d^* \geq 2N - 2$. From our argument it also follows that equality $d^* = 2N - 2$ is only possible for curves without hyperosculation points, i.e., for rational normal curves. □

Lower bounds for the class can be also obtained for smooth surfaces (in this case the defect is always equal to zero, cf. [Z1]).

Proposition 3.2. *Let* $X \subset \mathbb{P}^N$ *be a nondegenerate nonsingular surface of degree* d *and codegree* d^*.

(i) $d^* \geq d - 1$. *Furthermore,* $d^* = d - 1$ *if and only if* X *is the Veronese surface* $v_2(\mathbb{P}^2)$ *or its isomorphic projection to* $\mathbb{P}^4$, *and* $d^* = \operatorname{codeg} X = \deg X = d$ *if and only if* X *is a scroll over a curve.*

(ii) $d^* \geq N - 2$. *Furthermore,* $d^* = \operatorname{codeg} X = \operatorname{codim} X = N - 2$ *if and only if* X *is the Veronese surface* $v_2(\mathbb{P}^2)$, *and* $d^* = N - 1$ *if and only if* X *is either an isomorphic projection of the Veronese surface* $v_2(\mathbb{P}^2)$ *to* $\mathbb{P}^4$ *or a rational normal scroll.*

Proof. Assertion (i) is proved in [Z1, Proposition 3] (cf. also [M], [G1], [G2]).

Assertion (ii) follows from (i) since $d \geq N - 1$ for an arbitrary nondegenerate smooth surface X with equality holding if and only if X is either the Veronese surface $v_2(\mathbb{P}^2) \subset \mathbb{P}^5$ or a normal rational scroll (this is a special case of Theorem 3.4 below). □

Some work has been done in the direction of classification of smooth surfaces and, to a smaller degree, threefolds of small class (cf. [L], [LT], [LT2], [TV], but the corresponding results are based on computations involving Betti numbers or Chern classes and do not extend to higher dimensions.

Furthermore, it is not even clear which values of class should be considered "small". In most cases, in the existing literature class is compared to degree, which only seems reasonable for low-dimensional varieties (anyhow, the bounds obtained in [Bal] are very rough). On the other hand, in [TV] the authors give a classification of smooth surfaces whose class does not exceed twenty five, which does not seem to be a "small" number. It is clear that results of such type rely on classification of curves and surfaces and can hardly be generalized to higher dimensions.

To understand the situation better and pose the problem in a reasonable way, it makes sense to consider a similar question for varieties of low *degree.* There are two types of results concerning such varieties, viz. classification theorems for varieties whose degree is absolutely low and bounds and classification theorems for varieties whose degree is lowest possible (in terms of codimension).

An example of the first type is given by the following

Theorem 3.3. *Let $X^n \subset \mathbb{P}^N$ be a nondegenerate variety of degree d. Then*

(i) $d = 1$ *if and only if* $X = \mathbb{P}^N$;
(ii) $d = 2$ *if and only if X is a quadric hypersurface in* $\mathbb{P}^N$;
(iii) $d = 3$ *if and only if X is one of the following varieties*:
  I. *X is a cubic hypersurface in* $\mathbb{P}^N$;
  II. $X = \mathbb{P}^1 \times \mathbb{P}^2$ *is a Segre variety,* $n = \dim X = 3$, $N = 5$;
  II'_1. $X = \mathbb{F}_1$ *is a nonsingular hyperplane section of the variety* II, *i.e.,* $n = \dim X = 2$, $N = 4$, *and X is the scroll to which $\mathbb{P}^2$ is mapped by the linear system of conics passing through a point*;
  II'_2. $X = v_3(\mathbb{P}^1)$ *is a nonsingular hyperplane section of the variety* II'_1, *i.e.,* $n = \dim X = 1$, $N = 3$, *and X is a twisted cubic curve*;
  II''. *X is a cone over one of the varieties* II, II'_1, II'_2 (in this case $n = \dim X$ and $N = n + 2$ can be arbitrarily large).

It should be noted that, apart from hypersurfaces and cones, there exists only a finite number of varieties of degree three.

Proof. (i) and (ii) are obvious, and (iii) is proved in [X.X.X]. □

Swinnerton–Dyer [S-D] obtained a similar classification for varieties of degree four (apart from complete intersections, all such varieties are obtained from the Segre variety $\mathbb{P}^1 \times \mathbb{P}^3 \subset \mathbb{P}^7$ and the Veronese surface $v_2(\mathbb{P}^2) \subset \mathbb{P}^5$ by projecting, taking linear sections, and forming cones). Singularities create additional difficulties for higher degrees, but Ionescu [Io1], [Io2] gave a classification of all *nonsingular* varieties up to degree eight.

Clearly, one cannot hope to proceed much further in this way, and it is necessary to specify which degrees should be considered "low". To this end, there is a well known classical theorem dating back to Del Pezzo and Bertini.

Theorem 3.4. *Let $X^n \subset \mathbb{P}^N$ be a nondegenerate variety of degree d. Then*

(i) $d \geq N - n + 1$;
(ii) *If $d = N - n + 1$, then there are the following possibilities:*
  I. $X = \mathbb{P}^n$, $N = n$, $d = 1$;
  II. X *is a quadric hypersurface,* $N = n + 1$, $d = 2$;
  III. $X = v_2(\mathbb{P}^2) \subset \mathbb{P}^5$ *is a Veronese surface,* $n = 2$, $N = 5$, $d = 4$;
  IV. $X = \mathbb{P}^1 \times \mathbb{P}^{n-1} \subset \mathbb{P}^{2n-1}$ *is a Segre variety,* $N = 2n - 1$, $d = n$;
  IV′ X *is a linear section of* IV (the hyperplane sections of I and II have the same type as the original varieties, and the irreducible hyperplane sections of III are also linear sections of $\mathbb{P}^1 \times \mathbb{P}^3$). *It should be noted that varieties of this type are scrolls* (cf. [E-H]; in particular, if $n = 1$, then X is a rational normal curve);
  III″–IV″ X *is a cone over* III *or over one of the varieties described in* IV *and* IV′ (cones over I and II are of the same type as the original varieties).

Proof. Cf. [E-H]. □

A next step is to classify nondegenerate varieties $X^n \subset \mathbb{P}^N$ for which the difference $\Delta_X = d - (N - n + 1)$ is small (if X is linearly normal, then Δ_X is called the Δ-*genus* of X; cf. Remark 2.13, (iv)). Much is known about classification of varieties with small Δ-genus and a related problem of classification of varieties of small sectional genus (cf., e.g., [Fu]); we will not go into details here.

There is an analogue of Theorem 3.3 for codegree.

Theorem 3.5. *Let $X^n \subset \mathbb{P}^N$ be a nondegenerate variety of codegree d^*. Then*

(i) $d^* = 1$ *if and only if* $X = \mathbb{P}^N$;
(ii) $d^* = 2$ *if and only if X is a quadric hypersurface in* $\mathbb{P}^N$ (furthermore, $\operatorname{def} X$ is the dimension of singular locus (vertex) of X);
(iii) *If X is smooth, then $d^* = 3$ if and only if X is one of the following varieties:*
  I. $X = \mathbb{P}^1 \times \mathbb{P}^2 \subset \mathbb{P}^5$ *is a Segre variety. In this case* $n = 3$, $\operatorname{def} X = 1$, *and X^* is isomorphic to X;*
  I′ $X = \mathbb{F}_1 \subset \mathbb{P}^4$ *is a nonsingular hyperplane section of the Segre variety from* I, *and X^* is the projection of the dual Segre variety from the point corresponding to the hyperplane;*
  II. X *is a Severi variety* (cf. [Z2, Chapter IV]). *More precisely, in this case there are the following possibilities:*
   II.1. $X = v_2(\mathbb{P}^2) \subset \mathbb{P}^5$ *is a Veronese surface, and X^* is projectively isomorphic to X^2* (cf. Examples 1.5, 5) and 2.6, 5));

II.2. $X = \mathbb{P}^2 \times \mathbb{P}^2 \subset \mathbb{P}^8$ *is a Segre fourfold, and* X^* *is projectively isomorphic to* X^2 (cf. Examples 1.5, 1) and 2.6, 1));

II.3. $X = G(5,1) \subset \mathbb{P}^{14}$ *is an eight-dimensional Grassmann variety of lines in* $\mathbb{P}^5$*, and* X^* *is projectively isomorphic to* X^2 (cf. Examples 1.5, 6) and 2.6, 6));

II.4. $X = E \subset \mathbb{P}^{26}$ *corresponds to the orbit of highest weight vector of the lowest dimensional nontrivial representation of the group of type* E_6*,* $\dim X = 16$*, and* X^* *is projectively isomorphic to* X^2*;*

II′. *X is an isomorphic projection of one of the Severi varieties* $Y^n \subset \mathbb{P}^{\frac{3n}{2}+2}$ *from* II *to* $\mathbb{P}^{\frac{3n}{2}+1}$*,* $n = 2^i$*,* $1 \le i \le 4$*, and* X^* *is the intersection of the corresponding* Y^* *with the hyperplane corresponding to the center of projection* (as in II, here we obtain four cases II′.1–II′.4).

Proof. (i) and (ii) are almost obvious (they also follow from Theorem 2.2, Proposition 2.4 and Theorem 3.3 (i), (ii)), and (iii) is proved in [Z2, Chapter IV, §5, Theorem 5.2]. □

Remark 3.6. Comparing Theorems 3.3 (iii) and 3.5 (iii), one observes that classification of varieties of codegree three is already much harder than that of varieties of degree three and that there exist varieties of codegree three having large codimension (while the codimension of varieties of degree three is at most two). It is likely that one can classify all smooth varieties of codegree four. It is interesting that, unlike varieties of low degree, smooth varieties of low codegree tend to be homogeneous. They also have nice geometric properties. Thus, the "main series" II in Theorem 3.5 (iii) is formed by Severi varieties which are defined as varieties of lowest codimension that can be isomorphically projected (i.e., they do not have apparent double points) and the "main series" of varieties of codegree four is conjecturally formed by the homogeneous Legendrean varieties which, incidentally, have one apparent double point (i.e., their projection from a general point acquires only one singularity; cf. [Hw], [CMR], [LM], [Mu]).

In what follows we address the problem of finding a sharp lower bound for the codegree in terms of dimension and codimension, i.e., proving an analogue of Theorem 3.4 (i) and describing the varieties on the boundary, i.e., proving an analogue of Theorem 3.4 (ii).

4 Jacobian Systems and Hessian Matrices

Let $X \subset \mathbb{P}^N$, $\dim X = n$ be a nondegenerate variety, and let $d^* = \operatorname{codeg} X$. Applying Proposition 2.4 and Theorem 2.2, one can replace X by its generic linear section $X' = \mathbb{P}^{N'} \cap X$ such that $N' = N - \operatorname{def} X$, $n' = \dim X' = n - \operatorname{def} X$, $\operatorname{def} X' = 0$, and X'^* is the projection of X^* from the linear subspace ${}^{\perp}\mathbb{P}^{N'} \subset \mathbb{P}^{N*}$ of hyperplanes passing through $\mathbb{P}^{N'}$. It is clear that $\operatorname{codeg} X' = \operatorname{codeg} X = d^*$ and the dual variety X'^* is a hypersurface of degree d^* in $\mathbb{P}^{N'*}$

defined by vanishing of a form F of degree d^* in $N'+1 = N - \operatorname{def} X + 1$ variables $x_0, \dots, x_{N'}$.

Let $\mathcal{J} = \mathcal{J}_F$ be the Jacobian (or polar) linear system on $\mathbb{P}^{N'*}$ spanned by the partial derivatives $\frac{\partial F}{\partial x_i}$, $i = 0, \dots, N'$, and let $\phi = \phi_{\mathcal{J}} : \mathbb{P}^{N'*} \dashrightarrow \mathbb{P}^{N'}$ be the corresponding rational map. By Theorem 2.2, the Gauß map $\gamma = \phi\big|_{X'^*}$ associating to a smooth point $\alpha \in X'^*$ the point in $\mathbb{P}^{N'}$ corresponding to the tangent hyperplane $T_{X'^*,\alpha}$ maps X'^* to X'. Thus one gets a rational map $\gamma = \phi\big|_{X'^*} : X'^* \dashrightarrow X'$, and $n' = n - \operatorname{def} X$ is equal to the rank of the differential $\mathrm{d}_\alpha \gamma$ at a general point $\alpha \in X'^*$. Since γ is defined by the partial derivatives of F, it is not surprising that, at a smooth point $\alpha \in X'^*$, the rank n' of the differential $\mathrm{d}_\alpha\gamma$ can be expressed in terms of the Hessian matrix $H_\alpha = \left(\frac{\partial^2 F}{\partial x_i \partial x_j}(\alpha)\right)$, $0 \le i, j \le N'$.

We use the following general

Definition 4.1. Let M be a matrix over an integral ring A, and let $\mathfrak{a} \subset A$ be an ideal. We say that M *has rank* r *modulo* $\mathfrak{a}$ and write $\operatorname{rk}_{\mathfrak{a}} M = r$ if all minors of M of order larger than r are contained in $\mathfrak{a}$, but there exists a minor of order r not contained in $\mathfrak{a}$. If $\mathfrak{a} = (a)$ is a principal ideal, then we simply say that M *has rank* r *modulo* a and write $\operatorname{rk}_a M = r$.

Thus $\operatorname{rk}_e M = 0$ for every invertible element $e \in A$, $\operatorname{rk}_0 M = \operatorname{rk} M$ is the usual rank, and $0 \le \operatorname{rk}_{\mathfrak{a}} M \le \operatorname{rk} M$ for an arbitrary ideal $\mathfrak{a} \in A$.

It turns out that the rank of Hessian matrix modulo F determines the rank of the differential $\mathrm{d}_\alpha\gamma$ at a general point $\alpha \in X'^*$, which is equal to n'.

Proposition 4.2. *In the above notations,* $\operatorname{rk}_F H = n' + 2$.

Proof. This can be verified by an easy albeit tedious computation (cf. [S, Theorem 2]).

For a more direct geometric proof, we observe that, for a point $\xi \in \mathbb{P}^{N'*}$, the Hessian matrix $H_\xi = \left(\frac{\partial^2 F}{\partial x_i \partial x_j}(\xi)\right)$ is the matrix of the (projective) differential $\mathrm{d}_\xi \phi : \mathbb{P}^{N'*} \dashrightarrow \mathbb{P}^{N'}$. Since, for a general point $\alpha \in X'^*$,

$$\operatorname{rk}_F H = (N'+1) - \dim \operatorname{Ker} \mathrm{d}_\alpha \phi \tag{4.2.1}$$

and

$$\dim \operatorname{Ker} \mathrm{d}_\alpha \gamma = \operatorname{def} X'^* = N' - n' - 1 = N - n - 1, \tag{4.2.2}$$

to prove the proposition it suffices to verify that for $\alpha \in \operatorname{Sm} X'^*$ one has

$$\operatorname{Ker} \mathrm{d}_\alpha \phi = \operatorname{Ker} \mathrm{d}_\alpha \gamma. \tag{4.2.3}$$

Suppose to the contrary that

$$\operatorname{Ker} \mathrm{d}_\alpha \phi \supsetneq \operatorname{Ker} \mathrm{d}_\alpha \gamma. \tag{4.2.4}$$

Then

$$\operatorname{Im} \mathrm{d}_\alpha \phi = \operatorname{Im} \mathrm{d}_\alpha \gamma. \tag{4.2.5}$$

In particular, if $\nu = \left(\frac{\partial F}{\partial x_0}(\alpha), \ldots, \frac{\partial F}{\partial x_{N'}}(\alpha)\right)$ and $\bar{\nu}$ is the complex conjugate vector, then there exists a vector v such that

$$\nu^* \cdot v = 0, \qquad H_\alpha \cdot \bar{\nu} = H_\alpha \cdot v \tag{4.2.6}$$

(here and in what follows $\nu = \nu^{(N+1)\times 1}$ is a column vector and $\nu^* = \nu^{*\,1\times(N+1)}$ is a row vector). Denoting a vector corresponding to the point α by the same letter and recalling that H_α is a symmetric matrix, one sees that (4.2.6) yields the following chain of equalities:

$$\begin{aligned}(H_\alpha \cdot \alpha)^* \cdot v &= (\alpha^* \cdot H_\alpha) \cdot v = \alpha^* \cdot (H_\alpha \cdot v) \\ &= \alpha^* \cdot (H_\alpha \cdot \bar{\nu}) = (\alpha^* \cdot H_\alpha) \cdot \bar{\nu} = (H_\alpha \cdot \alpha)^* \cdot \bar{\nu}.\end{aligned} \tag{4.2.7}$$

To complete the proof, we recall that, by Euler's formula,

$$H_\alpha \cdot \alpha = (d^* - 1)\nu, \tag{4.2.8}$$

so that, in view of (4.2.6) and (4.2.7),

$$0 = (d^* - 1)\nu^* \cdot v = (H_\alpha \cdot \alpha)^* \cdot v = (H_\alpha \cdot \alpha)^* \cdot \bar{\nu} = (d^* - 1)\nu^* \cdot \bar{\nu}. \tag{4.2.9}$$

But from (4.2.9) it follows that $\nu = 0$, which is only possible if $\alpha \in \operatorname{Sing} X'^*$, contrary to the assumption that $\alpha \in X'^*$ is smooth. □

We now use the following result which generalizes the obvious fact that if all elements of a square matrix M of order m are multiples of p, then $\det M$ is a multiple of p^m.

Proposition 4.3. *Let M be a matrix over an integral ring A, let $\mathfrak{p} \subset A$ be a prime ideal, and let k be a natural number. Then $\operatorname{rk}_{\mathfrak{p}^k} M \leq \operatorname{rk}_{\mathfrak{p}} M + k - 1$.*

In particular, if M is a square matrix of order m, then, for each prime ideal $\mathfrak{p} \subset A$, $\det M \in \mathfrak{p}^{\operatorname{cork}_{\mathfrak{p}} M}$, where $\operatorname{cork}_{\mathfrak{p}} M = m - \operatorname{rk}_{\mathfrak{p}} M$ is the corank of M modulo $\mathfrak{p}$.

Proof. By an easy induction argument, it suffices to consider the case when M is a square matrix of order m and to show that

$$\nu_{\mathfrak{p}}(\det M) \geq \min_{1 \leq i,j \leq m} \nu_{\mathfrak{p}}(\det M_{ij}) + 1, \tag{4.3.1}$$

where, for an element $a \in A$, $\nu_{\mathfrak{p}}(a) = \max\{k \mid a \in \mathfrak{p}^k\}$ and M_{ij} is the $(m-1)\times(m-1)$-matrix obtained from M by removing the i-th row and the j-th column. To this end, consider the matrix $\operatorname{adj} M$ for which $(\operatorname{adj} M)_{ij} = (-1)^{i+j} \det M_{ji}$. It is well known and easy to check that

$$M \cdot \operatorname{adj} M = \det M \cdot I_m, \tag{4.3.2}$$

where I_m is the identity matrix of order m (in standard courses of linear algebra it is shown that if A is a field and $\det M \neq 0$, then M is invertible and $M^{-1} = (\det M)^{-1} \cdot \operatorname{adj} M$). If $\det M = 0$, then (4.3.1) is clear; otherwise, taking determinants of both sides of (4.3.2), we see that

$$\det(\operatorname{adj} M) = (\det M)^{m-1} \tag{4.3.3}$$

and so

$$m \cdot \min_{1 \leq i,j \leq m} \nu_{\mathfrak{p}}(\det M_{ij}) \leq \nu_{\mathfrak{p}}(\det(\operatorname{adj} M)) = (m-1) \cdot \nu_{\mathfrak{p}}(\det M). \tag{4.3.4}$$

Thus

$$\nu_{\mathfrak{p}}(\det M) \geq \frac{m}{m-1} \cdot \min_{1 \leq i,j \leq m} \nu_{\mathfrak{p}}(\det M_{ij}) > \min_{1 \leq i,j \leq m} \nu_{\mathfrak{p}}(\det M_{ij}), \tag{4.3.5}$$

which implies (4.3.1). □

If the ring A is nötherian, then, by Krull's theorem, $\bigcap_k \mathfrak{p}^k = 0$ and the numbers $\nu_{\mathfrak{p}}(a)$ are finite for each element $a \in A$, $a \neq 0$. However, we do not need this fact here.

Applying Proposition 4.3 to the Hessian matrix $H = \left(\frac{\partial^2 F}{\partial x_i \partial x_j}\right)$, $0 \leq i,j \leq N'$ and using Proposition 4.2, one gets the following

Corollary 4.4. $F^{N'-n'-1} | \det H$.

Since all the entries of H have degree $d^* - 2$, one has $\deg \det H = (N' + 1)(d^* - 2)$. On the other hand, $\deg F^{N'-n'-1} = d^*(N' - n' - 1)$, and so Corollary 4.4 yields the following

Corollary 4.5. *Suppose that* $h = \det H \neq 0$. *Then*

$$\operatorname{codeg} X = d^* \geq 2 \cdot \frac{N'+1}{n'+2} = 2 \cdot \frac{N - \operatorname{def} X + 1}{n - \operatorname{def} X + 2} \geq 2 \cdot \frac{N+1}{n+2}.$$

Corollary 4.5 gives the desired lower bound for codegree in terms of dimension and codimension. Clearly, unless X is a quadratic cone, this bound can be sharp only if $\operatorname{def} X = 0$, i.e., X^* is a hypersurface. Furthermore, Examples 2.6, 1), 5) and 6) (for r even) and Theorem 3.5 (iii), II.4 show that the bound given in Corollary 4.5 is indeed sharp.

The catch is that we do not know how to tell whether or not the determinant $h = h_F$ of the matrix H (called the *hessian* of F) vanishes identically for a given form F. The simplest example of this phenomenon is when, after a linear change of coordinates, the form F depends on fewer than $N' + 1$ variables (cf. [H1, Lehrsatz 2]). Geometrically, this means that X'^* is a cone, and therefore the variety X' is degenerate. But then X is also degenerate, contrary to our assumptions. Thus in our setup F always depends on all the variables.

Hesse claimed that the converse is also true, i.e., in our language, $h_F = 0$ if and only if X' is degenerate (cf. [H1, Lehrsatz 3]). Moreover, his paper [H2] written eight years later is, in his own words, devoted to "giving a stronger foundation to this result". However, as Gauß put it, "unlike with lawyers for whom two half proofs equal a whole one, with mathematicians half proof equals zero, and real proof should eliminate even a shadow of doubt". This was not the case here, and, twenty five years after the publication of [H1], Gordan and Nöther [GN] found that Hesse's claim was wrong. To wit, they did not produce an explicit example of form with vanishing hessian, but rather verified the existence of solutions of certain systems of partial differential equations yielding such forms. In particular, they checked that, while Hesse's claim is true for $N' = 2$ and $N' = 3$, it already fails for $N' = 4$ (i.e., for quinary forms).

Even though constructing examples of forms with vanishing hessian might not be evident from the point of view of differential equations, it is easy from geometric viewpoint. From the above it is clear that $h = h_F \equiv 0$ if and only if the Jacobian map $\phi = \phi_{\mathcal{J}} : \mathbb{P}^{N'*} \dashrightarrow \mathbb{P}^{N'}$ fails to be surjective.

Example 4.6. Let $X = \mathbb{P}^1 \times \mathbb{P}^2 \subset \mathbb{P}^5$ be the Segre variety considered in Theorem 3.5 (iii), I, so that $X^* \simeq X$ and $\deg X = \deg X^* = 3$, and let $X' = \mathbb{F}_1 \subset \mathbb{P}^4$ be a nonsingular hyperplane section of X. Then X'^* is the projection of X^* from the point corresponding to the hyperplane (the center of projection is not contained in X^*; cf. Theorem 3.5 (iii), I$'$). I claim that the hessian $h = h_F$ of the cubic form F defining X'^* is identically equal to zero (this was first discovered by Perazzo [P] who used the theory of polars).

In fact, it is clear that the entry locus of the center of projection in $\mathbb{P}^{5*}$ is a nonsingular quadric surface. The restriction of the projection on this quadric is a double covering of the image, a plane $\Pi \subset \mathbb{P}^{4*}$, ramified along a nonsingular conic $C \subset \Pi$. Thus $\operatorname{Sing} X'^* = \Pi$ and X'^* is swept out by a family of planes $\mathbb{P}^2_\alpha$, $\alpha \in C$ such that $\mathbb{P}^2_\alpha \cap \Pi = T_{C,\alpha}$, i.e., the planes of the family meet Π along the tangent lines to the conic. Furthermore, the pencil of hyperplanes passing through a fixed plane $\mathbb{P}^2_\alpha$ is parameterized by a fibre f_α of the scroll $X' = \mathbb{F}_1$, the corresponding hyperplane sections of X'^* are a union of $\mathbb{P}^2_\alpha$ and a nonsingular quadric, and these quadrics meet $\mathbb{P}^2_\alpha$ along a union of $T_{C,\alpha}$ and lines from the pencil defined by the point $\alpha \in \mathbb{P}^2_\alpha \cap C$. Finally, the hyperplane sections of X'^* corresponding to the points of the exceptional section s of the scroll $X' = \mathbb{F}_1$, i.e., those cut out by the hyperplanes in $\mathbb{P}^{4*}$ passing through Π, have the form $\mathbb{P}^2_\alpha + 2\Pi$.

In this example $\mathcal{J}$ is a linear system of quadrics with base locus Π. We need to check that $\dim \phi(\mathbb{P}^{4*}) < 4$, i.e., the fibres of ϕ are positive-dimensional. To this end, we observe that, by the above, a general point $\xi \in \mathbb{P}^{4*}$ is contained in a unique hyperplane of the form $\mathbb{P}^3_\alpha = \langle \mathbb{P}^2_\alpha, \Pi \rangle$. Restricted to $\mathbb{P}^3_\alpha$, the linear system $\mathcal{J}$ has fixed component Π. Thus $\phi\big|_{\mathbb{P}^3_\alpha}$ is a linear projection, and to prove our claim it suffices to recall that $\phi\big|_{X'^*} = \gamma$ and $\phi(X'^*) = X'$. In particular, ϕ blows down the plane $\mathbb{P}^2_\alpha$ to a line, hence it maps $\mathbb{P}^3_\alpha$ to a plane, and the fibres of ϕ are lines. Furthermore, $Z = \phi(\mathbb{P}^{4*}) \subset \mathbb{P}^4$ is a quadratic cone with vertex at the minimal section s of the scroll $X' = \mathbb{F}_1$, and $Z = T(s, X') = \bigcup_{x \in s} T_{X',x} = l \cdot X'$ is the join of s with X'. □

Some work has been done towards better understanding the condition of vanishing of the hessian (cf. [P], [Fr], [C], [Il], [PW]), but it does not seem to be helpful in dealing with our problem.

In what follows we show that Corollary 4.5 is true for an arbitrary nondegenerate variety X, even though the corresponding hessian may vanish.

Theorem 4.7. *In the above notations,*

$$\operatorname{codeg} X = d^* \geq 2 \cdot \frac{N' + 1}{n' + 2} = 2 \cdot \frac{N - \operatorname{def} X + 1}{n - \operatorname{def} X + 2} \geq 2 \cdot \frac{N + 1}{n + 2}.$$

In view of Proposition 4.2, Theorem 4.7 is equivalent to the following elementary statement not involving any algebraic geometry.

Theorem 4.8. *Let F be an irreducible form of degree d in m variables which cannot be transformed into a form of fewer variables by a linear change of coordinates, and let $H = \left(\dfrac{\partial^2 F}{\partial x_i \partial x_j}\right)$, $1 \leq i, j \leq m$ be its Hessian matrix. Then $\operatorname{rk}_F H \geq \dfrac{2m}{d}$.*

Before proceeding with the proof of Theorem 4.7, we need to study the rational map $\phi = \phi_{\mathcal{J}} : \mathbb{P}^{N'*} \dashrightarrow \mathbb{P}^{N'}$ in more detail. From the Euler formula it follows that ϕ is regular outside of $\Sigma = \operatorname{Sing} X'^*$, and it is clear that $Z = \phi(\mathbb{P}^{N'*}) \subset \mathbb{P}^{N'}$ is an irreducible variety.

Proposition 4.9. (i) $Z \supset X'$, $r = \dim Z = \operatorname{rk} H_\xi - 1$, *where $\xi \in \mathbb{P}^{N'*}$ is a general point, and $n' + 1 \leq r \leq N'$.*

(ii) *Let $z \in Z$ be a general point, and let $\mathcal{F}_z = \phi^{-1}(z)$ be the corresponding fibre. Then $\mathcal{F}_z$ is an $(N' - r)$-dimensional linear subspace passing through the $(N' - r - 1)$-dimensional linear subspace ${}^\perp T_{Z,z}$.*

(iii) *$Z^* \subset \Sigma$. Furthermore, if $z \in Z$ is a general point, then $\mathcal{F}_z \cap X'^* = \mathcal{F}_z \cap \Sigma = \mathcal{F}_z \cap Z^* = {}^\perp T_{Z,z}$.*

Proof. (i) Since the ground field has characteristic zero, $\phi = \phi_{\mathcal{J}}$ is generically smooth (or submersive), i.e., $r = \operatorname{rk} \mathrm{d}_\xi \phi$ for a general point $\xi \in \mathbb{P}^{N'*}$. On the other hand, it is clear that $\operatorname{rk} \mathrm{d}_\xi \phi = \operatorname{rk} H_\xi - 1$, which proves the first assertion of (i).

The second assertion easily follows from the observation that $\operatorname{rk} \mathrm{d}_\xi \phi$ is lower semicontinuous as a function of ξ and, by virtue of (4.2.1) and (4.2.3), $\operatorname{rk} \mathrm{d}_\alpha \phi = \operatorname{rk} \mathrm{d}_\alpha \gamma + 1$ for a general point $\alpha \in X'^*$.

(ii) Let $\xi \in \mathcal{F}_z \setminus \Sigma$, and let H_ξ be the Hessian matrix at the point ξ. Since the matrix H_ξ is symmetric and $\phi(\xi) = z$, one sees that, in the obvious notations,

$$T_{\mathcal{F}_z,\xi} = \langle \xi, \mathbb{P}(\operatorname{Ker} H_\xi) \rangle = \langle \xi, {}^\perp\mathbb{P}(\operatorname{Im} H_\xi) \rangle \supseteq \langle \xi, {}^\perp T_{Z,z} \rangle. \qquad (4.9.1)$$

For $\xi \notin \Sigma$, the Euler formula (4.2.8) shows that

$$\xi \notin {}^\perp\mathbb{P}(\operatorname{Im} H_\xi). \qquad (4.9.2)$$

If the fibre $\mathcal{F}_z$ is reduced, then, since the ground field has characteristic zero, from (4.9.1) it follows that $\mathcal{F}_z$ is a cone with vertex ${}^\perp T_{Z,z}$. If, moreover, $z \in \operatorname{Sm} Z$ and $\mathcal{F}_z$ is equidimensional of dimension $N' - r$, then (4.9.1) shows that $\mathcal{F}_\xi$ is a union of $(N'-r)$-dimensional linear subspaces of $\mathbb{P}^{N'*}$ meeting along ${}^\perp T_{Z,z}$.

Considering the graph of ϕ obtained, at a general point $z \in Z = (Z^*)^*$, by blowing up Z (so that z is replaced by the linear subspace ${}^\perp T_{Z,z}$ along which the hyperplane ${}^\perp z$ is tangent to Z^*) and restricting the projection corresponding to ϕ on the inverse image of Z, we see that, for a general point $\zeta \subset Z^*$, the fibre $\mathcal{F}_\zeta$ has only one component.

(iii) To prove that $Z^* \subset \Sigma$ it suffices to verify that, for a general point $z \in Z$, ${}^\perp T_{Z,z} \subset \Sigma$. Suppose that this is not so, and let $\xi \in {}^\perp T_{Z,z} \setminus \Sigma$. By (ii), ${}^\perp T_{Z,z} \subset \mathcal{F}_z$ and thus $\phi(\xi) = z$. But then ${}^\perp T_{Z,z} \subset {}^\perp\mathbb{P}(\operatorname{Im} H_\xi)$ and from (4.9.2) it follows that, contrary to our assumption, $\xi \notin {}^\perp T_{Z,z}$. The resulting contradiction shows that $Z^* \subset \Sigma$.

If $z \notin X'$, then, clearly, $\mathcal{F}_z \cap X'^* = \mathcal{F}_z \cap \Sigma$. From the first assertion of (iii) it follows that ${}^\perp T_{Z,z} \subset \mathcal{F}_z \cap Z^* \subset \mathcal{F}_z \cap \Sigma$. If, furthermore, z is general in the sense of (ii), then from (4.9.1) and (ii) it follows that $\mathcal{F}_z \cap \Sigma = {}^\perp T_{Z,z}$. □

Describing forms with vanishing hessian and the dual varieties of the corresponding hypersurfaces is an interesting geometric problem which will be dealt with elsewhere. In particular, one can obtain natural proofs and far going generalizations of the relevant results in [GN], [Fr], [C] and [Il]. However in the present paper we only need the properties listed in Proposition 4.9.

We proceed with showing that if the hessian $h = h_F$ vanishes identically on $\mathbb{P}^{N'*}$, then we get an even better bound for $d^* = \deg F$. To this end, we first recall some useful classical notions.

Let $G(m,l)$ denote the Grassmann variety of l-dimensional linear subspaces in $\mathbb{P}^m$, let $\mathcal{U}$ be the universal bundle over $G(m,l)$, and let $p : \mathcal{U} \to \mathbb{P}^m$ and $q : \mathcal{U} \to G(m,l)$ denote the natural projections. For a subvariety $\Theta \subset G(m,l)$, we denote by $\mathcal{U}_\Theta$ the restriction of $\mathcal{U}$ on Θ and put $p_\Theta = p\big|_{\mathcal{U}_\Theta}$, $q_\Theta = q\big|_{\mathcal{U}_\Theta}$.

Definition 4.10. An irreducible subvariety $\Theta \subset G(m,l)$ is called *congruence* if $\dim \Theta = m - l$.

The number of points in a general fibre of the map $p_\Theta : \mathcal{U}_\Theta \to \mathbb{P}^m$ (equal to $\deg p_\Theta$) is called the *order* of Θ (usually there is no danger of confusing this notion with the one introduced in Definition 1.3). In particular, the map p_Θ is birational if and only if Θ is a congruence of order one.

The subvariety $B \subset \mathbb{P}^m$ of points in which p_Θ fails to be biholomorphic is called the *focal* or *branch locus* of Θ. In particular, if Θ is a congruence of order one, then $B = \{w \in \mathbb{P}^m \mid \dim p_\Theta^{-1}(w) > 0\}$ and $\operatorname{codim}_{\mathbb{P}^m} B > 1$; in this case B is also called the *jump locus* of the congruence Θ.

Let $\Theta \subset G(m,l)$ be a congruence of order one, and let $L \subset \mathbb{P}^m$, $\dim L = m-l$ be a general linear subspace. Intersection with L gives rise to a rational map $\rho_L : \Theta \dashrightarrow \mathbb{P}^{m-l}$ with fundamental locus $B = B_L = \Theta \cap \Gamma_L$, where $\Gamma_L = \{\ell \in G(m,l) \mid \dim \ell \cap L > 0\}$.

Lemma 4.11. *Either Θ is the congruence formed by all the l-dimensional subspaces passing through a fixed $(l-1)$-dimensional linear subspace of $\mathbb{P}^m$ or $B \neq \varnothing$ and* $\operatorname{codim}_\Theta B = 2$.

Proof. Let $\widetilde{L} \subset \mathbb{P}^m$ be a general $(m-l+1)$-dimensional linear subspace containing L. Intersection with $\widetilde{L}$ gives rise to a rational map $\rho_{\widetilde{L}} : \Theta \dashrightarrow G(m-l+1,1)$, and it is clear that ρ_L factors through $\rho_{\widetilde{L}}$. Now it suffices to prove the claim for $\widetilde{\Theta} = \rho_{\widetilde{L}}(\Theta)$, which is clearly a congruence of lines of order one. In other words, it is enough to prove the lemma for $l=1$.

Suppose now that $l=1$ and $B=\varnothing$, so that $V = V_\Theta = \bigcup\limits_{\vartheta \in \Theta} {}^\perp\ell_\vartheta \subsetneq \mathbb{P}^{m*}$, where $\ell_\vartheta = p_\Theta(q_\Theta^{-1}(\vartheta))$. Since each ${}^\perp\ell_\vartheta$ is a linear subspace of codimension two in $\mathbb{P}^{m*}$, V is a hypersurface, and a general point $v \in V$ is contained in at least an $(m-2)$-dimensional family of ${}^\perp\ell_\vartheta$'s. Thus, for $m>2$, the tangent hyperplane $T_{V,v}^{m-1}$ meets V along a positive-dimensional family of linear subspaces of dimension $m-2$. Hence V is a hyperplane, and all lines ℓ_ϑ, $\vartheta \in \Theta$ pass through one and the same point $z = V^\perp \in \mathbb{P}^m$. The remaining assertions of the lemma, as well as the cases when $m \le 2$ are now clear. □

Returning to our setup, we recall that $h_F = 0$ if and only if $Z \subsetneq \mathbb{P}^{N'}$. Suppose that this is so, let $l = N'-r$, and let $\Theta = \Theta_F \subset G(N',l)$ be the corresponding congruence of order one formed by the fibres of the Jacobian map $\phi : \mathbb{P}^{N'*} \dashrightarrow Z$ (cf. Proposition 4.9, (ii)). Cutting X'^* with a general r-dimensional linear subspace $\Lambda \subset \mathbb{P}^{N'*}$ and restricting ϕ on Λ, we obtain a rational map $\Lambda \dashrightarrow \mathbb{P}^{N'}$. Let $\varpi = \varpi_\Lambda : \mathbb{P}^{N'} \dashrightarrow \mathbb{P}^r$ denote the projection with center at the $(N'-r-1)$-dimensional linear subspace ${}^\perp\Lambda$, and put $X'' = \varpi(X')$. Then, by Proposition 2.4, $X''^* = X'^* \cap \Lambda$. We denote by F' the restriction of F on Λ and by $\mathcal{J}'$ the Jacobian system of F'. Then $\mathcal{J}'$ defines a rational map $\phi' = \phi_{\mathcal{J}'} : \Lambda \dashrightarrow \mathbb{P}^r$ and $\phi' = \varpi \circ \phi\big|_\Lambda$. Furthermore, it is clear that ϕ' is dominant and the hessian $h_{F'}$ does not vanish identically on Λ. More precisely, we have the following

Proposition 4.12. *In the above notations* $(h_{F'}) \ge (r-n'-1)(F') + E_\Lambda + W_\Lambda$, *where parentheses denote the divisor of form,* $E_\Lambda = p_\Theta(q_\Theta^{-1}(B)) \cap \Lambda$ *is the*

exceptional divisor of ϕ' *outside of* X''^*, $W_\Lambda = (\phi|_\Lambda)^{-1}(R_\Lambda)$, *and* R_Λ *is the ramification divisor of the finite covering* $\varpi_\Lambda : Z \to \mathbb{P}^r$. *Furthermore,* $d^* = \operatorname{codeg} X = \operatorname{codeg} X' = \operatorname{codeg} X'' \geq \dfrac{2(r+1) + \deg E_\Lambda + \deg W_\Lambda}{n'+2}$.

Proof. The fact that the divisor $(h_{F'})$ contains $(r - n' - 1)(F')$ was proved in Corollary 4.4, and from the above discussion it is clear that $(h_{F'})$ also contains E_Λ and W_Λ. Computing the degrees, we get $(r+1)(d^* - 2) \geq (r - n' - 1)d^* + \deg E_\Lambda + \deg W_\Lambda$, so that $d^* \geq \dfrac{2(r+1) + \deg E_\Lambda + \deg W_\Lambda}{n'+2}$. □

Proposition 4.12 shows that to give a lower bound for the codegree it suffices to obtain lower bounds for $\deg W_\Lambda$ and $\deg E_\Lambda$.

Proof of Theorem 4.7. In view of Proposition 4.12, to prove Theorem 4.7 it suffices to show that $\deg E_\Lambda + \deg W_\Lambda \geq 2(N' - r)$. The case when $h_F \neq 0$, i.e., $Z = \mathbb{P}^{N'}$ was dealt with in Corollary 4.5; so, we may assume that $r < N'$.

We claim that under this assumption, i.e., when the hessian vanishes, one has a strict inequality

$$\deg E_\Lambda + \deg W_\Lambda > 2(N' - r), \tag{4.7.1}$$

which, by virtue of 4.12, yields a strict inequality

$$d^* > 2\frac{N+1}{n+2}. \tag{4.7.2}$$

To prove (4.7.1) we recall that from Lemma 4.11 it follows that $E_\Lambda \neq \varnothing$, and so

$$\deg E_\Lambda > 0 \tag{4.7.3}$$

(as a matter of fact, it is very easy to show that E_Λ is nonlinear, and so $\deg E_\Lambda \geq 2$). On the other hand, from Proposition 4.9, (i) it follows that Z is nondegenerate, and, applying Proposition 3.1, (ii), we conclude that

$$\deg W_\Lambda \geq 2(N' - r). \tag{4.7.4}$$

The inequality (4.7.1) (hence (4.7.2), hence Theorem 4.7) is an immediate consequence of the inequalities (4.7.3) and (4.7.4). □

Remarks 4.13. (i) The bounds in Theorems 2.7 and 4.7 appear to be quite different, and the methods used to prove these bounds are very different as well. However, there seems to be a connection between these bounds. To wit, one may conjecture that, at least for nondegenerate smooth varieties $X \subset \mathbb{P}^N$ of dimension n, one always has

$$\frac{N+1}{n+1} \leq \operatorname{ord} X \leq 2\frac{N+1}{n+2} \tag{4.13.1}$$

(we recall that the first inequality in (4.13.1) was proved in Proposition 1.8). Thus, at least for smooth varieties, Theorem 2.7 should follow from Theorem 4.7. The upper bound for order in (4.13.1) can be viewed as a generalization of the theorem on linear normality. Indeed, if the secant variety SX is a proper subvariety of $\mathbb{P}^N$ or, equivalently, $\operatorname{ord} X \geq 3$, then from (4.13.1) it follows that $N \geq \frac{3n}{2} + 2$, which coincides with the bound for linear normality (cf. [Z2, Chapter 2, §2]). Similarly, for $\operatorname{ord} X \geq 4$, (4.13.1) yields $N \geq 2n + 3$, which can be easily proven directly.

(ii) From (4.7.2) and Corollary 4.5 it follows that if the bound in Theorem 4.7 is sharp (i.e., $d^* = 2\frac{N+1}{n+2}$) and X is not a quadratic cone, then $\operatorname{def} X = 0$ and $h_F \neq 0$ (i.e., $r = N$); see Conjecture 4.15 for a more precise statement.

(iii) The lower bound in (4.7.1) can be considerably improved. To wit, as in the case of W_L, one can obtain a lower bound for $\deg E_L$ in terms of the codimension of Z. This yields a better lower bound for d^* in the case of vanishing hessian. However, we do not need it here.

(iv) The techniques of studying Jacobian maps and Hessian matrices that we started developing in this section can also be useful in studying other problems, such as classification of Jacobian (or polar) Cremona transformations (cf. [EKP], [Dolg]) or, more generally, classification of forms F for which the general fibre of the Jacobian map $\phi_{\mathcal{J}_F}$ is linear (i.e., classification of homaloidal and subhomaloidal forms). This topic will be dealt with elsewhere.

Proposition 4.14. *Let $X^n \subset \mathbb{P}^N$ be a nondegenerate variety. Then the following conditions are equivalent:*

(i) $d^* = \operatorname{codeg} X = 2\frac{N+1}{n+2}$;

(ii) *Either $N = n+1$, $h_F = 0$ and X is a quadratic cone with vertex $\mathbb{P}^{\operatorname{def} X - 1}$ or $\operatorname{def} X = 0$ and $h_F = F^{N-n-1}$, where F is a suitably chosen equation of the hypersurface X^*.*

Proof. In Remark 4.13, (ii) we already observed that, unless X is a quadratic cone, (i) implies $\operatorname{def} X = 0$ and $h_G \neq 0$, where G is an equation of X^*. Thus from Corollary 4.4 it follows that $G^{N-n-1} \mid h_G$. Since $\deg G^{N-n-1} = d^*(N - n - 1)$ and $\deg h_G = (N+1)(d^* - 2)$, (i) implies that $\deg G^{N-n-1} = \deg h_G$, and so $h_G = c \cdot G^{N-n-1}$, where $c \in \mathbb{C}$ is a nonzero constant. Hence, replacing G by $F = c^{-\frac{1}{n+2}}$, we get $h_F = F^{N-n-1}$. Thus (i)$\Rightarrow$(ii).

Conversely, in the non-conic case from (ii) it follows that $\deg F^{N-n-1} = \deg h_F$, hence $d^*(N - n - 1) = (N+1)(d^* - 2)$, whence (ii)$\Rightarrow$(i). □

Having obtained a bound for codegree in Theorem 4.7, it is natural to proceed with describing the varieties for which this bound is sharp, just as in Theorem 2.12 we classified the varieties on the boundary of Theorem 2.7. However, we have not proved classification theorem for these varieties yet.

Conjecture 4.15. A nondegenerate variety $X^n \subset \mathbb{P}^N$ satisfies the equivalent conditions of Proposition 4.14 if and only if X is one of the following varieties:

I. X is a quadric, $N = n + 1$, $d^* = 2$;

II. X is a Scorza variety (cf. [Z2, Chapter VI]). More precisely, in this case there are the following possibilities:

II.1. $X = v_2(\mathbb{P}^n)$ is a Veronese variety, $N = \frac{n(n+3)}{2}$, $d^* = n + 1$;

II.2. $X = \mathbb{P}^a \times \mathbb{P}^a$, $a \geq 2$ is a Segre variety, $n = 2a$, $N = a(a+2) = \frac{n(n+4)}{4}$, $d^* = a + 1$;

II.3. $X = G(2m + 1, 1)$ is the Grassmann variety of lines in $\mathbb{P}^{2m+1}$, $m \geq 2$, $n = 4m$, $N = m(2m + 3) = \frac{n(n+6)}{8}$, $d^* = m + 1$;

II.4. $X = E$ is the variety corresponding to the orbit of highest weight vector in the lowest dimensional nontrivial representation of the group of type E_6, $n = 16$, $N = 26$, $d^* = 3$.

Remarks 4.16. (i) As in Conjectured Theorem 2.12, it seems reasonable to argue by induction. Let $x \in X$ be a general point, and let $\mathcal{P}_x \subset X^*$ be the $(N-n-1)$-dimensional linear subspace of $\mathbb{P}^{N*}$ which is the locus of tangent hyperplanes to X at x. It is clear that the restriction of the Jacobian linear system $\mathcal{J}_F$ on $\mathcal{P}_x$ is a hypersurface in $\mathcal{P}_x$ defined by vanishing of $\frac{\partial F}{\partial x_i}$ for any $i = 0, \ldots, N$ (since $\mathcal{P}_x$ is a fibre of the Gauß map $\gamma : X^* \dashrightarrow X$, the nonvanishing partial derivatives are proportional to each other on $\mathcal{P}_x$). In this way one obtains a homogeneous polynomial $F' = \frac{\partial F}{\partial x_i}$ of degree $d^{*\prime} = d^* - 1$ on $\mathcal{P}_x = \mathbb{P}^{N-n-1\,*}$ and a map $\phi' : \mathbb{P}^{N-n-1\,*} \dashrightarrow \mathbb{P}^{N-n-1}$ defined by any row of the Hessian matrix of F restricted on $\mathcal{P}_x$.

(ii) It is worthwhile to observe that the varieties in Conjecture 4.15 are the same as those in Conjectured theorem 2.12, i.e., although the bounds in Theorem 2.7 and Theorem 4.7 are quite different, the extremal varieties are expected to be the same (and, in particular, Remark 2.13, (iii) should also apply to the varieties on the boundary of Theorem 4.7). This comes as no big surprise in view of Remark 4.13, (i). Assuming the bound (4.13.1), we observe that Conjecture 4.15 implies Theorem 2.12.

(iii) As in Remark 2.13, (iv), we observe that, having classified varieties of minimal codegree, it is natural to proceed with giving classification of varieties of "next to minimal codegree". However, this notion is not so easy to define. In particular, one should exclude projected varieties and varieties with positive defect. This having been said, we observe that, unlike the bound for minimal degree in Theorem 3.4 which is additive in $n = \dim X$ and $N = \dim \langle X \rangle$, the bound in Theorem 4.7 is of a multiplicative nature. Thus, varieties of next to minimal degree in the sense of Theorem 4.7 should have codegree $d^* = 2\frac{N+1}{n+1}$. An example of such varieties is given by the homogeneous Legendrean varieties $X^n \subset \mathbb{P}^{2n+1}$ mentioned in Remark 3.6, in which case $d^* = 4$.

References

[Bal] Ballico, E.: On the dual of projective varieties. Canad. Math. Bull. **34**(4) 433–439 (1991)

[C-J] Catalano–Johnson, M.: The homogeneous ideals of higher secant varieties. J. Pure Appl. Algebra **158**(2–3) 123–129 (2001)

[Ca] Cayley, A.: On the theory of linear transformations. Cambridge Math. J. **4** 1–16 (1845)

[Ch] Chaput, P.–E.: Scorza varieties and Jordan algebras. Indagationes Math. **14** 169–182 (2003)

[C] Ciliberto, C.: Ipersuperficie algebriche a punti parabolici e relative hessiane. Rend. Accad. Naz. Sci. XL Mem. Mat. **4** 25–42 (1979/80)

[CMR] Ciliberto, C., Mella, M., Russo, F.: Varieties with one apparent double point. J. Algebraic Geometry **13** (2004)

[Cl] Clebsch, A.: Über Kurven vierter Ordnung. J. reine angew. Math. **59** 125–145 (1861)

[Dim] Dimca, A.: Milnor numbers and multiplicities. Rev. Roumaine Math. Pures. Appl. **31**(6) 535–538 (1986)

[Dolg] Dolgachev I.: Polar Cremona transformations. Michigan Math. J. **48** 191–202 (2000)

[E-H] Eisenbud, D., Harris, J.: On varieties of minimal degree (a centennial account). In: Bloch, Spencer J. (ed.): *Algebraic Geometry.* Bowdoin, 1985 (Brunswick, Maine 1985), Part I, pp. 3–13. Proc. Sympos. Pure Math., **46** (Amer. Math. Soc., Providence, RI 1987)

[EKP] Etingof, P., Kazhdan, D., Polishchuk, A.: When is the Fourier transform of an elementary function elementary? Sel. Math., New ser. **8** 27–66 (2002)

[Fr] Franchetta, A.: Sulle forme algebriche di S_4 aventi l'hessiana indeterminata. Rend. Mat. e Appl. (5) **14** 252–257 (1954)

[Fu] Fujita, T.: *Classification Theories of Polarized Varieties.* London Mathematical Society Lecture Note Series, vol. 155 (Cambridge University Press, Cambridge 1990)

[Ful] Fulton, W.: *Intersection Theory.* Ergebn. Math. (3), **2** (Springer, Berlin, Heidelberg, New York 1984)

[G1] Gallarati, D.: Una proprietá caratteristica delle rigate algebriche. Rend. Accad. Naz. Lincei (8) **21** 55–56 (1956)

[G2] Gallarati, D.: Ancora sulla differenza tra la classe e l'ordine di una superficie algebrica. Ricerche Mat. **6** 111–124 (1957)

[GKZ] Gelfand, I.M., Kapranov, M.M., Zelevinsky, A.V.: *Discriminants, Resultants and Multidimensional Determinants.* Mathematics: Theory & Applications (Birkhäuser, Boston, Basel, Berlin 1994)

[Ge] Inverse systems of fat points: Waring's problem, secant varieties of Veronese varieties and parameter spaces for Gorenstein ideals. In: Geramita, A. (ed.): *The Curves Seminar at Queen's, X (1995)*, Queen's Papers in Pure and Appl. Math., **102** (Queen's Univ., Kingston, ON 1996)

[GN] Gordan, P., Nöther, M.: Über die algebraischen Formen, deren Hesse'sche Determinante identisch verschwindet. Math. Ann. **10** 547–568 (1876)

[Ha] Hartshorne, R. *Algebraic Geometry* (Springer, Berlin, Heidelberg, New York 1977)

[H1] Hesse, O.: Über die Bedingung, unter welcher eine homogene ganze Function von n unabhängigen Variabeln durch lineäre Substitutionen von n andern unabhängigen Variabeln auf eine homogene Function sich zurückführen läfst, die eine Variable weniger enthält. J. reine angew. Math. **42** 117–124 (1851)

[H2] Hesse, O.: Zur Theorie der ganzen homogenen Functionen. J. reine angew. Math. **56** 263–269 (1859)

[HT] Harris, J., Tu, L.W.: On symmetric and skew-symmetric determinantal varieties. Topology **23** 71–84 (1984)

[HR] Holme, A., Roberts, J.: Pinch-points and multiple locus of generic projections of singular varieties. Adv. Math. **33** 212–256 (1979)

[Hw] Hwang, J.–M.: Rigidity of homogeneous contact manifolds under Fano deformations. J. reine angew. Math. **486** 153–163 (1997)

[Il] Ilardi, G.: Sulle ipersuperfici algebriche ad hessiana indeterminata. Rend. Ist. Mat. Univ. Trieste **26**(1–2) 47–78 (1995)

[Io1] Ionescu, P.: Variétés projectives lisses de degrés 5 et 6. C. R. Acad. Sci. Paris, Ser. **I** Math. **293**(15) 685–687 (1981)

[Io2] Ionescu, P.: Embedded projective varieties of small invariants. III. In: Sommese, A. J., Biancofiore, A., and Livorni, E. L. (eds.): *Algebraic Geometry* (L'Aquila, 1988) pp. 138–154. Lecture Notes in Math., **1417** (Springer, Berlin, Heidelberg, New York 1990)

[LM] Landsberg, J., Manivel, L.: Representation theory and projective geometry. This volume, pp. 71–122

[L] Lanteri, A.: On the class of a projective algebraic surface. Arch. Math. **45** 79–85 (1985)

[LT] Lanteri, A., Turrini, C.: Projective threefolds of small class. Abh. Math. Sem. Univ. Hamburg **57** 103–117 (1987)

[LT2] Lanteri, A., Turrini, C.: Projective surfaces with class less than or equal to twice the degree. Math. Nachrichten **175** 199–207 (1995)

[M] Marchionna, E.: Sopra una disuguaglianza tra i caratteri proiettivi di una superficie algebraica. Boll. Un. Mat. Ital. (3) **10** 478–480 (1955)

[Mu] Mukai, Sh.: Simple Lie algebra and Legendre variety. Nagoya Sūri Forum **3** 1–12 (1996) (in Japanese). Revised translation into English: http://www.math.nagoya-u.ac.jp/~mukai/warwick15.pdf (1998) pp. 1–14

[Mum] Mumford, D.: *Algebraic Geometry* I. Complex Projective Varieties (Springer, Berlin, Heidelberg, New York 1976)

[P] Perazzo, U.: Sulle varietá cubiche la cui hessiana svanisce identicamente. Atti R. Acc. Lincei, 337–354 (1900)

[PW] Pinkus, A., Wajnryb, B.: A problem of approximation using multivariate polynomials. Uspekhi Mat. Nauk **50**(2) 89–110 (1995) (in Russian). English translation in: Russian Math. Surveys **50**(2) 319–340 (1995)

[S] Segre, B.: Bertini forms and Hessian matrices. J. London Math. Soc. **26** 164–176 (1951)

[S-D] Swinnerton–Dyer, H.P.F.: An enumeration of all varieties of degree 4. Amer. J. Math. **95** 403–418 (1973)

[Syl] Sylvester, J.J.: An essay on canonical forms supplemented by a sketch of a memoir on elimination, transformation, and canonical forms. In: Baker, H. F. (ed.): Collected Mathematical Papers, vol. 1 (Cambridge University Press, Cambridge 1904) pp. 184–216

[Tev] Tevelev, E.: *Projective Duality and Homogeneous Spaces.* Encyclopaedia of Mathematical Sciences, Subseries *Invariant Theory and Algebraic Transformation Groups* (Springer, Berlin, Heidelberg, New York) to appear

[TV] Turrini, C., Verderio, E.: Projective surfaces of small class. Geometria Dedicata **47**(1) 1–14 (1993)

[X.X.X] Serre, J.-P.: Correspondence. Amer. J. Math., 951–952 (1957)

[Z1] Zak, F.L.: On surfaces with zero Lefschetz cycles. Mat. Zametki, **13**, 869–880 (1973). English transl. in: Math. Notes **13** 520–525

[Z2] Zak, F.L.: *Tangents and Secants of Algebraic Varieties.* Translations of mathematical monographs, vol. 127 (American Mathematical Society, Providence, Rhode Island 1993)

[Z3] Zak, F.L.: Projective invariants of quadratic embeddings. Math. Ann. **313** 507–545 (1999)

[Å] Ådlansvik, B.: Varieties with constrained higher secant varieties. Preprint, University of Bergen (1987)

Printing: Strauss GmbH, Mörlenbach
Binding: Schäffer, Grünstadt